New Trends in Differential and Difference Equations and Applications

New Trends in Differential and Difference Equations and Applications

Special Issue Editors

Feliz Manuel Minhós
João Fialho

MDPI • Basel • Beijing • Wuhan • Barcelona • Belgrade

Special Issue Editors
Feliz Manuel Minhós
University of Évora
Portugal

João Fialho
British University Vietnam
Vietnam

Editorial Office
MDPI
St. Alban-Anlage 66
4052 Basel, Switzerland

This is a reprint of articles from the Special Issue published online in the open access journal *Axioms* (ISSN 2075-1680) from 2018 to 2019 (available at: https://www.mdpi.com/journal/axioms/special_issues/differential_difference_equations).

For citation purposes, cite each article independently as indicated on the article page online and as indicated below:

LastName, A.A.; LastName, B.B.; LastName, C.C. Article Title. *Journal Name* **Year**, *Article Number*, Page Range.

ISBN 978-3-03921-538-6 (Pbk)
ISBN 978-3-03921-539-3 (PDF)

Contents

About the Special Issue Editors vii

Preface to "New Trends in Differential and Difference Equations and Applications" ix

Feliz Minhós
Heteroclinic Solutions for Classical and Singular ϕ-Laplacian Non-Autonomous Differential Equations
Reprinted from: *Axioms* **2019**, *8*, 22, doi:10.3390/axioms8010022 1

Ravi Agarwal, Snezhana Hristova and Donal O'Regan
Lipschitz Stability for Non-Instantaneous Impulsive Caputo Fractional Differential Equations with State Dependent Delays
Reprinted from: *Axioms* **2019**, *8*, 4, doi:10.3390/axioms8010004 15

Mina Torabi and Mohammad-Mehdi Hosseini
A New Efficient Method for the Numerical Solution of Linear Time-Dependent Partial Differential Equations
Reprinted from: *Axioms* **2018**, *7*, 70, doi:10.3390/axioms7040070 32

Vladimir I. Semenov
The 3D Navier–Stokes Equations: Invariants, Local and Global Solutions
Reprinted from: *Axioms* **2019**, *8*, 41, doi:10.3390/axioms8020041 45

Shin Min Kang, Zain Iqbal, Mustafa Habib and Waqas Nazeer
Sumudu Decomposition Method for Solving Fuzzy Integro-Differential Equations
Reprinted from: *Axioms* **2019**, *8*, 74, doi:10.3390/axioms8020074 96

Mama Foupouagnigni and Salifou Mboutngam
On the Polynomial Solution of Divided-Difference Equations of the Hypergeometric Type on Nonuniform Lattices
Reprinted from: *Axioms* **2019**, *8*, 47, doi:10.3390/axioms8020047 114

Michael Gil'
Solution Estimates for the Discrete Lyapunov Equation in a Hilbert Space and Applications to Difference Equations
Reprinted from: *Axioms* **2019**, *8*, 20, doi:10.3390/axioms8010020 125

Jan Andres and Denis Pennequin
Note on Limit-Periodic Solutions of the Difference Equation $x_{t+1} - [h(x_t) + \lambda]x_t = r_t, \lambda > 1$
Reprinted from: *Axioms* **2019**, *8*, 19, doi:10.3390/axioms8010019 147

Tatyana V. Redkina, Robert G. Zakinyan, Arthur R. Zakinyan, Olesya B. Surneva and Olga S. Yanovskaya
Bäcklund Transformations for Nonlinear Differential Equations and Systems
Reprinted from: *Axioms* **2019**, *8*, 45, doi:10.3390/axioms8020045 157

João Fialho and Feliz Minhós
First Order Coupled Systems With Functional and Periodic Boundary Conditions: Existence Results and Application to an SIRS Model
Reprinted from: *Axioms* **2019**, *8*, 23, doi:10.3390/axioms8010023 175

About the Special Issue Editors

Feliz Minhós, Prof., has a Ph.D. and a Habilitation Degree, both in Mathematics, is the Coordinator of the Research Center on Mathematics and Applications, Director of the Ph.D. Program in Mathematics at the University of Évora, Portugal, has published a large number of papers, books, monographs, proceedings, ... and is a member of the Editorial Board of several international journals. Prof. Minhós' main research interests are in differential equations, boundary value problems with variational and topological methods, fixed point theory, impulsive problems, functional differential equations, nonlocal problems, degree theory, lower and upper solutions, Green's functions, among others.

Joao Fialho, Prof., is a Ph.D. in Mathematics at University of Évora, Portugal. He is currently Senior Lecturer and MBA Director at the British University of Vietnam. He has diverse practical working experience and teaching experience, having spent more than 5 years working as a business analyst for private companies in Lisbon, Portugal, and for the last 7 years he has been working in higher education, in several different countries, as Head of the Mathematics Department at College of the Bahamas, Nassau, Bahamas and as Assistant Professor of Mathematics at the American University of the Middle East, Kuwait, before joining the British University of Vietnam.

Dr. Fialho has a vast research record of international publications, a book, and several presentations in international conferences. His current research interests are in differential equations, boundary value problems, mathematical modelling, and big data analytics in sports and industry.

Preface to "New Trends in Differential and Difference Equations and Applications"

Differential and difference equations, their methods, their techniques, and their huge variety of applications have attracted interest in different fields of science in the last few years. Not only their solvability and the study of qualitative properties have been the aim of many research papers, but, also, their role in different types of boundary value problems have allowed the study of many real-world phenomena.

This Special Issue provides examples of some new methods and techniques on research topics, such as sufficient conditions to obtain heteroclinic solutions for phi-Laplacian equations, invariants, local and global solutions, stability theory (asymptotic, exponential, Lipsichtz, etc.), numerical methods for partial differential equations, fuzzy integro-differential equations, divided-difference equations, limit-periodic solutions for difference equations, Bäcklund transformations for nonlinear equations and systems, and coupled systems with functional boundary conditions that include the periodic case.

These topics, which encompass several areas of mathematical research, give the reader a comprehensive and quick overview of the trends and recent research results, which may be useful in their research or in future research topics.

Feliz Manuel Minhós, João Fialho
Special Issue Editors

axioms

MDPI

Article

Heteroclinic Solutions for Classical and Singular ϕ-Laplacian Non-Autonomous Differential Equations

Feliz Minhós

Departamento de Matemática, Escola de Ciências e Tecnologia, Centro de Investigação em Matemática e Aplicações (CIMA), Instituto de Investigação e Formação Avançada, Universidade de Évora. Rua Romão Ramalho, 59, 7000-671 Évora, Portugal; fminhos@uevora.pt

Received: 28 December 2018; Accepted: 11 February 2019; Published: 15 February 2019

Abstract: In this paper, we consider the second order discontinuous differential equation in the real line, $(a\,(t,u)\;\phi\,(u'))' = f\,(t,u,u')\,,\ a.e.t \in \mathbb{R}, u(-\infty) = \nu^-,\ \ u(+\infty) = \nu^+$, with ϕ an increasing homeomorphism such that $\phi(0) = 0$ and $\phi(\mathbb{R}) = \mathbb{R}$, $a \in C(\mathbb{R}^2, \mathbb{R})$ with $a(t,x) > 0$ for $(t,x) \in \mathbb{R}^2$, $f : \mathbb{R}^3 \to \mathbb{R}$ a L^1-Carathéodory function and $\nu^-, \nu^+ \in \mathbb{R}$ such that $\nu^- < \nu^+$. The existence and localization of heteroclinic connections is obtained assuming a Nagumo-type condition on the real line and without asymptotic conditions on the nonlinearities ϕ and f. To the best of our knowledge, this result is even new when $\phi(y) = y$, that is for equation$(a\,(t,u(t))\,u'(t))' = f\,(t,u(t),u'(t))\,,\ a.e.t \in \mathbb{R}$. Moreover, these results can be applied to classical and singular ϕ-Laplacian equations and to the mean curvature operator.

Keywords: ϕ-Laplacian operator; mean curvature operator; heteroclinic solutions; problems in the real line; lower and upper solutions; Nagumo condition on the real line; fixed point theory

2010 Mathematics Subject Classification: 34C37; 34B40; 34B15; 47H10

---

1. Introduction

In this paper, we study the second order non-autonomous half-linear equation on the whole real line,

$$\left(a\,(t,u)\,\phi\,(u')\right)' = f\,(t,u,u')\,,\ a.e.t \in \mathbb{R}, \tag{1}$$

with ϕ an increasing homeomorphism, $\phi(0) = 0$ and $\phi(\mathbb{R}) = \mathbb{R}$, $a \in C(\mathbb{R}^2, \mathbb{R})$ such that $a(t,x) > 0$ for $(t,x) \in \mathbb{R}^2$, and $f : \mathbb{R}^3 \to \mathbb{R}$ a L^1-Carathéodory function, together with the asymptotic conditions:

$$u(-\infty) = \nu^-,\ \ u(+\infty) = \nu^+, \tag{2}$$

with $\nu^+, \nu^- \in \mathbb{R}$ such that $\nu^- < \nu^+$. Moreover, an application to singular ϕ-Laplacian equations will be shown.

This problem (1) and (2) was studied in [1,2] . This last paper contained several results and criteria. For example, Theorem 2.1 in [2] guarantees the existence of heteroclinic solutions under, in short, the following main assumptions:

- ϕ grows at most linearly at infinity;
- $f(t,\nu^-,0) \le 0 \le f(t,\nu^+,0)$ for $a.e.t \in \mathbb{R}$;

- there exist constants $L, H > 0$, a continuous function $\theta : \mathbb{R}^+ \to \mathbb{R}^+$ and a function $\lambda \in L^p([-L, L])$, with $1 \le p \le \infty$, such that:

$$\begin{aligned} |f(t,x,y)| &\le \lambda(t)\,\theta\left(a(t,x)\,|y|\right), \text{ for a.e. } |t| \le L, \text{ every } x \in \left[\nu^-, \nu^+\right], \\ |y| &> H, \int^{+\infty} \frac{s^{1-\frac{1}{q}}}{\theta(s)} ds = +\infty; \end{aligned}$$

- for every $C > 0$, there exist functions $\eta_C \in L^1(\mathbb{R})$, $\Lambda_C \in L^1_{loc}([0, +\infty))$, null in $[0, L]$ and positive in $[L, +\infty)$, and $N_C(t) \in L^1(\mathbb{R})$ such that:

$$\begin{aligned} f(t,x,y) &\le -\Lambda_C(t)\phi\left(|y|\right), \\ f(-t,x,y) &\ge \Lambda_C(t)\phi\left(|y|\right), \text{ for a.e. } t \ge L, \text{ every } x \in \left[\nu^-, \nu^+\right], \\ |y| &\le N_C(t), \\ |f(t,x,y)| &\le \eta_C(t) \text{ if } x \in \left[\nu^-, \nu^+\right], |y| \le N_C(t), \text{ for } a.e. t \in \mathbb{R}. \end{aligned}$$

Motivated by these works, we prove, in this paper, the existence of heteroclinic solutions for (1) assuming a Nagumo-type condition on the real line and without asymptotic assumptions on the nonlinearities ϕ and f. The method follows arguments suggested in [3–5], applying the technique of [3] to a more general function a, with an adequate functional problem and to classical and singular ϕ-Laplacian equations. The most common application for ϕ is the so-called p-Laplacian, i.e., $\phi(y) = |y|^{p-2}p$, $p > 1$, and even in this particular case, verifying (4), the new assumption on ϕ.Moreover, this type of equation includes, for example, the mean curvature operator. On the other hand, to the best of our knowledge, the main result is even new when $\phi(y) = y$, that is for equation:

$$\left(a\,(t,u)\,u'\right)' = f\left(t,u,u'\right),\ a.e. t \in \mathbb{R}.$$

The study of differential equations and boundary value problems on the half-line or in the whole real line and the existence of homoclinic or heteroclinic solutions have received increasing interest in the last few years, due to the applications to non-Newtonian fluids theory, the diffusion of flows in porous media, and nonlinear elasticity (see, for instance, [6–16] and the references therein). In particular, heteroclinic connections are related to processes in which the variable transits from an unstable equilibrium to a stable one (see, for example, [17–24]); that is why heteroclinic solutions are often called transitional solutions.

The paper is organized in this way: Section 2 contains some notations and auxiliary results. In Section 3, we prove the existence of heteroclinic connections for a functional problem, which is used to obtain an existence and location theorem for heteroclinic solutions for the initial problem. Section 4 contains an example, to show the applicability of the main theorem. The last section applies the above theory to singular ϕ-Laplacian differential equations.

2. Notations and Auxiliary Results

Throughout this paper, we consider the set $X := BC^1(\mathbb{R})$ of the $C^1(\mathbb{R})$ bounded functions, equipped with the norm $\|x\|_X = \max\left\{\|x\|_\infty, \|x'\|_\infty\right\}$, where $\|y\|_\infty := \sup_{t\in\mathbb{R}} |y(t)|$.

By standard procedures, it can be shown that $(X, \|.\|_X)$ is a Banach space.

As a solution of the problem (1) and (2), we mean a function $u \in X$ such that $t \mapsto \left(a\,(t,u(t))\,\phi\left(u'(t)\right)\right) \in W^{1,1}(\mathbb{R})$ and satisfying (1) and (2).

The L^1-Carathéodory functions will play a key role throughout the work:

Definition 1. *A function $f : \mathbb{R}^3 \to \mathbb{R}$ is L^1-Carathéodory if it verifies:*

(*i*) *for each $(x,y) \in \mathbb{R}^2$, $t \mapsto f(t,x,y)$ is measurable on $\mathbb{R}$;*

(ii) *for almost every* $t \in \mathbb{R}$, $(x,y) \mapsto f(t,x,y)$ *is continuous in* $\mathbb{R}^2$;
(iii) *for each* $\rho > 0$, *there exists a positive function* $\varphi_\rho \in L^1(\mathbb{R})$ *such that, for* $\max\left\{\sup_{t\in\mathbb{R}}|x(t)|, \sup_{t\in\mathbb{R}}|y(t)|\right\} < \rho$,

$$|f(t,x,y)| \leq \varphi_\rho(t), \text{ a.e. } t \in \mathbb{R}. \tag{3}$$

The following hypothesis will be assumed:

(H_1) ϕ is an increasing homeomorphism with $\phi(0) = 0$ and $\phi(\mathbb{R}) = \mathbb{R}$ such that:

$$\left|\phi^{-1}(w)\right| \leq \phi^{-1}(|w|); \tag{4}$$

(H_2) $a \in C(\mathbb{R}^2, \mathbb{R})$ is a continuous and positive function with $a(t,x) \to +\infty$ as $|t| \to +\infty$.

To overcome the lack of compactness of the domain, we apply the following criterion, suggested in [25]:

Lemma 1. *A set* $M \subset X$ *is compact if the following conditions hold:*

1. *M is uniformly bounded in X;*
2. *the functions belonging to M are equicontinuous on any compact interval of* $\mathbb{R}$;
3. *the functions from M are equiconvergent at* $\pm\infty$, *that is, given* $\epsilon > 0$, *there exists* $T(\epsilon) > 0$ *such that:*

$$|f(t) - f(\pm\infty)| < \epsilon \text{ and } \left|f'(t) - f'(\pm\infty)\right| < \epsilon,$$

for all $|t| > T(\epsilon)$ *and* $f \in M$.

3. Existence Results

The first existence result for heteroclinic connections will be obtained for an auxiliary functional problem without the usual asymptotic or growth assumptions on ϕ or on the nonlinearity f.

Consider two continuous operators $A : X \to C(\mathbb{R})$, $x \longmapsto A_x$, with $A_x > 0$, $\forall x \in X$, and $F : X \to L^1(\mathbb{R})$, $x \longmapsto F_x$, the functional problem composed of:

$$\left(A_u(t)\, \phi\left(u'(t)\right)\right)' = F_u(t), \text{ a.e. } t \in \mathbb{R}, \tag{5}$$

and the boundary conditions (2).

Define, for each bounded set $\Omega \subset X$,

$$m(t) := \min_{x\in\Omega} A_x(t) \tag{6}$$

and for the above operators, assume that:

(F_1) For each $\eta > 0$, there is $\psi_\eta \in L^1(\mathbb{R})$, with $\psi_\eta(t) > 0$, *a.e.* $t \in \mathbb{R}$, such that $|F_x(t)| \leq \psi_\eta(t)$, *a.e.* $t \in \mathbb{R}$, whenever $\|x\|_X < \eta$.

(A_1) $A_x(t) \to +\infty$ as $|t| \to +\infty$ and:

$$\int_{-\infty}^{+\infty} \phi^{-1}\left(\frac{2\int_{-\infty}^{+\infty} \psi_\eta(r)dr}{m(s)}\right) ds < +\infty. \tag{7}$$

Theorem 1. *Assume that conditions (H_1), (F_1), and (A_1) hold and there is $R>0$ such that:*

$$\max\left\{\begin{array}{c} |\nu^-|+\int_{-\infty}^{+\infty}\phi^{-1}\left(\frac{2\int_{-\infty}^{+\infty}\psi_R(r)dr}{m(s)}\right)ds, \\ \sup\phi^{-1}\left(\frac{2\int_{-\infty}^{+\infty}\psi_R(r)dr}{m(t)}\right)\end{array}\right\}<R. \tag{8}$$

Then, there exists $u\in X$ such that $A_u\cdot(\phi\circ u')\in W^{1,1}(\mathbb{R})$, verifying (5) and (2), given by:

$$u(t)=\nu^-+\int_{-\infty}^{t}\phi^{-1}\left(\frac{\tau_u+\int_{-\infty}^{s}F_u(r)\,dr}{A_u(s)}\right)ds. \tag{9}$$

where τ_u is the unique solution of:

$$\int_{-\infty}^{+\infty}\phi^{-1}\left(\frac{\tau_u+\int_{-\infty}^{s}F_u(r)\,dr}{A_u(s)}\right)ds=\nu^+-\nu^-. \tag{10}$$

Moreover, for $R>0$ such that $\|x\|_X<R$,

$$\tau_u\in[w_1,\,w_2], \tag{11}$$

with:

$$w_1:=-\int_{-\infty}^{+\infty}\Psi_R(r)dr, \tag{12}$$

and:

$$w_2:=\int_{-\infty}^{+\infty}\Psi_R(r)dr. \tag{13}$$

Proof. For every $x\in X$, define the operator $T:X\to X$ by

$$T_x(t)=\nu^-+\int_{-\infty}^{t}\phi^{-1}\left(\frac{\tau_x+\int_{-\infty}^{s}F_x(r)\,dr}{A_x(s)}\right)ds$$

where $\tau_x\in\mathbb{R}$ is the unique solution of:

$$\int_{-\infty}^{+\infty}\phi^{-1}\left(\frac{\tau_x+\int_{-\infty}^{s}F_x(r)\,dr}{A_x(s)}\right)ds=\nu^+-\nu^-.$$

To show that τ_x is the unique solution of (10), consider the strictly-increasing function in $\mathbb{R}$:

$$G(y):=\int_{-\infty}^{+\infty}\phi^{-1}\left(\frac{y+\int_{-\infty}^{s}F_x(r)\,dr}{A_x(s)}\right)ds,$$

and remark that:

$$\lim_{y\to-\infty}G(y)=\int_{-\infty}^{+\infty}\phi^{-1}(-\infty)\,ds=-\infty,$$

and:

$$\lim_{y\to+\infty}G(y)=\int_{-\infty}^{+\infty}\phi^{-1}(+\infty)\,ds=+\infty. \tag{14}$$

Moreover, for w_1 given by (12) and w_2 given by (13), $G(w_1)$ and $G(w_2)$ have opposite signs, as:

$$G(w_1)=\int_{-\infty}^{+\infty}\phi^{-1}\left(\frac{w_1+\int_{-\infty}^{s}F_x(r)\,dr}{A_x(s)}\right)ds\le 0<\nu^+-\nu^-,$$

$$G(w_2) = \int_{-\infty}^{+\infty} \phi^{-1}\left(\frac{w_2 + \int_{-\infty}^{s} F_x(r)\, dr}{A_x(s)}\right) ds \geq 0.$$

As G is strictly increasing in $\mathbb{R}$, by (14), there is $k \geq 0$ such that $w_3 = w_2 + k$ and $G(w_3) \geq \nu^+ - \nu^-$. Therefore, the equation $G(y) = \nu^- - \nu^+$ has a unique solution τ_x, and by Bolzano's theorem, $\tau_x \in [w_1, w_2]$, when $\|x\|_X < R$, for some $R > 0$.

It is clear that if T has a fixed point u, then u is a solution of the problem (5) and (2).

To prove the existence of such a fixed point, we consider several steps:

Step 1. *$T : X \to X$ is well defined*

By the positivity of A and the continuity of A and F, then T_x and:

$$T_x'(t) = \phi^{-1}\left(\frac{\tau_x + \int_{-\infty}^{t} F_x(r)\, dr}{A_x(t)}\right)$$

are continuous on $\mathbb{R}$, that is $T_x \in C^1(\mathbb{R})$.

Moreover, by (H_1), (F_1), (A_1), and (10), T_x and T_x' are bounded. Therefore, $T_x \in X$.

Step 2. *T is compact.*

Let $B \subset X$ be a bounded subset, $x \in B$, and $\rho_0 > 0$ such that $\|x\|_X < \rho_0$. Consider $m(t)$ given by (6) with $\Omega = B$.

Claim: *TB is uniformly bounded in X.*

By (4), (11), and (A_1), we have:

$$\begin{aligned}
\|T_x\|_\infty &= \sup_{t\in\mathbb{R}} \left| \nu^- + \int_{-\infty}^{t} \phi^{-1}\left(\frac{\tau_x + \int_{-\infty}^{s} F_x(r)\, dr}{A_x(s)}\right) ds \right| \\
&\leq \sup_{t\in\mathbb{R}} \left(|\nu^-| + \int_{-\infty}^{t} \phi^{-1}\left(\left|\frac{\tau_x + \int_{-\infty}^{s} F_x(r)\, dr}{A_x(s)}\right|\right) ds \right) \\
&\leq \sup_{t\in\mathbb{R}} \left(|\nu^-| + \int_{-\infty}^{t} \phi^{-1}\left(\frac{|\tau_x| + \int_{-\infty}^{s} |F_x(r)|}{A_x(s)} dr\right) ds \right) \\
&\leq |\nu^-| + \int_{-\infty}^{+\infty} \phi^{-1}\left(\frac{|\tau_x| + \int_{-\infty}^{s} \Psi_{\rho_0}(r) dr}{A_x(s)}\right) ds \\
&\leq |\nu^-| + \int_{-\infty}^{+\infty} \phi^{-1}\left(\frac{2\int_{-\infty}^{+\infty} \Psi_{\rho_0}(r) dr}{m(s)}\right) ds < +\infty,
\end{aligned}$$

and:

$$\begin{aligned}
\|T_x'\|_\infty &= \sup_{t\in\mathbb{R}} \left| \phi^{-1}\left(\frac{\tau_x + \int_{-\infty}^{t} F_x(r)\, dr}{A_x(t)}\right) \right| \leq \sup_{t\in\mathbb{R}} \phi^{-1}\left(\frac{|\tau_x| + \int_{-\infty}^{t} |F_x(r)|\, dr}{A_x(t)}\right) \\
&\leq \sup_{t\in\mathbb{R}} \phi^{-1}\left(\frac{|\tau_x| + \int_{-\infty}^{+\infty} \Psi_{\rho_0}(r) dr}{A_x(t)}\right) \\
&\leq \sup_{t\in\mathbb{R}} \phi^{-1}\left(\frac{2\int_{-\infty}^{+\infty} \Psi_{\rho_0}(r) dr + k}{m(t)}\right) < +\infty.
\end{aligned}$$

Therefore, TB is uniformly bounded in X.

Claim: *TB is equicontinuous on X.*

For $M > 0$, consider $t_1, t_2 \in [-M, M]$, and without loss of generality, $t_1 < t_2$.

Then, by (4), (11) and (A_1),

$$
\begin{aligned}
|T_x(t_1) - T_x(t_2)| &= \left| \int_{-\infty}^{t_1} \phi^{-1}\left(\frac{\tau_x + \int_{-\infty}^{s} F_x(r)\,dr}{A_x(s)} \right) ds \right. \\
&\quad \left. - \int_{-\infty}^{t_2} \phi^{-1}\left(\frac{\tau_x + \int_{-\infty}^{s} F_x(r)\,dr}{A_x(s)} \right) ds \right| \\
&= \left| \int_{t_1}^{t_2} \phi^{-1}\left(\frac{\tau_x + \int_{-\infty}^{s} F_x(r)\,dr}{A_x(s)} \right) ds \right| \\
&\leq \int_{t_1}^{t_2} \phi^{-1}\left(\frac{|\tau_x| + \int_{-\infty}^{s} |F_x(r)|\,dr}{A_x(s)} \right) ds \\
&\leq \int_{t_1}^{t_2} \phi^{-1}\left(\frac{2\int_{-\infty}^{+\infty} \Psi_{\rho_0}(r)dr}{m(s)} \right) ds \\
&\longrightarrow 0, \text{ uniformly as } t_1 \to t_2,
\end{aligned}
$$

and:

$$
\begin{aligned}
|T'_x(t_1) - T'_x(t_2)| &= \left| \phi^{-1}\left(\frac{\tau_x + \int_{-\infty}^{t_1} F_x(r)\,dr}{A_x(t_1)} \right) \right. \\
&\quad \left. - \phi^{-1}\left(\frac{\tau_x + \int_{-\infty}^{t_2} F_x(r)\,dr}{A_x(t_2)} \right) \right| \\
&\longrightarrow 0, \text{ uniformly as } t_1 \to t_2.
\end{aligned}
$$

Therefore, TB is equicontinuous on X.

Claim: *TB is equiconvergent at $\pm\infty$.*

Let $u \in B$. As in the claims above:

$$
\begin{aligned}
\left| T_x(t) - \lim_{t\to-\infty} (T_x(t)) \right| &= \left| \int_{-\infty}^{t} \phi^{-1}\left(\frac{\tau_x + \int_{-\infty}^{s} F_x(r)\,dr}{A_x(s)} \right) ds \right| \\
&\leq \int_{-\infty}^{t} \phi^{-1}\left(\frac{2\int_{-\infty}^{+\infty} \Psi_{\rho_0}(r)dr}{m(s)} \right) ds \\
&\longrightarrow 0, \text{ as } t \to -\infty,
\end{aligned}
$$

and:

$$
\begin{aligned}
\left| T_x(t) - \lim_{t\to+\infty} (T_x(t)) \right| &= \left| \int_{-\infty}^{t} \phi^{-1}\left(\frac{\tau_x + \int_{-\infty}^{s} F_x(r)\,dr}{A_x(s)} \right) ds \right. \\
&\quad \left. - \int_{-\infty}^{+\infty} \phi^{-1}\left(\frac{\tau_x + \int_{-\infty}^{s} F_x(r)\,dr}{A_x(s)} \right) ds \right| \\
&= \left| \int_{t}^{+\infty} \phi^{-1}\left(\frac{\tau_x + \int_{-\infty}^{s} F_x(r)\,dr}{A_x(s)} \right) ds \right| \\
&\leq \int_{t}^{+\infty} \phi^{-1}\left(\frac{2\int_{-\infty}^{+\infty} \Psi_{\eta}(r)dr}{m(s)} \right) ds \\
&\longrightarrow 0, \text{ as } t \to +\infty.
\end{aligned}
$$

Moreover, by (A_1),

$$\begin{aligned}\left|T_x'(t)-\lim_{t\to-\infty}T_x'(t)\right| &= \left|\phi^{-1}\left(\frac{\tau_x+\int_{-\infty}^{t}F_x(r)\,dr}{A_x(t)}\right)-\phi^{-1}\left(\frac{\tau_x}{\lim\limits_{t\to-\infty}A_x(t)}\right)\right|\\ &\leq \left|\phi^{-1}\left(\frac{\tau_x+\int_{-\infty}^{t}\Psi_{\rho_0}(r)dr}{A_x(t)}\right)\right|\\ &\longrightarrow 0,\text{ as } t\to-\infty,\end{aligned}$$

and:

$$\begin{aligned}\left|T_x'(t)-\lim_{t\to+\infty}T_x'(t)\right| &= \left|\phi^{-1}\left(\frac{\tau_x+\int_{-\infty}^{t}F_x(r)\,dr}{A_x(t)}\right)\right.\\ &\quad\left.-\,\phi^{-1}\left(\frac{\tau_x+\int_{-\infty}^{+\infty}F_x(r)\,dr}{\lim\limits_{t\to-\infty}A_x(t)}\right)\right|\\ &\longrightarrow 0,\text{ as } t\to+\infty.\end{aligned}$$

Therefore, TB is equiconvergent at $\pm\infty$, and by Lemma 1, T is compact.

Step 3. *Let $D\subset X$ be a closed and bounded set. Then, $TD\subset D$.*

Consider $D\subset X$ defined as:

$$D=\{x\in X:\|x\|_X\leq\rho_1\},$$

with ρ_1 such that:

$$\rho_1:=\max\left\{|\nu^-|+\int_{-\infty}^{+\infty}\phi^{-1}\left(\frac{K}{m^*(s)}\right)ds,\ \sup_{t\in\mathbb{R}}\phi^{-1}\left(\frac{K}{m^*(t)}\right)\right\},$$

with:

$$K:=2\int_{-\infty}^{+\infty}\Psi_{\rho_1}(r)dr$$

and:

$$m^*(t):=\min_{x\in B}A_x(t).$$

Let $x\in D$. Following similar arguments as in the previous claims, with $m(t)$ given by (6) and $\Omega=D$,

$$\begin{aligned}\|T_x\|_\infty &= \sup_{t\in\mathbb{R}}|T_x(t)|\\ &\leq |\nu^-|+\int_{-\infty}^{+\infty}\phi^{-1}\left(\frac{|\tau_x|+\int_{-\infty}^{s}\Psi_{\rho_1}(r)dr}{A_x(s)}\right)ds\\ &\leq |\nu^-|+\int_{-\infty}^{+\infty}\phi^{-1}\left(\frac{2\int_{-\infty}^{+\infty}\Psi_{\rho_1}(r)dr}{m^*(s)}\right)ds<\rho_1,\end{aligned}$$

and:

$$\begin{aligned} \|T_x'\|_\infty &= \sup_{t\in\mathbb{R}} |T_x'(t)| \le \sup_{t\in\mathbb{R}} \phi^{-1}\left(\frac{|\tau_x| + \int_{-\infty}^{t} |F_x(r)|\,dr}{A_x(t)}\right) \\ &\le \sup_{t\in\mathbb{R}} \phi^{-1}\left(\frac{2\int_{-\infty}^{+\infty} \Psi_{\rho_1}(r)dr}{m^*(t)}\right) < \rho_1. \end{aligned}$$

Therefore, $TD \subset D$. By Schauder's fixed point theorem, T_x has a fixed point in X. That is, there is a heteroclinic solution of the problem (5) and (2). □

To make the relation between the functional problem and the initial one, we apply the lower and upper solution method, according to the following definition:

Definition 2. *A function $\alpha \in X$ is a lower solution of the problem (1) and (2) if $t \mapsto (a(t,\alpha(t))\ \phi(\alpha'(t))) \in W^{1,1}(\mathbb{R})$,*

$$(a(t,\alpha))\ \phi(\alpha'))' \ge f(t,\alpha,\alpha'),\ a.e.\ t \in \mathbb{R}, \tag{15}$$

and:

$$\alpha(-\infty) \le \nu^-,\ \ \alpha(+\infty) \le \nu^+. \tag{16}$$

An upper solution $\beta \in X$ of the problem (1) and (2) satisfies $t \mapsto (a(t,\beta(t))\ \phi(\beta'(t))) \in W^{1,1}(\mathbb{R})$ and the reversed inequalities.

To have some control on the first derivative, we apply a Nagumo-type condition:

Definition 3. *A L^1-Carathéodory function $f : \mathbb{R}^3 \to \mathbb{R}$ satisfies a Nagumo-type growth condition relative to $\alpha, \beta \in X$, with $\alpha(t) \le \beta(t)$, $\forall t \in \mathbb{R}$ if there are positive and continuous functions $\psi, \theta : \mathbb{R} \to \mathbb{R}^+$ such that:*

$$\sup_{t\in\mathbb{R}} \psi(t) < +\infty,\quad \int_0^{+\infty} \frac{|\phi^{-1}(s)|}{\theta(|\phi^{-1}(s)|)} ds = +\infty, \tag{17}$$

and:

$$|f(t,x,y)| \le \psi(t)\ \theta(|y|),\ \text{for a.e. } t \in \mathbb{R} \text{ and } \alpha(t) \le x \le \beta(t). \tag{18}$$

Lemma 2. *Let $f : \mathbb{R}^3 \to \mathbb{R}$ be a L^1-Carathéodory function $f : \mathbb{R}^3 \to \mathbb{R}$ satisfying a Nagumo-type growth condition relative to $\alpha, \beta \in BC(\mathbb{R})$, with $\alpha(t) \le \beta(t)$, $\forall t \in \mathbb{R}$. Then, there exists $N > 0$ (not depending on u) such that for every solution u of (1) and (2) with:*

$$\alpha(t) \le u(t) \le \beta(t),\ \text{for } t \in \mathbb{R}, \tag{19}$$

we have:

$$\|u'\|_\infty < N. \tag{20}$$

Proof. Let u be a solution of (1) and (2) verifying (19). Take $r > 0$ such that:

$$r > \max\left\{|\nu^-|, |\nu^+|\right\}. \tag{21}$$

If $|u'(t)| \le r, \forall t \in \mathbb{R}$, the proof would be complete by taking $N > r$.
Suppose there is $t_0 \in \mathbb{R}$ such that $|u'(t_0)| > N$.

In the case $u'(t_0) > N$, by (17), we can take $N > r$ such that:

$$\int_{a(t,u))\phi(r)}^{a(t,u))\phi(N)} \frac{\left|\phi^{-1}\left(\frac{s}{a(s,u(s))}\right)\right|}{\theta\left(\left|\phi^{-1}\left(\frac{s}{a(s,u(s))}\right)\right|\right)} ds > M\left(\sup_{t\in\mathbb{R}} \beta(t) - \inf_{t\in\mathbb{R}} \alpha(t)\right) \tag{22}$$

with $M := \sup_{t\in\mathbb{R}} \psi(t)$, which is finite by (17).

By (2), there are $t_1, t_2 \in \mathbb{R}$ such that $t_1 < t_2$, $u'(t_1) = N$, $u'(t_2) = r$, and $r \leq u'(t) \leq N, \forall t \in [t_1, t_2]$. Therefore, the following contradiction with (22) holds, by the change of variable $a(t,u)\phi(u'(t)) = s$ and (17):

$$\begin{aligned}
\int_{a(t,u)\phi(r)}^{a(t,u)\phi(N)} \frac{\left|\phi^{-1}\left(\frac{s}{a(s,u(s))}\right)\right|}{\theta\left(\left|\phi^{-1}\left(\frac{s}{a(s,u(s))}\right)\right|\right)} ds &= \int_{a(t,u)\phi(u'(t_2))}^{a(t,u)\phi(u'(t_1))} \frac{\left|\phi^{-1}\left(\frac{s}{a(s,u(s))}\right)\right|}{\theta\left(\left|\phi^{-1}\left(\frac{s}{a(s,u(s))}\right)\right|\right)} ds \\
&= \int_{t_2}^{t_1} \frac{u'(s)}{\theta(u'(s))} \left(\phi\left(u'(s)\right)\right)' ds \\
&= -\int_{t_1}^{t_2} \frac{f(s,u(s),u'(s))}{\theta(u'(s))} u'(s)\, ds \\
&\leq \int_{t_1}^{t_2} \frac{|f(s,u(s),u'(s))|}{\theta(u'(s))} u'(s)\, ds \\
&\leq \int_{t_1}^{t_2} \psi(s)\, u'(s)\, ds \leq M \int_{t_1}^{t_2} u'(s)\, ds \\
&\leq M\left(u(t_2) - u(t_1)\right) \\
&\leq M\left(\sup_{t\in\mathbb{R}} \beta(t) - \inf_{t\in\mathbb{R}} \alpha(t)\right).
\end{aligned}$$

Therefore, $u'(t) < N, \forall t \in \mathbb{R}$.

By similar arguments, it can be shown that $u'(t) > -N, \forall t \in \mathbb{R}$. Therefore, $\|u'\|_\infty < N, \forall t \in \mathbb{R}$. □

The next lemma, in [26], provides a technical tool to use going forward:

Lemma 3. *For $v, w \in C(I)$ such that $v(x) \leq w(x)$, for every $x \in I$, define:*

$$q(x,u) = max\{v, min\{u, w\}\}.$$

Then, for each $u \in C^1(I)$, the next two properties hold:

(a) $\frac{d}{dx}q(x,u(x))$ *exists for a.e.$x \in I$.*

(b) *If $u, u_m \in C^1(I)$ and $u_m \to u$ in $C^1(I)$, then:*

$$\frac{d}{dx}q(x,u_m(x)) \to \frac{d}{dx}q(x,u(x)) \text{ for a.e. } x \in I.$$

The main result will be given by the next theorem:

Theorem 2. *Suppose that $f : \mathbb{R}^3 \to \mathbb{R}$ is a L^1-Carathéodory function verifying a Nagumo-type condition and hypotheses (H_1), (H_2), and (8). If there are lower and upper solutions of the problem (1) and (2), α and β, respectively, such that:*

$$\alpha(t) \leq \beta(t),\ \forall t \in \mathbb{R},$$

then there is a function $u \in X$ with $t \mapsto (a(t,u(t))\,\phi(u'(t))) \in W^{1,1}(\mathbb{R})$, the solution of the problem (1) and (2) and:

$$\alpha(t) \leq u(t) \leq \beta(t),\ \forall t \in \mathbb{R}.$$

Proof. Define the truncation operator $Q : W^{1,1}(\mathbb{R}) \to X \subset W^{1,1}(\mathbb{R})$ given by:

$$Q(x) := Q_x(t) = \begin{cases} \beta(t), & x(t) > \beta(t) \\ x(t), & \alpha(t) \le x(t) \le \beta(t) \\ \alpha(t), & x(t) < \alpha(t). \end{cases}$$

Consider the modified equation:

$$\begin{aligned} \left(a(t, Q_u)) \phi \left(\frac{d}{dt} Q_u\right)\right)' &= f\left(t, Q_u(t), \frac{d}{dt} Q_u(t)\right) \\ &+ \frac{1}{1+t^2} \frac{u(t) - Q_u(t)}{1 + |u(t) - Q_u(t)|}, \end{aligned} \tag{23}$$

for *a.e.* $t \in \mathbb{R}$, which is well defined by Lemma 3.

Claim 1: *Every solution $u(t)$ of the problem (23) and (2) verifies:*

$$\alpha(t) \le u(t) \le \beta(t), \ \forall t \in \mathbb{R}.$$

Let u be a solution of the problem (23) and (2), and suppose, by contradiction, that there is t_0 such that $\alpha(t_0) > u(t_0)$. Remark that, by (16), $t_0 \neq \pm\infty$ as $u(\pm\infty) - \alpha(\pm\infty) \ge 0$.

Define:

$$\min_{t \in \mathbb{R}} (u(t) - \alpha(t)) := u(t_1) - \alpha(t_1) < 0.$$

Therefore, there is an interval $]t_2, t_1]$ such that $u(t) - \alpha(t) < 0$, for a.e. $t \in]t_2, t_1]$, and by (15), this contradiction is achieved:

$$\begin{aligned} (a(t,\alpha)\, \phi(\alpha'))' &= \left(a(t, Q_u(t))\, \phi\left(\frac{d}{dt} Q_u(t)\right)\right)' \\ &= f\left(t, Q_u(t), \frac{d}{dt} Q_u(t)\right) + \frac{1}{1+t^2} \frac{u(t) - Q_u(t)}{1 + |u(t) - Q_u(t)|} \\ &< f(t, \alpha(t), \alpha'(t)) \le (a(\alpha(t))\, \phi(\alpha'(t)))'. \end{aligned}$$

Therefore, $\alpha(t) \le u(t)$, $\forall t \in \mathbb{R}$. Following similar arguments, it can be proven that $u(t) \le \beta(t)$, $\forall t \in \mathbb{R}$.

Claim 2: *The problem (23) and (2) has a solution.*

Let $A : X \to C(\mathbb{R})$ and $F : X \to L^1(\mathbb{R})$ be the operators given by $A_x := a(t, Q_x(t))$ and:

$$F_x := f\left(t, Q_x(t), \frac{d}{dt} Q_x(t)\right) + \frac{1}{1+t^2} \frac{u(t) - Q_x(t)}{1 + |u(t) - Q_x(t)|}.$$

As, for:

$$\rho := \max\left\{ \|\alpha\|_\infty, \|\beta\|_\infty, \|\alpha'\|_\infty, \|\beta'\|_\infty, N \right\},$$

with N given by (20),

$$\begin{aligned} |F_x| &\le \left| f\left(t, Q_x(t), \frac{d}{dt} Q_x(t)\right) \right| + \frac{1}{1+t^2} \frac{|u(t) - Q_x(t)|}{1 + |u(t) - Q_x(t)|} \\ &\le \left| f\left(t, Q_x(t), \frac{d}{dt} Q_x(t)\right) \right| \le \varphi_\rho(t), \end{aligned}$$

then F_x verifies (F_1). Moreover, from:

$$a(t, Q_x(t)) \ge \min_{t \in \mathbb{R}} \{a(t,\alpha)), a(t,\beta)\},$$

we have that A satisfies (A_1) with $0 < m(t) \leq \min_{t\in\mathbb{R}} \{a(t,\alpha), a(t,\beta)\}$.

Therefore, by Schauder's fixed point theorem, the problem (23) and (2) has a solution, which, by Claim 1, is a solution of the problem (1) and (2). □

4. Example

Consider the boundary value problem, defined on the whole real line, composed by the differential equation:

$$\left[(t^2+1)^3\left((u)^4+1\right)(u')^3\right]' = \frac{1}{10000}\frac{\left[(u(t))^2-1\right]\ (u'(t))^2}{1+t^2},\ a.e.t \in \mathbb{R}, \tag{24}$$

coupled with the boundary conditions:

$$u(-\infty) = -1,\ u(+\infty) = 1. \tag{25}$$

Remark that the null function is not solution of the problem (24) and (25), which is a particular case of (1) and (2), with:

$$\begin{aligned} \phi(w) &= w^3, \\ a(t,x) &= (t^2+1)^3\left(x^4+1\right), \\ f(t,x,y) &= \frac{1}{10000}\frac{(x^2-1)\ y^2}{1+t^2}, \\ v^- &= -1, \text{ and } v^+ = 1. \end{aligned}$$

All hypotheses of Theorem 2 are satisfied. In fact:

- f is a L^1-Carathéodory function with:

$$\varphi_\rho(t) = \frac{1}{10000}\frac{(\rho^2+1)\ \rho^2}{1+t^2};$$

- $\phi(w)$ verifies (H_1), and function $a(t,x)$ satisfies (H_2);
- the constant functions $\alpha(t) \equiv -1$ and $\beta(t) \equiv k$, with $k \in]1,+\infty[$, are lower and upper solutions of the problem (24) and (25), respectively.
- $f(t,x,y)$ verifies (8) for $\rho > 1.54$ and satisfies a Nagumo-type condition for $-1 \leq x \leq k$ with:

$$\psi(t) = \frac{1}{10000}\frac{k}{1+t^2} \text{ and } \theta(y) = y^2.$$

Therefore, by Theorem 2, there is a heteroclinic connection u between two equilibrium points -1 and one of the problem (24) and (25), such that:

$$-1 \leq u(t) \leq k,\ \forall t \in \mathbb{R},\ k \geq 1.$$

5. Singular ϕ-Laplacian Equations

The previous theory can be easily adapted to singular ϕ-Laplacian equations, that is for equations:

$$\left(a\left(t,u\right)\phi\left(u'\right)\right)' = f\left(t,u,u'\right),\ a.e.t \in \mathbb{R}, \tag{1s}$$

where ϕ verifies:

(H_s) $\phi : (-b, b) \to \mathbb{R}$, for some $0 < b < +\infty$, is an increasing homeomorphism with $\phi(0) = 0$ and $\phi(-b, b) = \mathbb{R}$ such that:

$$\left|\phi^{-1}(w)\right| \leq \phi^{-1}(|w|);$$

In this case, a heteroclinic solution of (1s), that is a solution for the problem (1s) and (2), is a function $u \in X$ such that $u'(t) \in (-b, b)$, for $t \in \mathbb{R}$, and $t \mapsto (a\,(t, u)\,\phi\,(u')) \in W^{1,1}(\mathbb{R})$, satisfying (1s) and (2).

The theory for singular ϕ-Laplacian equations is analogous to Theorems 1 and 2, replacing the assumption (H_1) by (H_s).

As an example, we can consider the problem, for $n \in \mathbb{N}$ and $k > 0$,

$$\begin{cases} \left(\left(1+t^2\right)\left(1+(u)^{2n}\right)\frac{u'}{\sqrt{1-(u')^2}}\right)' = \frac{((u)^2-1)(|u'|+1)}{1000(1+t^2)}, \ a.e.t \in \mathbb{R}, \\ \\ u(-\infty) = -1, \ \ u(+\infty) = 1. \end{cases} \tag{26}$$

Clearly, Problem (26) is a particular case of (1) and (2), with:

$$\phi(w) = \frac{w}{\sqrt{1-w^2}}, \text{ for } w \in (-1, 1),$$

which models mechanical oscillations under relativistic effects,

$$a(t, x) = \left(1+t^2\right)\left(1+x^{2n}\right), \tag{27}$$

$$f(t, x, y) = \frac{\left(x^2-1\right)\ (|y|+1)}{1000\,(1+t^2)}, \tag{28}$$

$$\nu^- = -1, \text{ and } \nu^+ = 1.$$

Moreover, the nonlinearity f given by (28) is a L^1-Carathéodory function with:

$$\varphi_\rho(t) = \frac{\left(\rho^2+1\right)\ (\rho+1)}{1000\,(1+t^2)}.$$

The conditions of Theorem 2 are satisfied with (H_1) replaced by (H_s), as:

- the function $a(t, x)$, defined by (27), verifies (H_2) ;
- the constant functions $\alpha(t) \equiv -1$ and $\beta(t) \equiv 1$ are lower and upper solutions of Problem (26), respectively.
- $f(t, x, y)$ verifies (8) for $\rho \in [1.09, 5.91]$ and satisfies a Nagumo-type condition for $-1 \leq x \leq 1$ with:

$$\psi(t) = \frac{1}{1000} \text{ and } \theta(y) = |y| + 1.$$

Therefore, there is a heteroclinic connection u between two equilibrium points -1 and one, for the singular ϕ-Laplacian problem (26), such that:

$$-1 \leq u(t) \leq 1, \ \forall t \in \mathbb{R}.$$

6. Conclusions

As can be seen in the Introduction, sufficient conditions for the existence of heteroclinic solutions require strong assumptions on the nonlinearities. The goal of this paper is to weaken these conditions on the nonlinearity f, replacing them by assumptions on the inverse of the homeomorphism ϕ, following the ideas and methods suggested in [27,28].

7. Discussion

The present result guarantees the existence of heteroclinic solutions for a broader set of nonlinearities, without "asking too much" of the homeomorphism ϕ.

However, it is the author's feeling that Condition (8) can be improved, applying other techniques and method. These are, in my opinion, the next steps for the research in this direction.

Funding: This research received no external funding.

Conflicts of Interest: The author declares no conflict of interest.

References

1. Cupini, G.; Marcelli, C.; Papalini, F. On the solvability of a boundary value problem on the real line. *Bound. Value Probl.* **2011**, *2011*, 26. [CrossRef]
2. Marcelli, C. The role of boundary data on the solvability of some equations involving non-autonomous nonlinear differential operators. *Bound. Value Probl.* **2013**, *2013*, 252. [CrossRef]
3. Minhós, F. Sufficient conditions for the existence of heteroclinic solutions for *φ-Laplacian differential equations. Complex Var. Elliptic Equ.* **2017**, *62*, 123–134. [CrossRef]
4. Graef, J.; Kong, L.; Minhós, F. Higher order boundary value problems with *Φ-Laplacian and functional boundary conditions. Comput. Math. Appl.* **2011**, *61*, 236–249. [CrossRef]
5. Graef, J.; Kong, L.; Minhós, F.; Fialho, J. On the lower and upper solutions method for higher order functional boundary value problems. *Appl. Anal. Discret. Math.* **2011**, *5*, 133–146. [CrossRef]
6. Alves, C.O. Existence of heteroclinic solutions for a class of non-autonomous second-order equation. *Nonlinear Differ. Equ. Appl.* **2015**, *2*, 1195–1212. [CrossRef]
7. Bianconi, B.; Papalini, F. Non-autonomous boundary value problems on the real line. *Discret. Contin. Dyn. Syst.* **2006**, *15*, 759–776.
8. Calamai, A. Heteroclinic solutions of boundary value problems on the real line involving singular Φ-Laplacian operators. *J. Math. Anal. Appl.* **2011**, *378*, 667–679. [CrossRef]
9. Izydorek, M.; Janczewska, J. Homoclinic solutions for a class of the second order Hamiltonian systems. *J. Differ. Equ.* **2005**, *219*, 375–389. [CrossRef]
10. Liu, Y. Existence of Solutions of Boundary Value Problems for Coupled Singular Differential Equations on Whole Lines with Impulses. *Mediterr. J. Math.* **2015**, *12*, 697–716. [CrossRef]
11. Marcelli, C. Existence of solutions to boundary value problems governed by general non-autonomous nonlinear differential operators. *Electron. J. Differ. Equ.* **2012**, *2012*, 1–18.
12. Marcelli, C.; Papalini, F. Heteroclinic connections for fully non-linear non-autonomous second-order differential equations. *J. Differ. Equ.* **2007**, *241*, 160–183. [CrossRef]
13. Sun, J.; Chen, H.; Nieto, J.J. Homoclinic solutions for a class of subquadratic second-order Hamiltonian systems. *J. Math. Anal. Appl.* **2011**, *373*, 20–29. [CrossRef]
14. Tersian, S.; Chaparova, J. Periodic and Homoclinic Solutions of Extended Fisher–Kolmogorov Equations. *J. Math. Anal. Appl.* **2001**, *260*, 490–506. [CrossRef]
15. Wang, J.; Xu, J.; Zhang, F. Homoclinic orbits for a class of Hamiltonian systems with superquadratic or asymptotically quadratic potentials. *Commun. Pure Appl. Anal.* **2011**, *10*, 269–286. [CrossRef]
16. Smyrnelis, P. Minimal heteroclinics for a class of fourth order O.D.E. systems. *Nonlinear Anal.* **2018**, *173*, 154–163. [CrossRef]
17. Cabada, A.; Cid, J.Á. Heteroclinic solutions for non-autonomous boundary value problems with singular Φ-Laplacian operators. *Discret. Contin. Dyn. Syst.* **2009**, *2009*, 118–122.
18. Cabada, A.; Pouso, R.L. Existence results for the problem $(\phi(u'))' = f(t, u, u')$ with nonlinear boundary conditions. *Nonlinear Anal.* **1999**, *35*, 221–231. [CrossRef]
19. Fabry, C.; Manásevich, R. Equations with a p-Laplacian and an asymmetric nonlinear term. *Discret. Contin. Dyn. Syst.* **2001**, *7*, 545–557.
20. Ferracuti, L.; Papalini, F. Boundary value problems for strongly nonlinear multivalued equations involving different Φ-Laplacians. *Adv. Differ. Equ.* **2009**, *14*, 541–566.

21. Gilding, B.H.; Kersner, R. *Travelling Waves in Nonlinear Diffusion–Convection–Reaction*; Birkhäuser: Basel, Switzerland, 2004.
22. Maini, P.K.; Malaguti, L.; Marcelli, C.; Matucci, S. Diffusion-aggregation processes with mono-stable reaction terms. *Discret. Contin. Dyn. Syst. (B)* **2006**, *6*, 1175–1189.
23. Papageorgiou, N.S.; Papalini, F. Pairs of positive solutions for the periodic scalar p-Laplacian. *J. Fixed Point Theory* **2009**, *5*, 157–184. [CrossRef]
24. del Pino, M.; Elgueta, M.; Manásevich, R. A homotopic deformation along p of a Leray-Schauder degree result and existence for $(|u'|^{p-2}u')' + f(t,u) = 0$, $u(0) = u(T) = 0$, $p > 1$. *J. Differ. Equ.* **1989**, *80*, 1–13. [CrossRef]
25. Corduneanu, C. *Integral Equations and Stability of Feedback Systems*; Academic Press: New York, NY, USA, 1973.
26. Wang, M.X.; Cabada, A.; Nieto, J.J. Monotone method for nonlinear second order periodic boundary value problems with Carathéodory functions. *Ann. Polon. Math.* **1993**, 58, 221–235. [CrossRef]
27. Minhós, F.; de Sousa, R. Existence of homoclinic solutions for second order coupled systems. *J. Differ. Equ.* **2019**, *266*, 1414–1428. [CrossRef]
28. Minhós, F. On heteroclinic solutions for BVPs involving 3c6-Laplacian operators without asymptotic or growth assumptions. *Math. Nachr.* **2018**, 1–9. [CrossRef]

axioms

MDPI

Article

Lipschitz Stability for Non-Instantaneous Impulsive Caputo Fractional Differential Equations with State Dependent Delays

Ravi Agarwal [1,2], Snezhana Hristova [3,*] and Donal O'Regan [4]

1 Department of Mathematics, Texas A&M University-Kingsville, Kingsville, TX 78363, USA; agarwal@tamuk.edu
2 Distinguished University Professor of Mathematics, Florida Institute of Technology, Melbourne, FL 32901, USA
3 Department of Applied Mathematics and Modeling, University of Plovdiv "Paisii Hilendarski", 4000 Plovdiv, Bulgaria
4 School of Mathematics, Statistics and Applied Mathematics, National University of Ireland, H91 CF50 Galway, Ireland; donal.oregan@nuigalway.ie
* Correspondence: snehri@gmail.com

Received: 21 November 2018; Accepted: 25 December 2018; Published: 29 December 2018

Abstract: In this paper, we study Lipschitz stability of Caputo fractional differential equations with non-instantaneous impulses and state dependent delays. The study is based on Lyapunov functions and the Razumikhin technique. Our equations in particular include constant delays, time variable delay, distributed delay, etc. We consider the case of impulses that start abruptly at some points and their actions continue on given finite intervals. The study of Lipschitz stability by Lyapunov functions requires appropriate derivatives among fractional differential equations. A brief overview of different types of derivative known in the literature is given. Some sufficient conditions for uniform Lipschitz stability and uniform global Lipschitz stability are obtained by an application of several types of derivatives of Lyapunov functions. Examples are given to illustrate the results.

Keywords: non-instantaneous impulses; Caputo fractional derivative; differential equations; state dependent delays; lipschitz stability

AMS Subject Classifications: 34A37, 34K20, 34K37

1. Introduction

Many papers in the literature study stability of solutions of differential equations via Lyapunov functions. One type of stability, useful in real world problems, is the so-called Lipschitz stability and Dannan and Elaydi [1] introduced the notion of Lipschitz stability for ordinary differential equations. As noted in [1], this type of stability is important only for nonlinear problems since it coincides with uniform stability in linear systems. Based on theoretical results for Lipschitz stability in [1], the dynamic behavior of a spacecraft when a single magnetic torque-rod is used for achieving a pure spin condition is studied in [2]. Recently, stability properties of delay fractional differential equations without any type of impulse are considered and we refer the reader to [3] and the references therein.

In this paper, we study the Lipschitz stability for a nonlinear system of non-instantaneous impulsive fractional differential equations with state dependent delay (NIFrDDE). The impulses start abruptly at some points and their actions continue on given finite intervals. Non-instantaneous impulsive differential equations were introduced by Hernandez and O'Regan in 2013 (see, for example, [4]). The systematic description of solutions of both ordinary and Caputo fractional differential equations with non-instantaneous impulse and without delays is given in the monograph [5]. In addition, some results for non-instantaneous fractional equations without any type of delay are presented in [6–8]. In [9], Caputo fractional differential equations with time varying delays is considered (we note that the model had no impulses). However, in this paper, for the first time, we consider together

1. Lipschitz stability;
2. state dependent delays (note a special case is time varying delays); and
3. models with non-instantaneous impulses.

There are two different approaches in the literature for the interpretation of the solution of fractional differential equations with impulses (for more details, see [6] and Chapter 2 of the book [5]). In the first interpretation, the lower limit of the fractional derivative is one and the same on the whole interval of study and at each point of jump we consider a boundary value problem defined by the impulsive function. In the second interpretation, the lower limit of the fractional derivative changes at each time of jump with the idea of considering an initial value problem at each jump point.

In this paper, we use the second approach to study Lipschitz stability properties of nonlinear non-instantaneous impulsive delay differential equations. The delays are bounded and depend on both the time and the state. Note several stability properties are studied in the literature for Caputo fractional differential equations (for example, see [10] (without delays), [3] (with delays and no impulses), and [11] (with multiple discrete delays without impulses)). Our study is based on Lyapunov functions and the Razumikhin technique. A brief overview in the literature of different types of derivatives of Lyapunov functions among the studied fractional differential equation is given. Several sufficient conditions for uniform Lipschitz stability and global uniform Lipschitz stability are obtained by an application of these derivatives. Some examples illustrating the results are given.

2. Notes on Fractional Calculus

We give the main definition of fractional derivatives used in the literature (see, for example, [12–14]). We give these definitions for scalar functions. Throughout the paper, we assume $q \in (0,1)$.

- *Riemann–Liouville (RL) fractional derivative* :

$$ {}_{t_0}^{RL}D_t^q m(t) = \frac{1}{\Gamma(1-q)} \frac{d}{dt} \int_{t_0}^{t} (t-s)^{-q} m(s) ds, \; t \geq t_0 $$

  where $\Gamma(.)$ denotes the Gamma function.
- *Caputo fractional derivative*

$$ {}_{t_0}^{C}D_t^q m(t) = \frac{1}{\Gamma(1-q)} \int_{t_0}^{t} (t-s)^{-q} m'(s) ds, \quad t \geq t_0. $$

  Note that for a constant m the equality ${}_{t_0}^{C}D_t^q m = 0$ holds. However, for any given t^* ,we denote ${}_{t_0}^{C}D_t^q m(t^*) = {}_{t_0}^{C}D_t^q m(t)|_{t=t^*}$.
- *The Grünwald–Letnikov fractional derivative* is given by

$$ {}_{t_0}^{GL}D_t^q m(t) = \lim_{h \to 0} \frac{1}{h^q} \sum_{r=0}^{[\frac{t-t_0}{h}]} (-1)^r \, {}_qC_r \, m(t-rh), \quad t \geq t_0 $$

and the Grünwald–Letnikov fractional Dini derivative by

$$ {}_{t_0}^{GL}D_+^q m(t) = \limsup_{h\to 0+} \frac{1}{h^q} \sum_{r=0}^{[\frac{t-t_0}{h}]} (-1)^r \, {}_qC_r m(t-rh), \quad t \ge t_0, $$

where ${}_qC_r = \frac{q(q-1)\dots(q-r+1)}{r!}$ and $[\frac{t-t_0}{h}]$ denotes the integer part of the fraction $\frac{t-t_0}{h}$.

From the relation between the Caputo fractional derivative and the Grünwald–Letnikov fractional derivative using Equation (1), we define the Caputo fractional Dini derivative of a function as

$$ {}_{t_0}^{C}D_+^q m(t) = {}_{t_0}^{GL}D_+^q [m(t) - m(t_0)], $$

i.e.,

$$ {}_{t_0}^{C}D_+^q m(t) = \limsup_{h\to 0+} \frac{1}{h^q} \Big[m(t) - m(t_0) - \sum_{r=1}^{[\frac{t-t_0}{h}]} (-1)^{r+1} \binom{q}{r} \Big(m(t-rh) - m(t_0) \Big) \Big]. $$

The fractional derivatives for scalar functions could be easily generalized to the vector case by taking fractional derivatives with the same fractional order for all components.

3. Statement of the Problem and Basic Definitions

Let the positive constant r be given and the points $\{t_i\}_1^\infty, \{s_i\}_1^\infty$ be such that $0 < s_i < t_i < s_{i+1}$, $i = 1, 2, \dots$. Let $t_0 \ge 0$ be the given initial time. Without loss of generality, we can assume $t_0 \in [0, s_1)$.

Consider the space PC_0 of all functions $y : [-r, 0] \to \mathbb{R}^n$, which are piecewise continuous endowed with the norm $||y||_{PC_0} = \sup_{t\in[-r,0]}\{||y(t)|| : y \in PC_0\}$ where $||.||$ is a norm in $\mathbb{R}^n$.

The intervals (t_i, s_{i+1}), $i = 0, 1, 2, \dots$ are the intervals on which the fractional differential equations are given and on the intervals (s_i, t_i), $i = 1, 2, \dots$ the impulsive conditions are given.

The Caputo fractional derivative has a memory and it depends significantly on its lower derivative. This property as well as the meaning of impulses in the differential equation lead to two basic approaches to Caputo fractional differential equations with non-instantaneous impulses:

- Unchangeable lower limit of the Caputo fractional derivative: the lower limit of the fractional derivative is equal to the initial time t_0 on the whole interval of consideration.
- Changeable lower limit of the Caputo fractional derivative: the lower limit of the fractional derivative is equal to the left end t_i on the interval $(t_i, s_{i+1}), i = 0, 1, 2, \dots$ without impulses.

In this paper, we study the case of changeable lower limit of the Caputo fractional derivative.

Consider the initial value problem (IVP) for a nonlinear system of non-instantaneous impulsive fractional differential equations with state dependent delay (NIFrDDE) with $q \in (0, 1)$:

$$ \begin{aligned} &{}_{t_i}^{C}D_t^q x(t) = f(t, x(t), x_{\rho(t,x_t)}) \ \text{ for } t \in (t_i, s_{i+1}], i = 0, 1, 2, \dots, \\ &x(t) = \phi_i(t, x(s_i)), \ \ t \in (s_i, t_i], \ i = 1, 2, \dots, \\ &x(t + t_0) = \varphi(t) \ \text{ for } t \in [-r, 0], \end{aligned} \tag{1} $$

where $x \in \mathbb{R}^n$, ${}_{t_i}^{C}D_t^q x(t)$ denotes the Caputo fractional derivative with lower limit t_i for the state $x(t)$, the functions $f : [0, s_1] \bigcup_{i=1}^\infty [t_i, s_{i+1}] \times R^n \times PC_0 \to \mathbb{R}^n$; $\rho : [0, s_1] \bigcup_{i=1}^\infty [t_i, s_{i+1}] \times PC_0 \to \mathbb{R}$, $\varphi \in PC_0$; $\phi_i : [s_i, t_i] \times \mathbb{R}^n \to \mathbb{R}^n$, $i = 1, 2, \dots$. Here, $x_t(s) = x(t+s), s \in [-r, 0]$, i.e., represents the history of the state from time $t - r$ up to the present time t. Note that for any $t \ge 0$ we let $x_{\rho(t,x_t)} = x(\rho(t, x(t+s))), s \in [-r, 0]$, i.e., the function ρ determines the state-dependent delay. Note, the integer order differential equations with non-instantaneous impulses and state dependent delay are studied in [15].

Let $\mathcal{PC}[t_0, \infty)$ be the space of all functions $y : [t_0 - r, \infty) \to \mathbb{R}^n$ which are piecewise continuous on $[t_0 - r, \infty)$ with points of discontinuity s_i, $i = 1, 2, \dots$, the limits $y(s_i - 0) = \lim_{t\to s_i,\, t<s_i} y(t) = y(s_i)$

and $y(s_i+) = \lim_{t\to s_i,\, t>s_i} y(t)$ exist, for any $t \in (t_i, s_i]$ the Caputo fractional derivative ${}_{t_i}^{C}D_t^q y(t), i = 0,1,\dots,$ exists and it is endowed with the norm $||y||_{\mathcal{PC}} = \sup_{t\in[t_0-r,\infty)}\{||y(t)|| : y \in \mathcal{PC}[t_0,\infty)\}$ where $||.||$ is a norm in $\mathbb{R}^n$.

Define the set $PC^q[t_0,\infty) = \{y \in C(\bigcup_{i=0}^{\infty}(t_i,s_i],\mathbb{R}^n)$ such that for any $t \in (t_i,s_i] : \int_{t_i}^{t}(t-s)^{q-1}y(s)ds < \infty,\ i = 1,2,\dots\}$.

We introduce the assumptions:

A1. The function $f \in C([0,s_1]\bigcup_{i=1}^{\infty}[t_i,s_{i+1}] \times \mathbb{R}^n \times PC_0, \mathbb{R}^n)$ is such that for any $y \in \mathbb{R}^n, u \in PC_0$ the inclusion $f(.,y,u) \in PC^q[0,\infty)$ holds.

A2. The function $\rho \in C([0,s_1]\bigcup_{i=1}^{\infty}[t_i,s_{i+1}] \times PC_0, [-r,\infty))$ and for any $(t,y) \in \bigcup_{i=0}^{k}[t_i,s_{i+1}] \times PC_0$ the inequalities $t-r \le \rho(t,y) \le t$ holds.

A3. The functions $\phi_i \in C([s_i,t_i] \times \mathbb{R}^n, \mathbb{R}^n),\ i = 1,2,\dots$.

A4. The function $\varphi \in PC_0$.

A5. The function $f(t,0) = 0$ for $t \in [0,s_1]\bigcup_{i=1}^{\infty}[t_i,s_{i+1}]$ and $\phi_i(t,0) = 0$ for $t \in [s_i,t_i],\ i = 1,2,\dots$.

Remark 1. *Assumption A5 guarantees the existence of the zero solution of IVP for NIFrDDE (Equation (1)) with the zero initial function* $\varphi \equiv 0$.

Remark 2. *Assumption A2 guarantees the delay of the argument in Equation (1).*

Definition 1. *Let the conditions A1–A4 be satisfied. The function* $x \in \mathcal{PC}[t_0,\infty)$ *is a solution of the IVP in Equation (1) iff it satisfies the following integral-algebraic equation*

$$x(t) = \begin{cases} \varphi(t), & t \in [-r,0], \\ \varphi(0) + \frac{1}{\Gamma(q)}\int_0^t (t-s)^{q-1} f(s,x(s),x_{\rho(s,x_s)})ds, & t \in (0,s_1], \\ \phi_i(t,x(s_i)), & t \in (s_i,t_i],\ i = 1,2,\dots, \\ \phi_i(t_i,x(s_i)) + \frac{1}{\Gamma(q)}\int_{t_i}^t (t-s)^{q-1} f(s,x(s),x_{\rho(s,x_s)})ds, & t \in (t_i,s_{i+1}],\ i = 1,2,\dots. \end{cases} \quad (2)$$

Definition 2. *The functions* f,ρ *are defined only on the intervals without impulses on which the differential equation is given.*

We generalize Lipschitz stability ([1]) for ordinary differential equations to systems of Caputo fractional non-instantaneous impulsive differential equations with state dependent delay.

Definition 3. *The zero solution of NIFrDDE (Equation (1)) is said to be:*

- *Uniformly Lipschitz stable if there exists* $M \ge 1$ *and* $\delta > 0$ *such that, for any for any initial time* $t_0 \in [0,s_1]\bigcup_{k=1}^{\infty}[t_k,s_k]$ *and any initial function* $\varphi \in PC_0$, *the inequality* $||\varphi||_{PC_0} < \delta$ *implies* $||x(t;t_0,\varphi)|| \le M||\varphi||_{PC_0}$ *for* $t \ge t_0$ *where* $x(t;t_0,\varphi)$ *is a solution of Equation (1).*
- *Globally uniformly Lipschitz stable if there exists* $M \ge 1$ *such that, for any initial time* $t_0 \in [0,s_1]\bigcup_{k=1}^{\infty}[t_k,s_k]$ *and any initial function* $\varphi \in PC_0$, *the inequality* $||\varphi||_{PC_0} < \infty$ *implies* $||x(t;t_0,\varphi)|| \le M||\varphi||_{PC_0}$ *for* $t \ge t_0$.

Let $J \subset \mathbb{R}_+,\ 0 \in J, \rho > 0$. Consider the following sets:

$$\begin{aligned} M(J) &= \{a \in C[J,\mathbb{R}^+] : a(0) = 0, \text{ a(r) is strictly increasing in } J, \text{ and} \\ &\quad a^{-1}(\alpha r) \le r q_a(\alpha) \text{ for some function } q_a : q_a(\alpha) \ge 1, \text{ if } \alpha \ge 1\}, \\ K(J) &= \{a \in C[J,\mathbb{R}^+] : a(0) = 0, \text{ a(r) is strictly increasing in } J, \text{ and} \\ &\quad a(r) \le K_a r \text{ for some constant } K_a > 0\}, \\ S_\rho &= \{x \in \mathbb{R}^n : ||x|| \le \rho\}. \end{aligned}$$

Remark 3. *The function* $a(u) = K_1 u$, $K_1 > 0$ *is from the class* $K(\mathbb{R}_+)$ *with* $K_a = K_1$. *The function* $a(u) = K_2 u^2$, $K_2 \in (0,1]$ *is from the class* $M([1,\infty))$ *with* $q(u) = \sqrt{\frac{u}{K_2}} \geq 1$ *for* $u \geq 1$.

4. Lyapunov Functions and Their Derivatives among Nonlinear Non-Instantaneous Caputo Delay Fractional Differential Equations

One approach to study Lipschitz stability of solutions of Equation (1) is based on using Lyapunov-like functions. The first step is to define a Lyapunov function. The second step is to define its derivative among the fractional equation.

We use the class Λ of Lyapunov-like functions, defined and used for impulsive differential equations in [16].

Definition 4. *Let* $J \in \mathbb{R}_+$ *be a given interval, and* $\Delta \subset \mathbb{R}^n$ *be a given set. We say that the function* $V(t,x) : J \times \Delta \to \mathbb{R}_+$, *belongs to the class* $\Lambda(J,\Delta)$ *if*

- *The function* $V(t,x)$ *is continuous on* $J/\{s_k \in J\} \times \Delta$ *and it is locally Lipschitz with respect to its second argument.*
- *For each* $s_k \in J$ *and* $x \in \Delta$, *there exist finite limits*

$$V(s_k,x) = V(s_k - 0,x) = \lim_{t \uparrow s_k} V(t,x) \text{ and } V(s_k + 0,x) = \lim_{t \downarrow s_k} V(t,x).$$

In connection with the Caputo fractional derivative, it is necessary to define in an appropriate way the derivative of Lyapunov functions among the studied equation. We give a brief overview of the derivatives of Lyapunov functions among solutions of fractional differential equations known and used in the literature. There are mainly three types of derivatives of Lyapunov functions from the class $\Lambda(J,\Delta)$ used in the literature to study stability properties of solutions of Caputo fractional differential in Equation (1):

- First type: **the Caputo fractional derivative** of the function $V(t,x(t)) \in \Lambda([a,b),\Delta)$ defined by

$$ {}_{t_k}^{c}D^q V(t,x(t)) = \frac{1}{\Gamma(1-q)} \int_{t_k}^{t} (t-s)^{-q} \frac{d}{ds}\Big(V(s,x(s))\Big) ds, \quad t \in [t_k, s_{k+1}) \tag{3}$$

  where $x(t)$ is a solution of Equation (1).
- Second type: **Dini fractional derivative** of the Lyapunov function $V \in \Lambda([t_0,\infty),\mathbb{R}^n)$ among Equation (1): Let $\phi \in PC_0$ and $t \in (t_k, s_{k+1})$ for a non-negative integer k. Then,

$$D^+_{(1)} V(t,\phi(0),t_k,\phi) = \limsup_{h\to 0} \frac{1}{h^q}\Big[V(t,\phi(0)) - \sum_{r=1}^{[\frac{t-t_k}{h}]} (-1)^{r+1} {}_qC_r V(t-rh, \phi(0) - h^q f(t,\phi(0),\phi(\rho(t,\phi_0)-t)))\Big] \tag{4}$$

  where $\phi_0(s) = \psi(s)$ and $\phi(\rho(t,\phi_0) - t) = \phi(\rho(t,\phi(s)) - t)$ for any $s \in [-r,0]$. We note that, because of Assumption A2, the inequality $t - r < \rho(t,\phi(s)) < t$ holds ($-r < \rho(t,\phi(s)) - t < 0$), i.e $\phi(\rho(t,\phi(s)) - t)$ is well defined.

  The derivative of Equation (4) keeps the concept of fractional derivatives because it has a memory.

- <u>*Third type*</u>: **Caputo fractional Dini derivative** of a Lyapunov function $V \in \Lambda([t_0,\infty),\mathbb{R}^n)$ among Equation (1): Let the initial function $\varphi \in PC_0$ be given and the function $\phi \in PC_0$ and $t \in (t_k, s_{k+1})$ for a non-negative integer k. Then,

$$\begin{aligned} {}^{c}_{(1)}D^q_+V(t,\phi;t_k,\varphi(0)) &= \limsup_{h\to 0^+}\frac{1}{h^q}\Big\{V(t,\phi(0)) - V(t_k,\varphi(0)) \\ &- \sum_{r=1}^{[\frac{t-t_k}{h}]}(-1)^{r+1}\,{}_qC_r\Big(V(t-rh,\phi(0)-h^q f(t,\phi(0),\phi(\rho(t,\phi_0)-t))) - V(t_k,\varphi(0))\Big)\Big\}, \end{aligned} \tag{5}$$

or its equivalence

$$\begin{aligned} &{}^{c}_{(1)}D^q_+V(t,\phi;t_k,\varphi(0)) = \\ &\limsup_{h\to 0^+}\frac{1}{h^q}\Big\{V(t,\phi(0)) + \sum_{r=1}^{[\frac{t-t_k}{h}]}(-1)^r\,{}_qC_rV(t-rh,\phi(0)-h^q f(t,\phi(0),\phi(\rho(t,\phi_0)-t)))\Big\} \\ &- \frac{V(t_k,\varphi(0))}{(t-t_k)^q\Gamma(1-q)}. \end{aligned} \tag{6}$$

The derivative ${}^{c}_{(1)}D^q_+V(t,\phi;t_k,\varphi(0))$ given by Equation (6) depends significantly on both the fractional order q and the initial data (t_k,φ) of IVP for FrDDE (Equation (1)) and it makes this type of derivative close to the idea of the Caputo fractional derivative of a function.

Remark 4. *For any initial data $(t_k,\varphi) \in \mathbb{R}_+ \times PC_0$ of the IVP for NIFrDDE (Equation (1)) and any function $\phi \in PC_0$ and any point $t \in (t_k, s_{i+1})$ for a non-negative integer k the relations*

$${}^{c}_{(1)}D^q_+V(t,\phi;t_k,\varphi(0)) = D^+_{(1)}V(t,\phi(0),t_k,\phi) - {}^{RL}_{t_k}D^q\Big(V(t_k,\varphi(0))\Big),$$

$${}^{c}_{(1)}D^q_+V(t,\phi;t_k,\varphi(0)) = D^+_{(1)}V(t,\phi(0),t_k\phi), \quad \text{if } V(t_k,\varphi(0)) = 0 \tag{7}$$

$${}^{c}_{(1)}D^q_+V(t,\phi;t_k,\varphi(0)) < D^+_{(1)}V(t,\phi(0),t_k,\phi), \quad \text{if } V(t_k,\varphi(0)) > 0. \tag{8}$$

are satisfied.

Remark 5. *A derivative of $V(t,x) \in \Lambda(J,\Delta)$ among a system of Caputo fractional differential equations without delays was introduced by V. Lakshmikantham et al. [17] in 2009. Later, it was generalized for fractional equations with delays ([18–20]):*

$$D^+_{(1)}V(t,\phi(0),\phi) = \limsup_{h\to 0}\frac{1}{h^q}\Big[V(t,\phi(0)) - V(t-h,\phi(0)-h^q f(t,\phi))\Big], \quad t \ge t_0 \tag{9}$$

where $\phi \in C([-\tau,0],\Delta)$.

This definition is a direct generalization of the well known Dini derivative among differential equations with ordinary derivatives. However, for equations with fractional derivatives, it seems strange. It does not depend on the order q of the fractional derivative nor on the initial time t_0. The operator defined by Equation (9) has no memory, which is typical for the fractional derivative.

The derivative $D^+_{(1)}V(t,\phi(0),\phi)$ defined by Equation (9) is applied in [18] to study stability of fractional delay differential equations where in the proof of the main comparison result (Theorem 4.3 [18]) the derivative $D^+_{(1)}V(t,\phi(0),\phi)$ is incorrectly substituted by the Caputo fractional derivative (see Equations (20) and (30) in [18]). A similar situation occurs with the application of the derivative of Equation (9) in [20] for studying stability of impulsive fractional differential equations.

In the next example to simplify the calculations and to emphasize the derivatives and their properties, we consider the scalar case, i.e. $n = 1$.

Example 1. *(Lyapunov function depending directly on the time variable). Let* $V(t,x) = m(t)\ x^2$ *where* $m \in C^1(\mathbb{R}_+,\mathbb{R}_+)$.

Case 1. Caputo fractional derivative. Let x be a solution of NIFrDDE (Equation (1)). Then, the fractional derivative

$$ {}_{t_0}^{c}D^q V(t,x(t)) = {}_{t_0}^{c}D^q \Big(m(t)\ x^2(t) \Big) = \frac{1}{\Gamma(1-q)} \int_{t_0}^{t} \frac{m'(s)x^2(s) + 2m(s)x(s)x'(s)}{(t-s)^q} ds $$

is difficult to obtain in the general case for any solution of Equation (1). In addition, the solution $x(t)$ might not be differentiable on the intervals of impulses.

Case 2. Dini fractional derivative. Let $\phi \in PC_0$ and $t \in (t_k, s_{k+1})$ for a non-negative integer k. Then, applying Equation (4), we obtain

$$ \begin{aligned} & D_{(1)}^{+} V(t,\phi(0),t_k,\phi) \\ &= \limsup_{h\to 0} \frac{1}{h^q} \Big[m(t)\ (\phi(0))^2 - \sum_{r=1}^{[\frac{t-t_k}{h}]} (-1)^{r+1}\ {}_qC_r m(t-rh)(\phi(0) - h^q f(t,\phi(0),\phi(\rho(t,\phi_0)-t))))^2 \Big] \\ &= \limsup_{h\to 0} \frac{1}{h^q} \Big[m(t) \Big((\phi(0))^2 - (\phi(0) - h^q f(t,\phi(0),\phi(\rho(t,\phi_0)-t))))^2 \Big) \\ &\quad + (\phi(0) - h^q f(t,\phi(0),\phi_0))^2 \sum_{r=0}^{[\frac{t-t_k}{h}]} (-1)^r\ {}_qC_r m(t-rh) \Big] \\ &= \phi(0)\ m(t) f(t,\phi(0),\phi(\rho(t,\phi_0)-t))) + (\phi(0))^2\ {}_{t_k}^{RL}D^q \Big(m(t) \Big). \end{aligned} $$

Case 2. Caputo fractional Dini derivative. Let $\varphi,\ \phi \in PC_0$ and $t \in (t_k, s_{k+1})$ for a non-negative integer k. Then, we use Equation (6) and obtain

$$ \begin{aligned} & {}_{(1)}^{c}D_+^q V(t,\phi;t_k,\varphi(0)) \\ &= \limsup_{h\to 0^+} \frac{1}{h^q} \Big\{ \phi(0)^2 m(t) - \sum_{r=1}^{[\frac{t-t_k}{h}]} (-1)^{r+1}\ {}_qC_r m(t-rh)\ (\phi(0) - h^q f(t,\phi(0),\phi(\rho(t,\phi_0)-t))))^2 \Big\} \\ &\quad - (\varphi(0))^2 m(t_k) \frac{(t-t_k)^{-q}}{\Gamma(1-q)} \\ &= 2\phi(0)m(t)f(t,\phi(0),\phi(\rho(t,\phi_0)-t))) + (\phi(0))^2\ {}_{t_k}^{RL}D^q \Big(m(t) \Big) - (\varphi(0))^2 m(t_k) \frac{(t-t_k)^{-q}}{\Gamma(1-q)} \\ &= D_{(1)}^{+} V(t,\phi(0),t_k,\phi) - V(t_k,\varphi(0)) \frac{(t-t_k)^{-q}}{\Gamma(1-q)}. \end{aligned} $$

□

5. Comparison Results

Lemma 1. [17]. *Let* $v \in C([a,b],\mathbb{R})$ *be such that* $(t-a)^{1-q}v \in C([a,b],\mathbb{R})$ *and there exists a point* $\tau \in (a,b]$: $v(\tau) = 0$ *and* $v(t) \le 0$ *for* $t \in [a,\tau]$. *Then,* ${}_a^C D_\tau^q v(\tau) \ge 0$.

We use the following comparison scalar fractional differential equation with non-instantaneous impulses:

$$\begin{aligned} &{}_{t_i}^{C}D_t^q u(t) = g(t, u(t)) \text{ for } t \in (t_i, s_{i+1}], i = 0, 1, 2, \dots, \\ &u(t) = \psi_i(t, u(s_i - 0)), \ t \in (s_i, t_i], \ i = 1, 2, \dots, \\ &u(t_0) = u_0, \end{aligned} \tag{10}$$

where $u, u_0 \in \mathbb{R}, g : [0, s_1] \cup_{k=1}^{\infty} [t_k, s_k] \times \mathbb{R} \to \mathbb{R}, \psi_k : [s_k, t_{k+1}] \times \mathbb{R} \to \mathbb{R} \ (k = 1, 2, 3, \dots)$.

We obtain some comparison results. Note some comparison results for fractional time delay differential equations are obtained in [18] by applying the derivative defined by Equation (9) and substituting it incorrectly as a Caputo fractional derivative (see Remark 5).

We introduce the following conditions:

A6. The function $g(t, u) \in C([0, s_1] \bigcup_{k=1}^{\infty} [t_k, s_{k+1}] \times \mathbb{R}_+, \mathbb{R})$ is strictly decreasing with respect to its second argument, and for any $k = 1, 2, \dots$ the functions $\psi_k : [s_k, t_k] \times \mathbb{R}_+ \to \mathbb{R}_+$ are nondecreasing with respect to their second argument.

A7. The function $g(t, 0) = 0$ for $t \in [0, s_1] \bigcup_{k=1}^{\infty} [t_k, s_{k+1}]$ and for any $k = 1, 2, \dots$ the function $\psi_k(t, 0) = 0$ for $t \in [s_k, t_k]$.

A8. For all $k = 1, 2, \dots$, the functions ψ_k satisfies $\psi_k(t, u) \le u, \ t \in [s_k, t_k], \ u \in \mathbb{R}$.

In our main results, we use the Lipschitz stability of the zero solution of the scalar comparison non-instantaneous impulsive fractional differential in Equation (10).

Example 2. *Let $t_k = 2k, k = 0, 1, \dots$ and $s_k = 2k - 1, \ k = 1, 2, \dots$. Consider the scalar non-instantaneous impulsive fractional differential equation*

$$\begin{aligned} &{}_{t_i}^{C}D_t^{0.25} u(t) = u(t) \ \text{for } t \in (t_i, s_{i+1}], i = 0, 1, 2, \dots, \\ &u(t) = \psi_k(t, u(s_k - 0)), \ t \in (s_i, t_i], \ i = 1, 2, \dots, \\ &u(0) = u_0, \end{aligned} \tag{11}$$

where $u, u_0 \in \mathbb{R}$.

Case 1. Suppose for all natural numbers $k = 1, 2, \dots$ the equality $\psi_k(t, u) = \frac{u}{2t}, \ u \in \mathbb{R}, t \in [s_k, t_k]$ holds. Then, the solution of Equation (11) is given by

$$u(t) = \begin{cases} u_0 E_{0.25}(t^{0.25}), & t \in (0, 1], \\ \frac{u_0 (E_{0.25}(1))^k}{2t \prod_{i=1}^{k-1} (4i)}, & t \in (2k - 1, 2k], \ k = 1, 2, \dots, \\ \frac{u_0 (E_{0.25}(1))^k}{\prod_{i=1}^{k} (4i)} E_{0.25}((t - 2k)^{0.25}), & t \in (2k, 2k + 1], \ k = 1, 2, \dots. \end{cases} \tag{12}$$

The solution of Equation (11) is uniformly Lipschitz stable with $M_1 = 30$ (see Figure 1 for the graph of the solutions with various initial values).

Case 2. Suppose for all natural numbers $k = 1, 2, \dots$ the equality $\psi_k(t, u) = tu, \ u \in \mathbb{R}, t \in [s_k, t_k]$ holds. Then, the solution of Equation (11) is given by

$$u(t) = \begin{cases} u_0 E_{0.25}(t^{0.25}), & t \in (0, 1], \\ u_0 (E_{0.25}(1))^k \prod_{i=1}^{k-1} (2i) \ t, & t \in (2k - 1, 2k], \ k = 1, 2, \dots, \\ u_0 (E_{0.25}(1))^k \prod_{i=1}^{k} (2i) E_{0.25}((t - 2k)^{0.25}), & t \in (2k, 2k + 1], \ k = 1, 2, \dots. \end{cases} \tag{13}$$

The solution of Equation (11) is unbounded (see Figure 2 for the graph of the solution).

Therefore, for $\psi_k(t, u) = \frac{u}{2t} \le u$ the solution is Lipschitz stable but for $\psi_k(t, u) = \frac{u}{2t} \ge u$ it is not (compare with condition (A8)). □

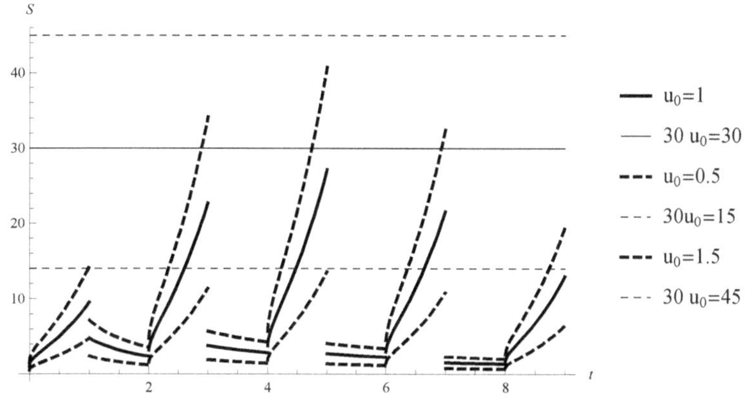

Figure 1. Example 2. Graph of the solution of Equation (11) with $\psi_k(t,u) = \frac{u}{2t}$ for various initial values.

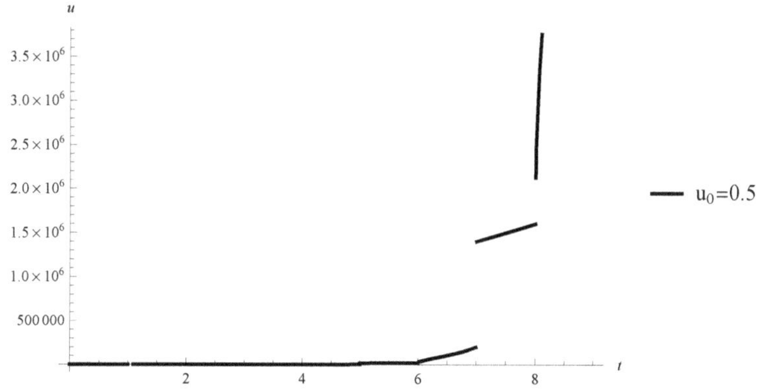

Figure 2. Example 2. Graph of the solution of Equation (11) with $\psi_k(t,u) = tu$.

In our study, we use some comparison results. When the Caputo fractional derivative is used, then the comparison result is:

Lemma 2. *(Caputo fractional derivative). Assume the following conditions are satisfied:*

1. *Assumptions A1–A4 and A6 are satisfied.*
2. *The function $x^*(t) = x(t;t_0,\varphi) : [t_0,T) \to \Delta, x^* \in PC^q([t_0,T))$ is a solution of Equation (1) where $\Delta \subset \mathbb{R}^n, 0 \in \Delta, T \leq \infty$.*
3. *The function $V \in \Lambda([t_0,T),\Delta)$ is such that*
   - (i) *For any $i = 0,1,2,\cdots : (t_i,s_{i+1}) \cap [t_0,T) \neq \emptyset$ and for $t \in (t_i,s_{i+1}) \cap [t_0,T)$, the inequality*

$$ {}_{t_i}^{C}D_t^q V(t,x^*(t)) \leq g(t,V(t,x^*(t))) $$

   holds.
   - (ii) *For all $i = 1,2,3,\cdots : (s_i,t_i) \cap [t_0,T) \neq \emptyset$ the inequality*

$$ V(t,\phi_i(t,x^*(s_i-0))) \leq \psi_i(t,V(s_i-0,x^*(s_i-0))) \quad for \ t \in (s_i,t_i] \cap [t_0,T) $$

   holds.

If $\sup_{s\in[-r,0]} V(t_0, \varphi(s)) \le u_0$, *then the inequality* $V(t, x^*(t)) \le r(t)$ *for* $t \in [t_0, T)$ *holds, where* $r(t) = r(t; t_0, u_0)$ *is the maximal solution on* $[t_0, T)$ *of Equation (10) with* $u_0 \ge 0$.

Proof. We use induction with respect to the intervals to prove Lemma 2. Let $m(t) = V(t, x^*(t)), t \ge t_0$. We prove

$$m(t) \le u(t), \quad t \ge t_0. \tag{14}$$

Let $t \in [t_0, s_1]$. Let $\varepsilon > 0$ be an arbitrary number. We prove

$$m(t) < u(t) + \varepsilon, \quad t \ge [t_0, s_1]. \tag{15}$$

Note $m(t_0) = V(t_0, \varphi(0)) \le \sup_{s\in[-r,0]} V(t_0, \varphi(s)) \le u_0$, i.e. the inequality in Equation (15) holds for $t = t_0$. If the inequality in Equation (15) is not true, then there exists a point $t^* \in (t_0, s_1]$ such that $m(t^*) = u(t^*) + \varepsilon, \ m(t) < u(t) + \varepsilon, \ t \in [t_0, t^*)$.

From Lemma 1 with $a = t_0, \ b = s_1, \ \tau = t^*$ and $v(t) = m(t) - u(t) - \varepsilon$ the inequality ${}_{t_0}^{C}D_t^q m(t^*) \ge {}_{t_0}^{C}D_t^q u(t^*) = g(t^*, u(t^*))$ holds.

From Assumption A6 and Condition 3(i), the inequality ${}_{t_0}^{C}D_t^q m(t^*) \le g(t^*, m(t^*)) = g(t^*, u(t^*) + \varepsilon) < g(t^*, u(t^*))$ holds. The contradiction proves the validity of Equation (15). Since ε is an arbitrary positive number, we obtain the inequality in Equation (14) for $t \in [t_0, s_1]$.

Let $t \in (s_1, t_1]$. Then, from the impulsive equality in Equation (1), Condition 3(ii), Assumption A6 and the inequality in Equation (14) for $t = s_1 - 0$, we obtain $m(t) = V(t, x^*(t)) = V(t, \phi_1(t, x^*(s_1 - 0))) \le \psi_1(t, V(s_1 - 0, x^*(s_1 - 0))) = \psi_1(t, m(s_1 - 0)) \le \psi_1(t, u(s_1 - 0)) = u(t)$, i.e. Equation (14) holds on $(s_1, t_1]$.

Let $t \in (t_1, s_2]$. Let $\varepsilon > 0$ be an arbitrary number. We prove Equation (15) for $t \in [t_1, s_2]$. Note that Equation (15) is true for $t = t_1$. If the inequality in Equation (15) is not true, then there exists a point $t^* \in (t_1, s_2]$ such that $m(t^*) = u(t^*) + \varepsilon, \ m(t) < u(t) + \varepsilon, \ t \in [t_1, t^*)$.

From Lemma 1 with $a = t_1, \ b = s_2, \ \tau = t^*$ and $v(t) = m(t) - u(t) - \varepsilon$, the inequality ${}_{t_1}^{C}D_t^q m(t^*) \ge {}_{t_1}^{C}D_t^q u(t^*) = g(t^*, u(t^*))$ holds.

From Assumption A6 and Condition 3(i), the inequality ${}_{t_1}^{C}D_t^q m(t^*) \le g(t^*, m(t^*)) = g(t^*, u(t^*) + \varepsilon) < g(t^*, u(t^*))$ holds. The contradiction proves the validity of Equation (15) and the inequality in Equation (14) for $t \in (t_1, s_2]$. Continuing this process and an induction argument prove Equation (14) and Lemma 2. □

Lemma 3. *[10] Let* $m \in C([t_0, T], \mathbb{R})$ *and there exists* $\tau \in (t_0, T]$, *such that* $m(\tau) = 0$ *and* $m(t) < 0$ *for* $t \in [t_0, \tau)$. *Then, the inequality* ${}_{t_0}^{GL}D_+^q m(\tau) > 0$ *holds.*

When the Dini fractional derivative defined by Equation (4) or Caputo fractional Dini derivative defined by Equation (5) is used then the comparison result is:

Lemma 4. *(Dini fractional derivative/Caputo fractional Dini derivative). Assume:*

1. *Assumptions A1–A4 and A6 are satisfied.*
2. *The function* $x^*(t) = x(t; t_0, \varphi) : [t_0, T) \to \Delta, x^* \in PC^q([t_0, T))$ *is a solution of Equation (1) where* $\Delta \subset \mathbb{R}^n, 0 \in \Delta, T \le \infty$.
3. *The function* $V \in \Lambda([t_0, T), \Delta)$ *is such that*

   (i) *For any* $i = 0, 1, 2, \cdots : \ (t_i, s_{i+1}) \cap [t_0, T) \ne \emptyset$ *and for* $t \in (t_i, s_{i+1}) \cap [t_0, T)$, *the inequality*

$$\mathcal{D}_{(1)} V(t, \phi, t_i) \le g(t, V(t, \phi(0)))$$

holds where $\phi(\Theta) = x^*(t+\Theta)$, $\Theta \in [-r,0]$, *and* $\mathcal{D}_{(1)}V(t,\phi,t_i)$ *is one of the following two derivatives: the Dini fractional derivative* $D^+_{(1)}V(t,\phi(0),t_i,\phi)$ *defined by Equation (4) or the Caputo fractional Dini derivative* ${}^c_{(1)}D^q_+V(t,\phi;t_i,\varphi(0))$ *defined by Equation (5).*

(ii) *For all* $i = 1,2,3,\cdots: (s_i,t_i)\cap[t_0,T) \neq \emptyset$, *the inequality*

$$V(t,\phi_i(t,x^*(s_i-0))) \le \psi_i(t,V(s_i-0,x^*(s_i-0))) \quad \text{for } t \in (s_i,t_i]\cap[t_0,T)$$

holds.

If $\sup_{s\in[-r,0]} V(t_0,\varphi(s)) \le u_0$, *then the inequality* $V(t,x^*(t)) \le r(t)$ *for* $t \in [t_0,T)$ *holds, where* $r(t) = r(t;t_0,u_0)$ *is the maximal solution on* $[t_0,T)$ *of Equation (10) with* $u_0 \ge 0$.

Proof. The proof is similar to the one in Lemma 2 where instead of the Caputo fractional derivative of the Lyapunov function, we use the Dini fractional derivative or the Caputo fractional Dini derivative which are less restrictive with respect to the properties of Lyapunov functions (for example, differentiability is not required). We sketch the proof emphasizing the differences with Lemma 2.

Case 1. Let $\mathcal{D}_{(1)}V(t,\phi,t_i) = {}^c_{(1)}D^q_+V(t,\phi;t_i,\varphi(0)), i = 1,2,\dots$ in Condition 3(i) of Lemma 4.

We use induction with respect to the intervals to prove Lemma 4. We prove the inequality in Equation (14).

Case 1.1. Let $t \in [t_0,s_1]$. We prove Equation (15) with $\varepsilon > 0$ an arbitrary number. Note that Equation (15) holds for $t = t_0$. If the inequality in Equation (15) is not true, then there exists a point $t^* \in (t_0,s_1]$ such that $p(t^*) = 0$ and $p(t^*) < 0$ for $t \in [t_0,t^*)$ where $p(t) = m(t) - u(t) - \varepsilon$. From Lemma 3 with $\tau = t^*$ we get the inequality

$${}^{GL}_{t_0}D^q_+m(t^*) >{}^{GL}_{t_0} D^q_+u(t^*) +{}^{GL}_{t_0} D^q_+\varepsilon.$$

Thus

$$\begin{aligned} {}^C_{t_0}D^q_t m(t^*) &={}^{GL}_{t_0} D^q_+(m(t^*) - m(t_0)) ={}^{GL}_{t_0} D^q_+m(t^*) - {}^{GL}_{t_0}D^q_+m(t_0) \\ &> {}^{GL}_{t_0}D^q_+(u(t^*) - u_0) ={}^C_{t_0} D^q_t u(t^*). \end{aligned} \tag{16}$$

Following the proof of Lemma 3 [3] from the choice of the point t^*, the definition of the function $m(t)$, the definition of the derivative ${}^c_{(1)}D^q_+V(\tau,\phi(0);t_0,\varphi(0))$, Assumption A2 and $x(t+s) = \phi(s)$, $x_{\rho(t,x_t)} = x_{\rho(t,\phi_0)} = x(\rho(t,\phi_0)) = x(t+(\rho(t,\phi_0)-t)) = \phi(\rho(t,\phi_0)-t)$, Assumption A6 and Condition 3(i) of Lemma 4, we obtain the inequality

$$\begin{aligned} {}^C_{t_0}D^q_t m(\tau) &= {}^{GL}_{t_0}D^q_+(m(t^*) - m(t_0)) \le {}^c_{(1)}D^q_+V(t^*,\phi(0);t_0,\varphi(0)) \\ &\le g(t^*,V(t^*,\phi(0)) = g(t^*,m(t^*)) = g(t^*,u(t^*)+\varepsilon) \le g(t^*,u(t^*)) \\ &={}^C_{t_0} D^q_t u(t^*) \end{aligned} \tag{17}$$

with $\phi(\Theta) = x(\tau+\Theta)$, $\Theta \in [-\tau,0]$.

The inequality in Equation (17) contradicts the inequality in Equation (16). The contradiction proves the validity of Equation (15) and, therefore, the validity of Equation (14) on $[t_0,s_1]$.

Case 1.2. Let $t \in (s_1,t_1]$. From the impulsive equality in Equation (1), Condition 3(ii) of Lemma 4, Assumption A6 and the inequality in Equation (14) for $t = s_1 - 0$, we obtain for $t \in (s_1,t_1]$ the inequalities $m(t) = V(t,x^*(t)) = V(t,\phi_1(t,x^*(s_1-0))) \le \psi_1(t,V(s_1-0,x^*(s_1-0))) = \psi_1(t,m(s_1-0)) \le \psi_1(t,u(s_1-0)) = u(t)$, i.e. Equation (14) holds on (s_1,t_1).

Case 1.3. Let $t \in (t_1,s_2]$. The proof of the inequality in Equation (15) for $t \ge (t_1,s_2]$ is similar to the one in Case 1.1 by replacing t_0 with t_1.

Case 2. Let $\mathcal{D}_{(1)}V(t,\phi,t_i)$ in Condition 3(i) of Lemma 4 be the Dini fractional derivative $D^{+}_{(1)}V(t,\phi(0),t_i,\varphi)$ defined by Equation (4). Then, based on the proof in Case 1 and Remark 4, we establish Lemma 4. □

6. Main Results

Theorem 1. (Caputo fractional derivative) *Let the following conditions be satisfied:*

1. *Assumptions A1–A8 are fulfilled.*
2. *There exist a function $V \in \Lambda(\mathbb{R}_+,\mathbb{R}^n)$ and*

   (i) *The inequalities*
   $$b(||x||) \le V(t,x) \le a(||x||), \quad x \in \mathbb{R}^n, t \in \mathbb{R}_+$$
   holds, where $a \in K([0,\rho])$, $b \in M([0,\rho])$, $\rho > 0$;

   (ii) *For any initial data and any solution $x(t)$ of Equation (1) defined on $[t_0,\infty)$ such that for any $\tau \in (t_k,s_{k+1})$, k is a non-negative integer, such that $x(t) \in S_\rho$, $t \in [t_0,\tau]$ and $V(\tau,x(\tau)) \ge V(s,x(s))$ for $s \in [t_0,\tau]$ the inequality*
   $${}_{t_i}^{C}D_t^q V(t,x(t)) \le g(t,V(t,x(t))), \quad t \in (t_i,s_{i+1}] \cap [t_0,\tau], i = 0,1,2,\dots,k$$
   holds.

   (iii) *For any $k = 0,1,2,\dots$ and $t \in (s_k,t_{k+1}]$, $y \in S_\rho$ the inequality*
   $$V(t,\phi_k(t,y)) \le \psi_k(t,V(s_k-0,y))$$
   holds.
3. *The zero solution of Equation (10) is uniformly Lipschitz stable (uniformly globally Lipschitz stable).*

Then, the zero solution of Equation (1) is uniformly Lipschitz stable (uniformly globally Lipschitz stable).

Proof. Let the zero solution of Equation (10) be uniformly Lipschitz stable. Let $t_0 \ge 0$ be an arbitrary. Without loss of generality, we assume $t_0 \in [0,s_1)$. From Condition 3, there exist $M \ge 1$, $\delta_1 > 0$ such that for any $u_0 \in \mathbb{R}: \ |u_0| < \delta_1$ the inequality

$$|u(t;t_0,u_0)| \ \le M\,|u_0| \text{ for } t \ge t_0 \tag{18}$$

holds, where $u(t;t_0,u_0)$ is a solution of Equation (10) with the initial data (t_0,u_0).

From the inclusions $a \in K([0,\rho])$ and $b \in M([0,\rho])$, there exist a function $q_b(u)$ and a positive constant K_a. Without loss of generality, we can assume $K_a \ge 1$. Choose the constant M_1 such that $M_1 > \max\{1, q_b(K_a), q_b(M)K_a\}$ and $\delta_2 \le \frac{\rho}{2M_1}$. Therefore, $2M_1\delta_2 \le \rho$.

Let $\delta = \min\left\{\delta_1,\delta_2,\frac{\delta_1}{K_a}\right\}$. Choose the initial function $\varphi \in PC_0([-r,0])$ such that $||\varphi||_{PC_0} < \delta$. Therefore, $||\varphi||_{PC_0} < \delta \le \delta_2 \le \rho$, i.e. $\varphi(s) \in S_\rho$ for $s \in [-r,0]$. Consider the solution $y(t) = y(t;t_0,\varphi)$ of the system in Equation (1) for the chosen initial data (t_0,φ).

Let $u_0^* = \sup_{s\in[-r,0]} V(t_0,\varphi(s))$. From the choice of φ and the properties of the function $a(u)$ applying condition 2(i) we get $u_0^* = V(t_0,\varphi(\xi)) \le a(||\varphi(\xi)||) \le a(||\varphi||_{PC_0}) \le K_a||\varphi||_{PC_0} < K_a\delta \le \delta_1$. Therefore, the function $u^*(t)$ satisfies Equation (18) for $t \ge t_0$ with $u_0 = u_0^*$, where $u^*(t) = u(t;t_0,u_0^*)$ is a solution of Equation (10) with initial data (t_0,u_0^*).

Let $\varepsilon \in (0, M_1\delta]$ be an arbitrary number. We prove

$$V(t,y(t)) < b(M_1||\varphi||_{PC_0} + \varepsilon), \quad t \ge t_0. \tag{19}$$

For $t = t_0$, we get $V(t_0,y(t_0)) = V(t_0,\varphi(0)) \le a(||\varphi(0)||) \le a(||\varphi||_{PC_0} \le K_a||\varphi||_{PC_0} \le b(q_b(K_a)||\varphi||_{PC_0}) \le b(M_1||\varphi||_{PC_0}) < b(M_1||\varphi||_{PC_0} + \varepsilon)$, i.e. the inequality in Equation (19) holds.

Assume Equation (19) is not true.

Case 1. There exists a point $T > t_0$, $T \in \bigcup_{k=0}^{\infty}(t_k, s_{k+1}]$ such that $V(t, y(t)) < b(M_1||\varphi||_{PC_0} + \varepsilon)$ for $t \in [t_0, T)$, $V(T, y(T)) = b(M_1||\varphi||_{PC_0} + \varepsilon)$, i.e. $V(s, y(s)) \leq V(T, y(T))$ for $s \in [t_0, T]$. Then, from Condition 2(i), we obtain the inequalities $||y(t)|| \leq b^{-1}(V(t, y(t)) \leq M_1||\varphi||_{PC_0} + \varepsilon < 2M_1\delta \leq 2M_1\delta_2 \leq \rho$ for $t \in [t_0, T]$, i.e., $y(t) \in S_\rho$ for $t \in [t_0, T]$ and, according to Condition 2(ii) of Theorem 1 with $\tau = T$, it follows that Condition 3(i) of Lemma 2 is satisfied for the solution $y(t)$ on the interval $[t_0, T]$ and $\Delta = S_\rho$.

According to Lemma 2, we get

$$V(t,\ y(t)) \leq u^*(t) \text{ for } t \in [t_0, T]. \tag{20}$$

From the inequality in Equation (20) and Condition 2(i), we obtain

$$\begin{aligned} M_1||\varphi||_{PC_0} &= b^{-1}(V(T, y(T))) \leq b^{-1}(u^*(T)) \\ &\leq b^{-1}(M\,|u_0^*|) = b^{-1}(MV(t_0, \varphi(\xi))) \leq q_b(M)V(t_0, \varphi(\xi)) \\ &\leq q_b(M)a(||\varphi(\xi)||) \leq q_b(M)a(||\varphi||_0) \leq q_b(M)K_a||\varphi)||_{PC_0} < M_1||\varphi||_{PC_0}. \end{aligned} \tag{21}$$

The contradiction proves the validity of Equation (19). From the inequality in Equation (19) and Condition 2(i), we have Theorem 1.

Case 2. There exists a point $T > t_0$, $T \in \bigcup_{k=1}^{\infty}(s_k, t_k)$ such that $V(t, y(t)) < b(M_1||\varphi||_{PC_0} + \varepsilon)$ for $t \in [t_0, T)$, $V(T, y(T)) = b(M_1||\varphi||_{PC_0} + \varepsilon)$. Then, as in Case 1 we get $y(t) \in S_\rho$ for $t \in [t_0, T]$. Let $T \in (s_j, t_{j+1})$ for a natural number j. According to Condition 2(iii) of Theorem 1, we obtain $b(M_1||\varphi||_{PC_0} + \varepsilon) = V(T, y(T)) = V(T, \phi_j(T, y(s_j - 0))) \leq \psi_j(T, V(s_j - 0, y(s_j - 0))) \leq \psi_j(T, b(M_1||\varphi||_{PC_0})) < \psi_j(T, b(M_1||\varphi||_{PC_0}) + \varepsilon)$. The contradiction proves this case is not possible.

Case 3. There exists a natural number k such that $V(t, y(t)) < b(M_1||\varphi||_0 + \varepsilon)$ for $t \in [t_0, s_k]$ and $V(s_k + 0, y(s_k + 0)) > b(M_1||\varphi||_0 + \varepsilon)$. Therefore, $\psi_k(s_k, b(M_1||\varphi||_0)) \geq \psi_k(s_k, V(s_k, y(s_k))) \geq V(s_k + 0, \phi_k(s_k, y(s_k - 0))) = V(s_k + 0, y(s_k + 0)) > b(M_1||\varphi||_0)$. The contradiction proves this case is not possible.

The proof of globally uniformly Lipschitz stability is analogous so we omit it. □

Theorem 2. *Let the conditions of Theorem 1 be satisfied where Condition 2(i) is replaced by:*

$2^*(i)$ *the inequalities* $\lambda_1(t)||x||^2 \leq V \leq \lambda_2(t)||x||^2$, $x \in S_\rho, t \in \mathbb{R}^+$ *holds, where* $\lambda_1, \lambda_2 \in C(\mathbb{R}_+, (0, \infty))$ *and there exists positive constant* $A_1, A_2 : A_1 < A_2$ *such that* $\lambda_1(t) \geq A_1$, $\lambda_2(t) \leq A_2$ *for* $t \geq 0$, *and* $\rho > 0$.

If the zero solution of Equation (10) is uniformly Lipschitz stable (uniformly globally Lipschitz stable), then the zero solution of Equation (1) is uniformly Lipschitz stable (uniformly globally Lipschitz stable).

Proof. The proof is similar to the one in Theorem 1 where $M_1 = \sqrt{M\frac{A_2}{A_1}}$.

Theorem 3. (Dini fractional derivative/ Caputo fractional Dini derivative) *Let the following conditions be satisfied:*

1. *Assumptions A1–A8 are fulfilled.*
2. *There exist a function* $V(t, x) \in \Lambda(\mathbb{R}_+, \mathbb{R}^n)$, $\rho > 0$ *and*

   (i) *The inequalities*

   $$b(||x||) \leq V(t, x) \leq a(||x||), \quad x \in \mathbb{R}^n, t \in \mathbb{R}_+$$

   holds, where $a \in K([0, \rho])$, $b \in M([0, \rho])$.

(ii) *For any function* $\phi \in PC_0 :\ \phi(s) \in S_\rho$ *for* $s \in [-r, 0]$ *such that for any* $t :\ t \in (t_k, s_{k+1})$, *k is a non-negative integer, such that* $V(t+s, \phi(s)) \leq V(t, \phi(0))$, $s \in [-r, 0]$ *the inequality*

$$\mathcal{D}_{(1)} V(t, \phi, t_k) \leq g(t, V(t, \phi(0)))$$

holds where $\mathcal{D}_{(1)} V(t, \phi, t_k)$ *is one of the following two derivatives: the Dini fractional derivative* $D^+_{(1)} V(t, \phi(0), t_k, \phi)$ *defined by Equation (4) or the Caputo fractional Dini derivative* ${}^{c}_{(1)}D^q_+ V(t, \phi; t_k, \varphi(0))$ *defined by Equation (5) and* $\rho > 0$.

(iii) *For any* $k = 0, 1, 2, \dots$ *and* $t \in (s_k, t_{k+1}]$, $y \in S_\rho$ *the inequality*

$$V(t, \phi_k(t, y)) \leq \psi_k(t, V(s_k - 0, y))$$

holds.

3. *The zero solution of Equation (10) is uniformly Lipschitz stable (uniformly globally Lipschitz stable).*

Then, the zero solution of Equation (1) is uniformly Lipschitz stable (uniformly globally Lipschitz stable).

The proof of Theorem 3 is similar to the one in Theorem 1 where Lemma 4 is applied instead of Lemma 2.

Example 3. *Let* $t_k = 2k, k = 0, 1, 2 \dots$ *and* $s_k = 2k - 1$, $k = 1, 2, \dots$. *Consider the non-instantaneous impulsive fractional differential equations*

$$\begin{aligned}
&{}^{C}_{t_i}D^{0.25}_t x_1(t) = 0.25 x_1(t) - x_2(t) + 0.25 x_1(t) (x_{\rho(t,x_t)})_2^2, \\
&{}^{C}_{t_i}D^{0.25}_t x_2(t) = 0.25 x_2(t) + x_1(t) + 0.25 x_2(t) (x_{\rho(t,x_t)})_1^2 \\
&\qquad \text{for } t \in (t_i, s_{i+1}], i = 0, 1, 2, \dots, \\
&x_1(t) = \frac{x_1(s_i - 0)}{\sqrt{2it}}, \quad x_2(t) = \frac{x_2(s_i - 0)}{\sqrt{2it}}, \quad t \in (s_i, t_i],\ i = 1, 2, \dots,
\end{aligned} \tag{22}$$

where $x = (x_1, x_2)$, $\rho(t, u) = t - \sin^2(u) : t - 1 \leq \rho(t, u) \leq t$, $x_{\rho(t,x_t)} = ((x_{\rho(t,x_t)})_1, (x_{\rho(t,x_t)})_2)$ *and* $(x_{\rho(t,x_t)})_i = x_i(t - \sin^2(x_i(t+s)))$, $s \in [-1, 0], i = 1, 2$.

Let $V(t, x) = x_1^2 + x_2^2$, $x = (x_1, x_2)$.

Let $x(t)$ be a solution of Equation (22). Let the point $\tau \in (t_k, s_{k+1}]$, k is a non-negative integer, be such that $x(t) \in S_1$, $t \in [0, \tau]$ and $x_1(\tau)^2 + x_2(\tau)^2 \geq x_1(s)^2 + x_2(s)^2$, $s \in [0, \tau]$. Using the notation $x_{\rho(\tau,x_\tau)}$ and Assumption A2, it follows that $\rho(\tau, x_j(\tau + \Theta)) \in [\tau - r, \tau], j = 1, 2$, $\Theta \in [-r, 0]$ and therefore $(x_{\rho(\tau,x_\tau)})_1^2 + (x_{\rho(\tau,x_\tau)})_2^2 \leq x_1(\tau)^2 + x_2(\tau)^2$ or

$$x_1^2(t)(x_{\rho(t,x_t)})_2^2 \leq x_1^2(t)(x_{\rho(t,x_t)})_1^2 + x_1^2(t)(x_{\rho(t,x_t)})_2^2 \leq x_1^2(t)(x_1(\tau)^2 + x_2(\tau)^2) \leq x_1(t)^2$$

and

$$x_2^2(t)(x_{\rho(t,x_t)})_1^2 \leq x_2^2(t)(x_{\rho(t,x_t)})_1^2 + x_2^2(t)(x_{\rho(t,x_t)})_2^2 \leq x_2^2(t)(x_1(\tau)^2 + x_2(\tau)^2) \leq x_2(t)^2.$$

Then, for all $i = 0, 1, 2, \dots, k]$ and $t \in (t_i, s_{i+1}] \cap [0, \tau]$, we get the inequality

$$\begin{aligned}
{}_{t_i}^{C}D_t^{0.25} V(t, x(t)) &= {}_{t_i}^{C} D_t^{0.25} x_1^2(t) + {}_{t_i}^{C} D_t^{0.25} x_2^2(t) \\
&\le 2x_1(t)\, {}_{t_i}^{C}D_t^{0.25} x_1(t) + 2x_2(t)\, {}_{t_i}^{C}D_t^{0.25} x_2(t) \\
&= 2x_1(t)\Big(0.25x_1(t) - x_2(t) + 0.25x_1(t)(x_{\rho(t,x_t)})_2^2\Big) \\
&\quad + 2x_2(t)\Big(0.25x_2(t) + x_1(t) + 0.25x_2(t)(x_{\rho(t,x_t)})_1^2\Big) \\
&\le V(t, x(t)).
\end{aligned} \tag{23}$$

In addition, for any natural number i, $x \in S_1 \subset \mathbb{R}^2$ and $t \in [s_i, t_i] = [2i-1, 2i$, we get $V(t, \frac{x}{\sqrt{2it}}) = \left(\frac{x_1}{\sqrt{2it}}\right)^2 + \left(\frac{x_2}{\sqrt{2it}}\right)^2 = \frac{1}{2t}\left(\frac{x_1^2}{i} + \frac{x_2^2}{i}\right) \le \frac{x_1^2 + x_2^2}{2t} = \frac{V(s_i, x)}{2t} = \psi_i(t, V(s_i, x))$ with $\psi_i(t, u) = \frac{u}{2t}$.

According to Example 2, Case 1 and Theorem 1, the zero solution of Equation (22) is uniformly Lipschitz stable. □

Example 4. *Let $t_k = 2k, k = 0, 1, 2\dots$ and $s_k = 2k-1$, $k = 1, 2, \dots$. Consider the non-instantaneous impulsive fractional differential equations*

$$\begin{aligned}
&{}_{t_i}^{C}D_t^{0.25} x_1(t) = 0.5x_1(t) - x_2(t) + 0.5x_1(t)(x_{\rho(t,x_t)})_2^2 - x_1(t)\frac{{}_{t_i}^{RL}D^q(\cos(0.5\pi(t + t_i + 1)) + 1.1)}{\cos(0.5\pi(t + t_i + 1)) + 1.1}, \\
&{}_{t_i}^{C}D_t^{0.25} x_2(t) = 0.5x_2(t) + x_1(t) + 0.5x_2(t)(x_{\rho(t,x_t)})_1^2 - x_2(t)\frac{{}_{t_i}^{RL}D^q(\cos(0.5\pi(t + t_i + 1)) + 1.1)}{\cos(0.5\pi(t + t_i + 1)) + 1.1} \\
&\qquad \text{for } t \in (t_i, s_{i+1}], i = 0, 1, 2, \dots, \\
&x_1(t) = \frac{x_1(s_i - 0)}{\sqrt{2it}}, \quad x_2(t) = \frac{x_2(s_i - 0)}{\sqrt{2it}}, \quad t \in (s_i, t_i], \ i = 1, 2, \dots,
\end{aligned} \tag{24}$$

where $x = (x_1, x_2)$, $\rho(t, u) = t - \sin^2(u)$, $t - 0.5 \le \rho(t, u) \le t$, $x_{\rho(t,x_t)} = ((x_{\rho(t,x_t)})_1, (x_{\rho(t,x_t)})_2)$ and $(x_{\rho(t,x_t)})_i = x_i(t - 0.5\sin^2(x_i(t + s)))$, $s \in [-0.5, 0], i = 1, 2$, $p(t) = \cos(0.5\pi(t + t_k + 1)) + 1.1$ for $t \in [t_k, s_{k+1}]$.

Note that, for any $t \in [t_k, s_{k+1}]$, the inequality $p(t) \le p(t + s), s \in [-0.5, 0]$ holds.

In this case, the quadratic function and Theorem 1 does not work (as it did in Example 3) because ${}_{t_i}^{C}D_t^{0.25} V(t, x(t)) \le 2V(t, x(t))(1 - \frac{{}_{t_k}^{RL}D^q p(t)}{p(t))}) \le 2V(t, x(t))(1 - \frac{10}{11}\, {}_{t_k}^{RL}D^q(p(t))$ and the solution of the comparison Equation (10) with $g(t, u) = 2u(1 - \frac{10}{21}\, {}_{t_k}^{RL}D^q(\cos(0.5\pi(t + t_k + 1)) + 1.1)$ is difficult to obtain.

Consider the Lyapunov function $V(t, x) = p(t)(x_1^2 + x_2^2)$, $x = (x_1, x_2)$.

Let the function $\phi \in PC_0$, $r = 0.5$ be such that $\phi(s) \in S_1$ for $s \in [-0.5, 0]$. Let $t:\ t \in (t_k, s_{k+1})$, k is a non-negative integer, be such that $p(t + s)(\phi_1(s)^2 + \phi_2(s)^2) \le p(t)(\phi_1(0)^2 + \phi_2(0)^2)$, $s \in [-1, 0]$. From the definition of the function ρ, it follows that $\rho(t, \phi_j(s)) - t = -0.5\sin^2(\phi_j(s)) \in [-0.5, 0]$ for $s \in [-1, 0], j = 1, 2$ and therefore $p(t + s)((\phi_1(\rho(t, \phi_1(s)) - t))^2 + (\phi_2(\rho(t, \phi_2(s)) - t))^2) \le p(t)(\phi_1(0)^2 + \phi_2(0)^2)$, $s \in [-0.5, 0]$. Then

$$\begin{aligned}
p(t)(\phi_2(\rho(t, \phi_0) - t))^2 &\le p(t + s)(\phi_2(\rho(t, \phi_2(s)) - t))^2 \\
&\le p(t + s)\Big((\phi_1(\rho(t, \phi_1(s)) - t))^2 + (\phi_2(\rho(t, \phi_2(s)) - t))^2\Big) \\
&\le p(t)\Big(\phi_1(0)^2 + \phi_2(0)^2\Big), \quad s \in [-0.5, 0].
\end{aligned} \tag{25}$$

Similarly, we get $p(t)(\phi_1(\rho(t, \phi_0) - t))^2 \le p(t)\Big(\phi_1(0)^2 + \phi_2(0)^2\Big)$.

Then, using Example 1, Case 2 and the notations $x_t = \phi_0$, $x_{\rho(t,x_t)} = \phi(\rho(t,\phi_0) - t)$, i.e., $f(t, x(t), x_{\rho(t,x_t)}) = f(t, \phi(0), \phi(\rho(t,\phi_0) - t))$, we get the inequality

$$\begin{aligned} D^+_{(24)} V(t,\phi(0),t_k,\phi) &= \phi_1(0)p(t)f_1(t,\phi(0),\phi(\rho(t,\phi_0)-t)) + \phi_2(0)p(t)f_2(t,\phi(0),\phi(\rho(t,\phi_0)-t)) \\ &+ \left(\phi_1^2(0)+\phi_2^2(0)\right)\,{}^{RL}_{t_k}D^q p(t) \\ &= \phi_1(0)p(t)\Big(0.5\phi_1(0) - \phi_2(0) + 0.5\phi_1(0)(\phi_2(\rho(t,\phi_0)-t))^2 - \phi_1(0)\frac{{}^{RL}_{t_k}D^q p(t)}{p(t)}\Big) \\ &+ \phi_2(0)p(t)\Big(0.5\phi_2(0) + \phi_1(0) + 0.5\phi_2(0)(\phi_1(\rho(t,\phi_0)-t))^2 - \phi_2(0)\frac{{}^{RL}_{t_k}D^q p(t)}{p(t)}\Big) \\ &+ \left(\phi_1^2(0)+\phi_2^2(0)\right)\,{}^{RL}_{t_k}D^q p(t) \\ &\leq V(t,\phi(0)). \end{aligned}$$

In addition, for any natural number i, $x \in S_1 \subset \mathbb{R}^2$ and $t \in [s_i, t_i] = [2i-1, 2i$, we get $V(t, \frac{x}{\sqrt{2it}}) = \left(\frac{x_1}{\sqrt{2it}}\right)^2 + \left(\frac{x_2}{\sqrt{2it}}\right)^2 = \frac{1}{2t}\left(\frac{x_1^2}{i} + \frac{x_2^2}{i}\right) \leq \frac{x_1^2+x_2^2}{2t} = \frac{V(s_i,x)}{2t} = \psi_i(t, V(s_i,x))$ with $\psi_i(t,u) = \frac{u}{2t}$.

According to Example 2, Case 1 and Theorem 3, the zero solution of Equation (24) is uniformly Lipschitz stable.

Author Contributions: Conceptualization, A.R., S.H. and D.O.; methodology, A.R., S.H. and D.O.; formal analysis, A.R., S.H. and D.O.

Funding: This research received no external funding.

Conflicts of Interest: The authors declare no conflict of interest.

References

1. Dannan, F.M.; Elaydi, S. Lipschitz stability of nonlinear systems of differential equations. *J. Math. Anal. Appl.* **1986**, *113*, 562–577. [CrossRef]
2. Zavoli, A.; Giulietti, F.; Avanzini, G.; De Matteis, G. Spacecraft dynamics under the action of Y-dot magnetic control. *Acta Astronaut.* **2016**, **122**, 146–158. [CrossRef]
3. Agarwal, R.; Hristova, S.; O'Regan, D. Lyapunov Functions and Stability of Caputo Fractional Differential Equations with Delays. *Differ. Equ. Dyn. Syst.* 2018, 1–22. [CrossRef]
4. Hernandez, E.; O'Regan, D. On a new class of abstract impulsive differential equations. *Proc. Am. Math. Soc.* **2013**, *141*, 1641–1649. [CrossRef]
5. Agarwal, R.; Hristova, S.; O'Regan, D. *Non-Instantaneous Impulses in Differential Equations*; Springer: Berlin, Germany, 2017.
6. Agarwal, R.; Hristova, S.; O'Regan, D. Non-instantaneous impulses in Caputo fractional differential equations. *Fract. Calc. Appl. Anal.* **2017**, **20**, 595–622. [CrossRef]
7. Bai, L.; Nieto, J.J. Variational approach to differential equations with not instantaneous impulses. *Appl. Math. Lett.* **2017**, *73*, 44–48. [CrossRef]
8. Nieto, J.J.; Uzal, J.M. Pulse positive periodic solutions for some classes of singular nonlinearities. *Appl. Math. Lett.* **2018**, *86*, 134–140. [CrossRef]
9. Agarwal, R.; Hristova, S.; O'Regan, D. Lyapunov Functions to Caputo Fractional Neural Networks with Time-Varying Delays. *Axioms* **2018**, *7*, 30. [CrossRef]
10. Agarwal, R.P.; O'Regan, D.; Hristova, S. Stability of Caputo fractional differential equations by Lyapunov functions. *Appl. Math.* **2015**, *60*, 653–676. [CrossRef]
11. Zhang, H.; Wu, D.; Cao, J. Asymptotic Stability of Caputo Type Fractional Neutral Dynamical Systems with Multiple Discrete Delays. *Abstr. Appl. Analys.* **2014**, *2014*, 138124. [CrossRef]
12. Das, S. *Functional Fractional Calculus*; Springer: Berlin/Heidelberg, Germany, 2011.

13. Diethelm, K. *The Analysis of Fractional Differential Equations*; Springer: Berlin/Heidelberg, Germany, 2010.
14. Podlubny, I. *Fractional Differential Equations*; Academic Press: San Diego, CA, USA, 1999.
15. Pandey, D.N.; Das, S.; Sukavanam, N. Existence of solutions for a second order neutral differential equation with state dependent delay and not instantaneous impulses. *Intern. J. Nonlinear Sci.* **2014**, *18*, 145–155.
16. Lakshmikantham, V.; Bainov, D.; Simeonov, P. *Theory of Impulsive Differential Equations*; World Scientific: Singapore, 1989.
17. Lakshmikantham, V.; Leela, S.; Vasundhara, Devi, J. *Theory of Fractional Dynamic Systems*; CSP: Cambridge, UK, 2009; 170p.
18. Sadati, S.J.; Ghaderi, R.; Ranjbar A. Some fractional comparison results and stability theorem for fractional time delay systems. *Rom. Rep. Phy.* **2013**, *65*, 94–102.
19. Stamova, I.; Stamov, G. Lipschitz stability criteria for functional differential systems of fractional order. *J. Math. Phys.* **2013**, *54*, 043502. [CrossRef]
20. Stamova, I.; Stamov, G. *Functional and Impulsive Differential Equations of Fractional Order: Qualitative Analysis and Applications*; CRC Press: Boca Raton, FL, USA, 2016; 276p.

axiomS

MDPI

Article

A New Efficient Method for the Numerical Solution of Linear Time-Dependent Partial Differential Equations

Mina Torabi and **Mohammad-Mehdi Hosseini** *

Department of Applied Mathematics, Faculty of Mathematics, Yazd University, P. O. Box 89195-741, Yazd, Iran; m_torabi@stu.yazd.ac.ir
* Correspondence: mhosseini@uk.ac.ir

Received: 5 August 2018; Accepted: 28 September 2018; Published: 1 October 2018

Abstract: This paper presents a new efficient method for the numerical solution of a linear time-dependent partial differential equation. The proposed technique includes the collocation method with Legendre wavelets for spatial discretization and the three-step Taylor method for time discretization. This procedure is third-order accurate in time. A comparative study between the proposed method and the one-step wavelet collocation method is provided. In order to verify the stability of these methods, asymptotic stability analysis is employed. Numerical illustrations are investigated to show the reliability and efficiency of the proposed method. An important property of the presented method is that unlike the one-step wavelet collocation method, it is not necessary to choose a small time step to achieve stability.

Keywords: Legendre wavelets; collocation method; three-step Taylor method; asymptotic stability; time-dependent partial differential equations

MSC: 35K05, 41A30, 65M70

1. Introduction

In recent years, many kinds of wavelet bases have been utilized to solve functional equations; for example, Shannon wavelets [1], Daubechies wavelets [2] and Chebyshev wavelets [3,4]. In this paper, we utilize Legendre wavelets. Legendre wavelets are derived from Legendre polynomials [5]. These wavelets have been used in solving different kinds of functional equations such as integral equations [6,7], fractional equations [8,9], ordinary differential equations [5], partial differential equations [10,11], etc.

In solving time-dependent problems, Legendre wavelets are often used for spatial discretization. Different techniques are implemented for time discretization. In some articles, Legendre wavelets are also applied for time discretization. Therefore, the collocation points should be defined for both time and spatial variables. Also in this technique, multi-dimensional wavelets should be used to approximate required functions, which deal with large matrices and require large storage space. For example, readers can refer to [9].

There are many contexts that use collocation methods in solving functional equations. For example, Luo et al. [12] presented three collocation methods based on a family of barycentric rational interpolation functions for solving a class of nonlinear parabolic partial differential equations. Furthermore, for solving a class of fractional subdiffusion equation, Luo et al. in 2016 [13] used the quadratic spline collocation method.

Another path for time discretization uses a finite difference method. Islam et al. [10] used a fully implicit scheme, which is based on the first-order Taylor expansion. Yin et al. [11] employed the θ-weighted scheme for nonlinear Klein–Sine–Gordon equations. Stability is the important point in using finite difference methods. Thus, methods that are first-order accurate in time might be inappropriate.

Here, we exploit the three-step finite element method for time discretization [14–16]. For the suitable differentiable function $F(t)$, these three steps are defined as follows:

$$F(t+\frac{\Delta t}{3}) = F(t) + \frac{\Delta t}{3}\frac{\partial F}{\partial t}(t), \tag{1}$$

$$F(t+\frac{\Delta t}{2}) = F(t) + \frac{\Delta t}{2}\frac{\partial F}{\partial t}(t+\frac{\Delta t}{3}), \tag{2}$$

$$F(t+\Delta t) = F(t) + \Delta t\frac{\partial F}{\partial t}(t+\frac{\Delta t}{2}). \tag{3}$$

It can be shown that the above equations are equivalent to the third-order Taylor expansion. Therefore, this method is third-order accurate in t. The first idea of using these three steps has been demonstrated by Jiang and Kawahara [14]. Equations (1)–(3) are usually accompanied by the Galerkin finite element method, which is known as the three-step Taylor–Galerkin method [17]. Kumar and Mehra [2] proposed a three-step wavelet Galerkin method based on the Daubechies wavelets for solving partial differential equations subject to periodic boundary conditions. In this paper, motivated and inspired by the ongoing research, we develop a new effective method, which combines the Legendre wavelets collocation method for spatial discretization and the mentioned three steps for time discretization in the numerical solution of a linear time-dependent partial differential equation subject to the Dirichlet boundary conditions. We call this method the three-step wavelet collocation method. Furthermore, we explain the asymptotic stability of the proposed method.

The organization of this paper is as follows. In Section 2, fundamental properties of the Legendre wavelets are described. The three-step wavelet collocation method is presented in Section 3. The analysis of asymptotic stability is performed in Section 4. Some numerical examples are presented in Section 5. Finally, Section 6 provides the conclusions of the study.

2. Basic Properties of Legendre Wavelets

Legendre wavelets are defined on the interval $[0,1]$ as follows [5]:

$$\begin{cases} \psi_{l,m}(x) = \sqrt{m+\frac{1}{2}}2^{\frac{k+1}{2}} L_m(2^{k+1}x-(2l+1)), & \frac{l}{2^k} \le x < \frac{l+1}{2^k} \\ 0, \quad otherwise & \end{cases}$$

where k can assume any positive integer, $m = 0,1,\cdots,M$, $l = 0,1,\cdots,2^k-1$ and $L_m(x)$ are the well-known Legendre polynomials of order m.

A function $f(x)$ defined over $[0,1]$ can be approximated in terms of Legendre wavelets as:

$$f(x) \simeq \sum_{l=0}^{2^k-1}\sum_{m=0}^{M} c_{l,m}\psi_{l,m}(x) = C^T\Psi(x), \tag{4}$$

where:

$$\Psi(x) = [\psi_{0,0},\psi_{0,1},\cdots,\psi_{0,M},\psi_{1,0},\psi_{1,1},\cdots,\psi_{2^k-1,0},\psi_{2^k-1,1}\cdots,\psi_{2^k-1,M}]^T,$$

and $c_{l,m} =< f(x),\psi_{l,m} >$, in which $< .,. >$ denotes the inner product.

The derivative of the vector $\Psi(x)$ can be expressed by:

$$\frac{d\Psi(x)}{dx} = D\Psi(x),$$

where D is the $2^k(M+1)$ operational matrix. Mohammadi and Hosseini obtained D and the operational matrix for the n-th derivative:

$$\frac{d^n\Psi(x)}{dx^n} = D^n\Psi(x), \tag{5}$$

in [5].

3. Three-Step Wavelet Collocation Method

In this section, we explain the main structure of the three-step wavelet collocation method.

3.1. Time Discretization

Consider the following linear time-dependent partial differential equation:

$$\frac{\partial u}{\partial t} = \nu(\frac{\partial^2 u}{\partial x^2}) + \mu u + f(x,t), \tag{6}$$

with the initial condition:

$$u(x,0) = g(x), \quad 0 \leq x \leq 1 \tag{7}$$

and boundary conditions:

$$u(0,t) = h_0(t), \tag{8}$$

$$u(1,t) = h_1(t), \quad t \geq 0. \tag{9}$$

Assume that $n \geq 0$ and Δt denote the time step such that $t_n = n\Delta t,\ n = 0,1,\cdots N_t$. By using the Taylor expansion, the value of the function $u(x,t)$ at the time t_{n+1} can be expressed as follows:

$$u^{n+1} = u^n + \Delta t(\frac{\partial u}{\partial t})^n + \frac{(\Delta t)^2}{2}(\frac{\partial^2 u}{\partial t^2})^n + \frac{(\Delta t)^3}{6}(\frac{\partial^3 u}{\partial t^3})^n + \mathbf{o}[(\Delta t)^4], \tag{10}$$

where the symbols u^n and $(\frac{\partial u}{\partial t})^n$ represent $u(x,t_n)$ and $\frac{\partial u}{\partial t}(x,t_n)$, respectively.

We can use the first-order Taylor expansion for time discretization and Legendre wavelets for spatial discretization [10]. We call this method the one-step wavelet collocation method. In addition, the time derivative in the given differential equations is approximated by Euler's formula:

$$(\frac{\partial u}{\partial t})^n = \frac{u(x,t_{n+1}) - u(x,t_n)}{\Delta t} + \mathbf{o}[(\Delta t)],$$

and therefore, we have semi-discrete equation:

$$u^{n+1} = u^n + \Delta t(\frac{\partial u}{\partial t})^n.$$

The three-step Taylor method for time discretization is derived by applying a factorization process to the right side of Equation (10) as follows:

$$\Big(\mathbf{I} + \Delta t\frac{\partial}{\partial t}[\mathbf{I} + \frac{\Delta t}{2}\frac{\partial}{\partial t}[\mathbf{I} + \frac{\Delta t}{3}\frac{\partial}{\partial t}]]\Big)u^n = u^n + \Delta t\frac{\partial}{\partial t}[u^n + \frac{\Delta t}{2}\frac{\partial}{\partial t}[u^n + \frac{\Delta t}{3}(\frac{\partial u}{\partial t})^n]]. \tag{11}$$

where the symbol $\mathbf{I}$ is the identity operator.

Now, using Equation (11) and employing a new notation, the three-step Taylor method is obtained as follows:

$$u^{n+\frac{1}{3}} = u^n + \frac{\Delta t}{3}(\frac{\partial u}{\partial t})^n \tag{12}$$

$$u^{n+\frac{1}{2}} = u^n + \frac{\Delta t}{2}(\frac{\partial u}{\partial t})^{n+\frac{1}{3}} \tag{13}$$

$$u^{n+1} = u^n + \Delta t(\frac{\partial u}{\partial t})^{n+\frac{1}{2}}. \tag{14}$$

It should be noted that $u^{n+\frac{1}{3}}$, $u^{n+\frac{1}{2}}$ and u^{n+1} represent the computed solution at time level $(t_n + \frac{\Delta t}{3})$, $(t_n + \frac{\Delta t}{2})$ and $(t_n + \Delta t)$, respectively.

3.2. Spatial Discretization

After time discretization, the spatial derivatives of $u(x,t)$ are approximated by Legendre wavelets. The collocation method is utilized in this part. Let the unknown solution $u(x,t_n)$ be expanded by:

$$u(x,t_n) \simeq u^n = \sum_{l=0}^{2^k-1} \sum_{m=0}^{M} c_{l,m}^n \psi_{l,m}(x) = (C^n)^T \Psi(x). \tag{15}$$

According to Equation (15), we use only one-dimensional Legendre wavelets to approximate the solution. The solution dependence on the time variable is specified by the coefficient $c_{l,m}^n$. In other words, the vector coefficient C^n is calculated at time t_n. Therefore, the approximation solution at time $t_{n+\frac{1}{3}}$ can be written as follows:

$$u^{n+\frac{1}{3}} = (C^{n+\frac{1}{3}})^T \Psi(x). \tag{16}$$

We can also approximate $f(x,t)$ at time t_n as:

$$f(x,t_n) \simeq f^n = (F^n)^T \Psi(x), \tag{17}$$

where the vector $(F^n)^T$ is given by Equation (4) at time t_n.

Substituting Equation (6) into Equation (12) results in:

$$u^{n+\frac{1}{3}} = u^n + \frac{\Delta t}{3}\Big(\nu(\frac{\partial^2 u}{\partial x^2})^n + \mu u^n + f(x,t_n)\Big). \tag{18}$$

Now, by using the operation matrix of the derivative and Equation (5), we have:

$$\frac{\partial^2 u}{\partial x^2}(x,t_n) \simeq \sum_{l=0}^{2^k-1} \sum_{m=0}^{M} c_{l,m}^n \frac{\partial^2 \psi_{l,m}}{\partial x^2}(x) = (C^n)^T D^2 \Psi(x). \tag{19}$$

Then, substituting Equations (15)–(17) and (19) into Equation (18) yields:

$$(C^{n+\frac{1}{3}})^T \Psi(x) = (C^n)^T \Psi(x) + \frac{\Delta t}{3}\Big(\nu (C^n)^T D^2 \Psi(x) + \mu (C^n)^T \Psi(x) + (F^n)^T \Psi(x)\Big). \tag{20}$$

By using boundary conditions and Equation (16), the following equalities are satisfied:

$$h_0(t_{n+\frac{1}{3}}) = u(0,t_{n+\frac{1}{3}}) \simeq (C^{n+\frac{1}{3}})^T \Psi(0), \tag{21}$$

$$h_1(t_{n+\frac{1}{3}}) = u(1,t_{n+\frac{1}{3}}) \simeq (C^{n+\frac{1}{3}})^T \Psi(1). \tag{22}$$

Considering the initial Condition (7), we have:

$$g(x) = u(x,t_0) \simeq u^0 = (C^0)^T\Psi(x) \tag{23}$$

By substituting Equation (20) in $(2^k(M+1)-2)$ Gauss–Legendre points $\{x_i\}_{i=1}^{2^k(M+1)-2}$ and using Equations (21) and (22), we can obtain a linear system of equations with $2^k(M+1)$ unknown variables, $c_{l,m}^{n+\frac{1}{3}}$, which can be written in matrix form:

$$AC^{n+\frac{1}{3}} = B, \tag{24}$$

where A and B are $2^k(M+1) \times 2^k(M+1)$ and $2^k(M+1) \times 1$ matrices, respectively. Since the vector C^0 is obtained from Equation (23), all entries of B are known.

The above system of linear equations can be solved by numerical methods. Here, for square matrix A, we use LU decomposition to solve the linear System (24) with partial pivoting. In this method, the square matrix A can be decomposed into two square matrices L and U such that $A = LU$, where U is an upper triangular matrix formed as a result of applying the Gauss elimination method on A, and L is a lower triangular matrix with diagonal elements being equal to one. Solving the system $AC^{n+\frac{1}{3}} = B$ is then equivalent to solving the two simpler systems $Ly = B$ and $UC^{n+\frac{1}{3}} = y$. The first system can be solved by forward substitution, and the second system can be solved by backward substitution. Solving the linear system with triangular matrices makes it easy to do calculations in the process of finding the solution. Since the Gaussian elimination can produce bad results for small pivot elements, we adopt the partial pivoting strategy. In this strategy, when we are choosing the pivot element on the diagonal at position a_{ii}, locate the element in column i at or below the diagonal that has the largest absolute value, and make it as the pivot at that step by interchanging two rows. Applying this strategy to our matrix avoids any distortion due to the pivots being small. For more details, readers can refer to [18,19].

After solving this system and determining $C^{n+\frac{1}{3}}$, we exploit Equation (13) to find $C^{n+\frac{1}{2}}$. In addition, there is a similar process that results in:

$$(C^{n+\frac{1}{2}})^T\Psi(x_i) = (C^n)^T\Psi(x_i) + \frac{\Delta t}{2}\Big(\nu(C^{n+\frac{1}{3}})^T D^2\Psi(x_i) + \mu(C^{n+\frac{1}{3}})^T\Psi(x_i) + (F^{n+\frac{1}{3}})^T\Psi(x_i)\Big), \tag{25}$$

and:

$$h_0(t_{n+\frac{1}{2}}) = u(0,t_{n+\frac{1}{2}}) \simeq (C^{n+\frac{1}{2}})^T\Psi(0), \tag{26}$$

$$h_1(t_{n+\frac{1}{2}}) = u(1,t_{n+\frac{1}{2}}) \simeq (C^{n+\frac{1}{2}})^T\Psi(1), \tag{27}$$

where $\{x_i\}_{i=1}^{2^k(M+1)-2}$ are the same collocation points used in the previous step.

A matrix form of Equations (25)–(27) can be displayed as:

$$PC^{n+\frac{1}{2}} = Q,$$

where the dimension of the square matrix P and column vector Q is $2^k(M+1)$. Since the vector $C^{n+\frac{1}{3}}$ is obtained from the previous step, all entries of Q are known. Similarly, we use LU decomposition to solve the above system.

Finally, by implementing similar analysis in the two previous steps, using Equation (14) with boundary conditions and exploiting $C^{n+\frac{1}{2}}$, the vector C^{n+1} can be specified in each step for $n = 0,1,2,\cdots$. Therefore, we can obtain the numerical solution, $u(x,t_n)$, in any time $t = t_n$.

4. Stability Analysis

For stability analysis, we use the asymptotic (or absolute) stability of a numerical method, which is defined in [20]. In a numerical scheme, when we fix the final time $t = n\Delta t$ and let $n \to \infty$, we want the corresponding numerical solution to remain bounded; a scheme satisfying this property is called stable. Therefore, a stability analysis needs a restriction on the mesh size Δt. In practice, we can only choose a finite and proper mesh size. It is then important to study the region of absolute stability in order to to choose the proper mesh size in practical computation.

Let us start from the typical evolution equation:

$$\frac{\partial u}{\partial t} = f(u,t), \quad t > 0$$
$$u(0) = 0,$$

where the non-linear operator f contains the spatial part of the partial differential equation. Let us abbreviate $u(x_j, t_n)$ by u_j^n. We shall approximate u_j^n by U_j^n. Following the general formulation of the proposed method, the semi-discrete version is:

$$Q_j \frac{du_j}{dt} = Q_j f_j(u_j),$$

where u_j is the spectral approximation to u, f_j denotes the spectral approximation to the operator f and Q_j is the projection operator, which characterizes the scheme. Let us set $U(t) = Q_j u_j(t)$. Then, the previous discrete problem can be written in the form:

$$\frac{dU}{dt} = F(U). \tag{28}$$

As is often done, we confine our discussion of time-discretizations to the linearized version of (28):

$$\frac{dU}{dt} = LU \tag{29}$$

where L is the diagonalizable matrix resulting from the implementation of spectral method on the spatial variable of the partial differential equation.

According to different contexts, the time discretization is said to be stable if U^n, the computed solution at the time $t_n = n\Delta t$, has been upper bounded, i.e., there exists a constant M such that:

$$\|U^n\| \leqslant M. \tag{30}$$

In many problems, the solution is bounded in some norm for all $t > 0$. In these cases, a method that produces the exponential growth allowed by Estimate (30) is not practical for long-time integrations. For such problems, the notion of asymptotic (or absolute) stability is useful.

Definition 1. *The region of absolute stability of a numerical method is defined for the scalar model problem:*

$$\frac{dU}{dt} = \lambda U$$

to be the set of all $\lambda\Delta t$ such that $\|U^n\|$ is bounded as $t \to \infty$ [20].

Finally, we say that a numerical method is asymptotically stable for a particular problem if, for sufficiently small Δt, the product of Δt times every eigenvalue of L lies within the region of absolute stability. In the following items, we summarize some remarkable characteristics of absolute stability [21]:

1. An absolutely stable method is one that generates a solution u^n that tends to zero as t_n tends to infinity,
2. A method is said to be A-stable, if it is absolutely stable for any possible choice of the time-step, Δt, otherwise a method is called conditionally stable.
3. Absolutely stable methods keep the perturbation controlled,
4. The analysis of absolute stability for the linear model problem can be exploited to find stability conditions on the time step when considering some nonlinear problems.

Since the three-step Equations (12)–(14) are equivalent to the third-order Taylor expansion, to demonstrate the stability region and achieve the stability condition, we use Equation (10). For simplicity, consider Equation (6), where $\mu = 0$ and $f(x,t) = 0$. Then, successive differentiations of the obtained equation indicate that:

$$\frac{\partial^2 u}{\partial t^2} = \nu^2 \frac{\partial^4 u}{\partial x^4}, \tag{31}$$

$$\frac{\partial^3 u}{\partial t^3} = \nu^2 \frac{\partial^4}{\partial x^4}\left(\frac{u^{n+1} - u^n}{\Delta t}\right) + \mathbf{o}[(\Delta t)]. \tag{32}$$

In Equation (32), we use Euler's formula to avoid the third-order space derivatives, as it is used in the finite element context [22]. By rearranging Equation (10) and substitution of Equations (31) and (32), we have the semi-discrete equation:

$$\left(I - \frac{\nu^2 \Delta t^2}{6} \frac{\partial^4}{\partial x^4}\right)\left(\frac{u^{n+1} - u^n}{\Delta t}\right) = \nu\left(\frac{\partial^2 u}{\partial x^2}\right)^n + \frac{(\nu^2 \Delta t)}{2}\left(\frac{\partial^4 u}{\partial x^4}\right)^n. \tag{33}$$

After applying the wavelet collection method, Equation (33) transforms into the following equation:

$$\left(\frac{dC}{dt}\right)^n \simeq \left(\frac{C^{n+1} - C^n}{\Delta t}\right) = A^{-1} B C^n, \tag{34}$$

where:

$$A = \Big((I - \frac{\nu^2 \Delta t^2}{6} D^4)\, (\Psi(x_i))^T \Big),$$

$$B = \Big((\nu D^2 + \frac{\nu^2 \Delta t}{2} D^4)\, (\Psi(x_i))^T \Big),$$

and $\{x_i\}_{i=1}^{2^k(M+1)}$ are the collocation and boundary points. Here, the matrix L, which is introduced in Equation (29), is defined as $L = A^{-1}B$.

There is a similar process to the one-step method. Lambert provided an explanation for how to draw the stability region. Readers can refer to [23], Chapter 3. Briefly, we can plot the region of absolute stability, R_L, by the meaning of the first and second characteristic polynomials. If we set, $\hat{h} = \lambda \Delta t$, the region of absolute stability is a function of the method and the complex parameter $\hat{h}$ only, so that we are able to plot the region R_L in the complex $\hat{h}$-plane.

First of all, we can write Equation (34) as a usual linear multi-step method given by:

$$\sum_{j=0}^{k} \alpha_j C^{n+j} = \Delta t \sum_{j=0}^{k} \beta_j L(C^{n+j}), \tag{35}$$

where k is the number of steps required for the method, and α_j and β_j are constants subject to the conditions:

$$\alpha_k = 1, \quad |\alpha_0| + |\beta_0| \neq 0.$$

According to Equations (34) and (35), we have:

$$k = 1, \quad \alpha_0 = -1, \quad \alpha_1 = 1, \quad \beta_0 = 1, \quad \beta_1 = 0. \tag{36}$$

Afterward, the first and second characteristic polynomials are defined as follows, respectively:

$$\rho(\xi) = \sum_{j=0}^{k} \alpha_j \xi^j,$$
$$\sigma(\xi) = \sum_{j=0}^{k} \beta_j \xi^j,$$

where $\xi \in \mathbb{C}$ is a dummy variable. Using the values of k and $\{\alpha_j, \beta_j\}_{j=0}^{1}$ in (36), for the proposed method, we have:

$$\rho(\xi) = \xi - 1, \quad \sigma(\xi) = 1.$$

Then, we plot the boundary of R_L, which consists of the contour ∂R_L. The contour ∂R_L in the complex $\hat{h}$-plane is defined by the requirement that for all $\hat{h} \in \partial R_L$, one of the roots of $\pi(r, \hat{h}) := \rho(r) - \hat{h}\sigma(r)$ has modulus one, that is, it is of the form $r = \exp(i\theta)$. Thus, for all $\hat{h}$ in ∂R_L, the identity:

$$\pi(\exp(i\theta), \hat{h}) = \rho(\exp(i\theta)) - \hat{h}\sigma(\exp(i\theta)) = 0,$$

must hold. This equation is readily solved for $\hat{h}$, and we have that the locus of ∂R_L is given by:

$$\hat{h} = \hat{h}(\theta) = \frac{\rho(\exp(i\theta))}{\sigma(\exp(i\theta))} = \exp(i\theta) - 1. \tag{37}$$

Finally, we use (37) to plot $\hat{h}(\theta)$ for a range of $\theta \in [0, 2\pi]$ and link consecutive plotted points by straight lines to get a representation of ∂R_L.

Therefore, according to Lambert's book and the above explanations, the stability region of the three-step and one-step wavelet collocation methods is the circle with center $(-1, 0)$ and radius one. Therefore, these methods will be stable if the eigenvalues of the corresponding system and Δt satisfy $Re(\lambda_j \Delta t) \in [-2, 0]$.

5. Numerical Examples

In this section, some numerical examples in the form of Equation (6) with initial and boundary Conditions (7)–(9) are discussed. The error function is defined as the maximum error L_∞:

$$L_\infty = \max_{1 \leq i \leq 2^k(M+1)} |u_{exact}(x_i, t_n) - \overline{u}(x_i, t_n)|,$$

where $\overline{u}$ is the approximate solution, which is obtained by the proposed method and $\{x_i\}_{i=1}^{2^k(M+1)}$ are the Gauss–Legendre and boundary points. All programs have been performed in MATLAB 2016.

In general, the numerical results are sensitive to the selection of parameters such as time step Δt, final time t and parameters of wavelet order M and k. In the following examples, we choose $t = 0.5$. Although the one-step wavelet collocation method needs less calculation, the three-step wavelet collocation method is more successful in finding the numerical solution. Furthermore, we compare our method with the three-step method proposed in [17]. They used the finite element method with standard linear interpolation functions for spatial discretization and the same three-step formula for time discretization. Numerical results show that utilizing Legendre wavelets with these three steps gives higher accuracy.

Example 1. *Consider Equation (6) with* $v = 1/\pi^2$, $\mu = -3$, $f(x,t) = 3$, $g(x) = 1 + \sin(\pi x)$, $h_0(t) = 1$ *and* $h_1(t) = 1$. *The exact solution of this problem is* $u_{exact}(x,t) = 1 + e^{-4t}\sin(\pi x)$ [3].

Numerical results for $M = 6$ *and* $M = 8$ *are reported in Tables 1 and 2, respectively. The first two rows of Table 1 show the obtained results for* $k = 2$. *As can be seen from these rows, a change in the length of the time step makes the one-step method fail to find an approximate solution. In other words, the one-step method is unstable for* $\Delta t = 0.01$, *while the three-step method gives high accuracy. There is a similar analysis for other parameters.*

The exact and three-step approximate solutions of $u(x,t)$ *are shown in Figure 1, where* $M = 8$ *and* $k = 2$. *Figure 2 shows the absolute stability region based on the three-step and one-step methods with the position of* $\lambda_j \Delta t$, *where* $\{\lambda_j\}_{j=1}^{2k(M+1)}$ *are the eigenvalues of corresponding matrix L. This figure is drawn for* $M = 6$ *and* $\Delta t = 0.01$. *As can be seen in this figure, there are some eigenvalues for the one-step method with* $Re(\lambda_j \Delta t) \notin [-2, 0]$; *however, the stability region of the three-step method includes all* $\{\lambda_j \Delta t\}_{j=1}^{2k(M+1)}$. *Therefore, the one-step method is not stable for* $\Delta t = 0.01$, *while the three-step method is stable.*

Table 1. The L_∞ error of Example 1 in $M = 6$.

k	Δt	Method in [17]	One-Step Method	Three-Step Method
2	0.001	2.1847×10^{-3}	1.8061×10^{-4}	1.5183×10^{-4}
2	0.01	unstable	unstable	1.5179×10^{-4}
3	0.0001	1.6291×10^{-3}	1.3751×10^{-4}	1.0575×10^{-4}
3	0.003	unstable	unstable	1.0510×10^{-4}

Table 2. The L_∞ error of Example 1 in $M = 8$.

k	Δt	Method in [17]	One-Step Method	Three-Step Method
1	0.001	2.3446×10^{-3}	9.0287×10^{-5}	2.1781×10^{-5}
1	0.006	2.3325×10^{-3}	unstable	1.5179×10^{-4}
2	0.001	1.1569×10^{-3}	6.5330×10^{-5}	2.4159×10^{-5}
2	0.005	1.1531×10^{-3}	unstable	1.8234×10^{-5}

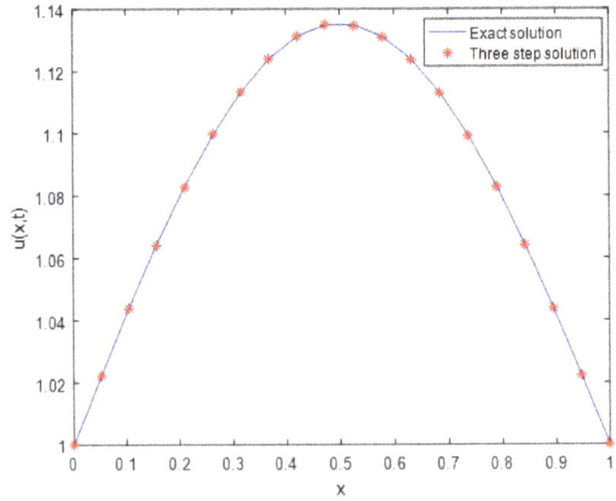

Figure 1. The exact and approximate solutions of Example 1 in the case $M = 8, k = 2$ and $\Delta t = 0.005$.

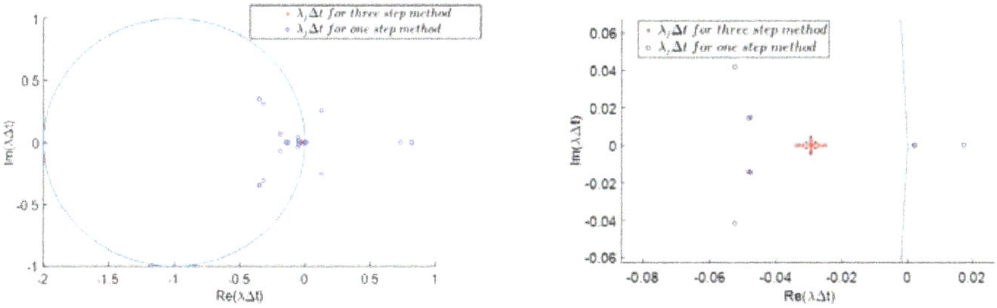

(**a**) Stability region (**b**) Magnification of (a)

Figure 2. Stability region of Example 1 with the position of $\lambda_j \Delta t$ for $M = 6$, $K = 2$ and $\Delta t = 0.01$. (**b**) is obtained from the magnification of (**a**).

Example 2. *In this example, we consider Equation (6) with* $\nu = 1/\pi^2$, $\mu = -4$, $f(x,t) = 0$, $g(x) = \sin(\pi x)$, $h_0(t) = 0$ *and* $h_1(t) = 0$. *The exact solution of this problem is* $u_{exact}(x,t) = e^{-5t}\sin(\pi x)$ [3].

Table 3 gives the comparison between the three-step wavelet collocation method and one-step wavelet collocation method for $M = 8$. *We can see from this table that the one-step wavelet collocation method tends to be unstable with a small change in time length. However, the three-step method keeps its stability for bigger* Δt.

The exact and approximate solutions for the three-step wavelet collocation method are shown in Figure 3. The stability region for both three-step and one-step methods by choosing $M = 8, k = 1$ *and* $\Delta t = 0.006$ *is shown in Figure 4. As can be seen from this figure, there are some eigenvalues in the system of the one-step collocation method with* $Re(\lambda_j \Delta t) \notin [-2, 0]$. *Therefore, this method is not stable for* $\Delta t = 0.006$. *In general, for* $k = 1$, *the one-step collocation method shows a stable and accurate result if* $\Delta t \leqslant 0.001$, *while the three-step collocation method is stable for* $\Delta t \leqslant 0.006$.

Table 3. The L_∞ error of Example 2 in $M = 8$.

k	Δt	Method in [17]	One-Step Method	Three-Step Method
1	0.001	8.6091×10^{-4}	7.3255×10^{-5}	3.0210×10^{-5}
1	0.006	8.5713×10^{-4}	unstable	3.0028×10^{-5}
2	0.001	7.2449×10^{-4}	5.2636×10^{-5}	1.6575×10^{-5}
2	0.0016	7.2280×10^{-4}	unstable	1.6528×10^{-5}

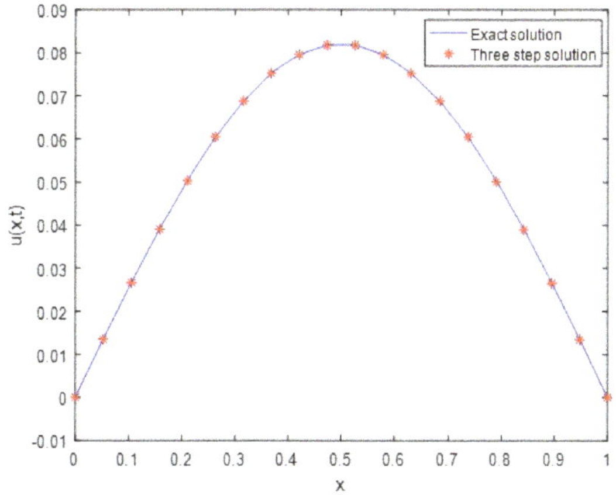

Figure 3. The exact and approximate solutions of Example 2 in the case $M = 8, k = 2$ and $\Delta t = 0.0016$.

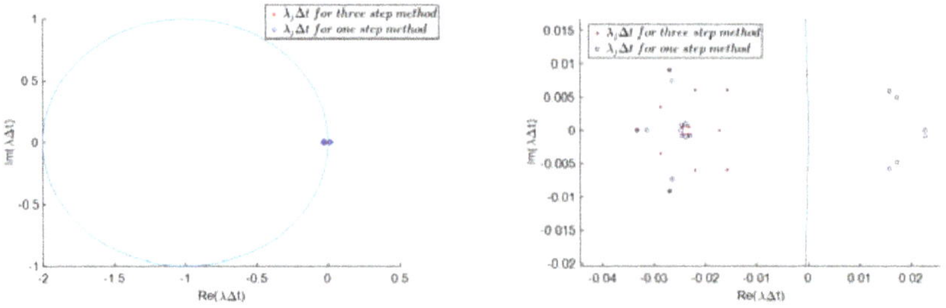

(**a**) Stability region (**b**) Magnification of (a)

Figure 4. Stability region of Example 2 with the position of $\lambda_j \Delta t$ for $M = 8$, $K = 1$ and $\Delta t = 0.006$. (**b**) is obtained from the magnification of (**a**).

Example 3. *For the last example, consider Equation (6) with* $\nu = 1/\pi^2$, $\mu = 0$, $f(x,t) = \sin(\pi x)$, $g(x) = \sin(\pi x) + \cos(\pi x)$, $h_0(t) = e^{-t}$ *and* $h_1(t) = -e^{-t}$. *The exact solution of this problem is* $u_{exact}(x,t) = \sin(\pi x) + e^{-t}\cos(\pi x)$ [3].

Table 4 shows the maximum error for some different values using the one-step and three-step wavelet collocation methods. It is clear from this table that the one-step wavelet collocation method is unstable, while the three-step wavelet collocation method has a more precise response.

Figure 5 shows the three-step approximate solution and the exact solution. The stability region for both three-step and one-step methods by choosing $M = 8, k = 1$ *and* $\Delta t = 0.006$ *is shown in Figure 6.*

Table 4. The L_∞ error of Example 3 in $M = 8$.

k	Δt	Method in [17]	One-Step Method	Three-Step Method
1	0.001	5.4341×10^{-3}	3.7505×10^{-4}	3.7552×10^{-4}
1	0.006	6.3181×10^{-3}	unstable	3.6961×10^{-4}
2	0.001	4.5033×10^{-3}	1.7750×10^{-4}	1.6859×10^{-4}
2	0.0016	4.2377×10^{-3}	unstable	1.6772×10^{-4}

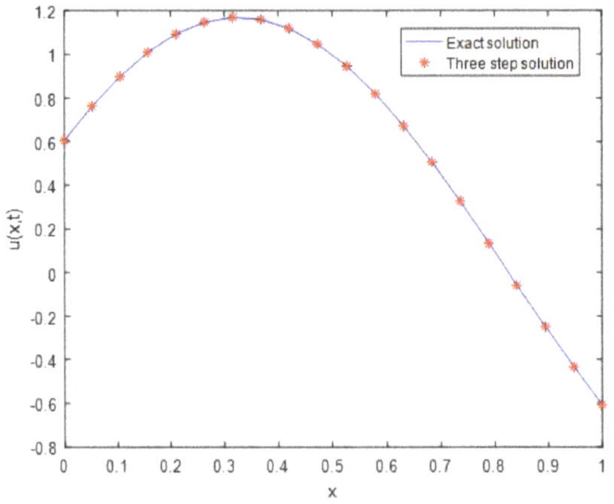

Figure 5. The exact solution of Example 3 in the case $M = 8, k = 2$ and $\Delta t = 0.0016$.

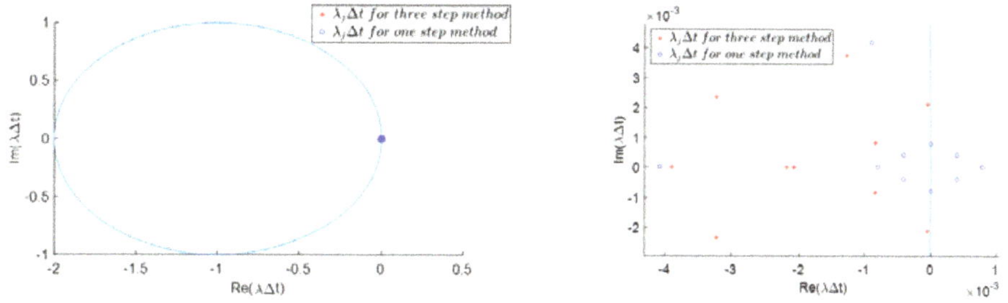

(**a**) Stability region (**b**) Magnification of (a)

Figure 6. Stability region of Example 3 with the position of $\lambda_j \Delta t$ for $M = 8$, $K = 1$ and $\Delta t = 0.006$. (**b**) is obtained from the magnification of (**a**).

6. Conclusions

In this paper, we proposed a new numerical method for a linear time-dependent partial differential equation. We called this method the three-step wavelet collocation method. In this method, time discretization was performed prior to the spatial discretization. These steps are equivalent to the third-order Taylor expansion; therefore, this method is third-order accurate in time. For spatial discretization, Legendre wavelets were used, which resulted in good spatial accuracy and spectral resolution. A comparison between the proposed method and other methods was presented. The theoretical aspect of absolute stability was discussed. This stability is based on $\lambda_j \Delta t$, where $\{\lambda_j\}$ are the eigenvalues of the corresponding system. Numerical performance shows that the three-step method leads to an effective time-accurate scheme with an improved stability property.

The proposed method can be easily implemented for other cases of time-dependent partial differential equations. For example, extending our results with Legendre wavelets in solving nonlinear partial differential equations or fractional equations is worthwhile for future contribution.

Author Contributions: All authors contributed significantly to the study and preparation of the article. Conceptualization, M.T. Formal analysis, M.T. and M.-M.H. Investigation, M.T. and M.-M.H. Supervision, M.-M.H.; Writing, original draft, M.T.

Funding: This research received no external funding.

Acknowledgments: The authors wish to thank Seyed Mehdi Karbassi for his useful comments.

Conflicts of Interest: The authors declare no conflict of interest.

References

1. Nouria, K.; Siavashani, N.B. Application of Shannon wavelet for solving boundary value problems of fractional differential equations. *Wavelet Linear Algebra* **2014**, *1*, 33–42.
2. Kumar, B.V.R.; Mehra, M. A three-step wavelet Galerkin method for parabolic and hyperbolic partial differential equations. *Int. J. Comput. Math.* **2006**, *83*, 143–157. [CrossRef]
3. Iqbal, M.A.; Ali, A.; Din, S.T.M. Chebyshev wavelets method for heat equations. *Int. J. Mod. Appl. Phys.* **2014**, *4*, 21–30.
4. Zhou, F.; Xu, X. Numerical solution of the convection diffusion equations by the second kind Chebyshev wavelets. *Appl. Math. Comput.* **2014**, *247*, 353–367. [CrossRef]
5. Mohammadi, F.; Hosseini, M.M. A new Legendre wavelet operational matrix of derivative and its applications in solving the singular ordinary differential equations. *J. Frankl. Inst.* **2011**, *348*, 1787–1796. [CrossRef]
6. Maleknejad, K.; Sohrabi, S. Numerical solution of Fredholm integral equations of the first kind by using Legendre wavelets. *J. Appl. Math. Comput.* **2007**, *186*, 836–843. [CrossRef]
7. Sahu, P.K.; Ray, S.S. Legendre wavelets operational method for the numerical solutions of nonlinear Volterra integro-differential equations system. *Appl. Math. Comput.* **2015**, *256*, 715–723. [CrossRef]

8. Chena, Y.M.; Weia, Y.Q.; Liub, D.Y.; Yu, H. Numerical solution for a class of nonlinear variable order fractional differential equations with Legendre wavelets. *Appl. Math. Lett.* **2015**, *43*, 83–88. [CrossRef]
9. Heydari, M.H.; Maalek Ghaini, F.M.; Hooshmandasl, M.R. Legendre wavelets method for numerical solution of time-fractional heat equation. *Wavelet Linear Algebra* **2014**, *1*, 15–24.
10. Islam, S.U.; Aziz, I.; Al-Fhaid, A.S.; Shah, A. A numerical assessment of parabolic partial differential equations using Haar and Legendre wavelets. *Appl. Math. Model.* **2013**, *37*, 9455–9481. [CrossRef]
11. Yin, F.; Tian, T.; Song, J.; Zhu, M. Spectral methods using Legendre wavelets for nonlinear Klein\Sine-Gordon equations. *J. Comput. Appl. Math.* **2015**, *275*, 321–324. [CrossRef]
12. Luo, W.H.; Huang, T.Z.; Gu, X.M.; Liu, Y. Barycentric rational collocation methods for a class of nonlinear parabolic partial differential equations. *Appl. Math. Lett.* **2017**, *68*, 13–19. [CrossRef]
13. Luo, W.H.; Huang, T.Z.; Wu, G.C.; Gu, X.M. Quadratic spline collocation method for the time fractional subdiffusion equation. *Appl. Math. Comput.* **2016**, *276*, 252–265. [CrossRef]
14. Jiang, C.B.; Kawahara, M. A three-step finite element method for unstesdy incompressible flows. *Comput. Mech.* **1993**, *11*, 355–370. [CrossRef]
15. Kashiyama, K.; Ito, H.; Behr, M.; Tezduyar, T. Three-step explicit finite element computation of shallow water flows on a massively parallel. *Int. J. Numer. Meth. Fluids* **1995**, *21*, 885–900. [CrossRef]
16. Quartapelle, L. *Numerical Solution of the Incompressible Navier-Stokes Equations*; Springer Basel AG: Basel, Switzerland, 1993.
17. Kumar, B.V.R.; Sangwan, V.; Murthy, S.V.S.S.N.V.G.K.; Nigam, M. A numerical study of singularly perturbed generalized Burgers-Huxley equation using three-step Taylor-Galerkin method. *Comput. Math. Appl.* **2011**, *62*, 776–786. [CrossRef]
18. Ford, W. *Numerical Linear Algebra with Applications Using MATLAB*; Elsevier: Waltham, MA, USA, 2015.
19. Toledo, S. Locality of reference in LU decomposition with partial pivoting. *SIAM J. Matrix. Anal. Appl.* **1997**, *18*, 1065–1081. [CrossRef]
20. Canuto, C.; Quarteroni, A.; Hussaini, M.Y.; Zang, T.A. *Spectral Methods in Fluid Dynamics*; Springer: Berlin, Germany, 1988.
21. Quarteroni , A.; Saleri, F. *Scientific Computing with MATLAB and Octave*; Springer: Berlin/Heidelberg, Germany, 2006.
22. Kumar, B.V.R.; Mehra, M. A wavelet-taylor galerkin method for parabolic and hyperbolic partial differential equations. *Int. J. Comput. Meth.* **2005**, *2*, 75–97. [CrossRef]
23. Lambert, J.D. *Numerical Methods For Ordinary Differential Systems*; John Wiley & Sons Ltd.: Hoboken, NJ, USA, 1991.

axioms

MDPI

Article

The 3D Navier–Stokes Equations: Invariants, Local and Global Solutions

Vladimir I. Semenov

Institute Physics, Mathematics and Computer Sciences, Baltic Federal University, Kaliningrad 236016, Russia; visemenov@rambler.ru

Received: 31 January 2019; Accepted: 1 April 2019; Published: 7 April 2019

Abstract: In this article, I consider local solutions of the 3D Navier–Stokes equations and its properties such as an existence of global and smooth solution, uniform boundedness. The basic role is assigned to a special invariant class of solenoidal vector fields and three parameters that are invariant with respect to the scaling procedure. Since in spaces of even dimensions the scaling procedure is a conformal mapping on the Heisenberg group, then an application of invariant parameters can be considered as the application of conformal invariants. It gives the possibility to prove the sufficient and necessary conditions for existence of a global regular solution. This is the main result and one among some new statements. With some compliments, the rest improves well-known classical results.

Keywords: Navier–Stokes equations; global solutions; regular solutions; a priori estimates; weak solutions; kinetic energy; dissipation

---

1. Introduction

During the last century, the Navier–Stokes equations attracted very much attention. The first essential steps in this way were offered by C. Oseen [1], F. K. G. Oldquist [2], J. Leray [3–5], and E. Hopf [6]. Later, the Cauchy problem and the boundary value problem were actively studied by many authors (see, for example, [7,8], the review [9–17] and etc.). The main objects and tools of these works were weak solutions or fix points of integral operators. Here, a special case is connected with the existence problem of a global and regular solution in the 3D Cauchy problem. In response to the new setting of this task by Ch. Fefferman in 2000 (see [18]), O.A. Ladyzhenskaya wrote in her review [9] that she would put the main question otherwise: "Do or don't the Navier–Stokes equations give, together with initial and boundary dates, the deterministic description of fluid dynamics?"

Then, this problem is more difficult and more interesting from the physical point of view. Therefore, I introduced some invariants for studying solutions properties. At least, it is natural for applications because invariants are very important and strong tools. Moreover, these invariants didn't apply earlier.

Let us describe them now. The first invariant connected with the Cauchy problem that provided initial data belongs to a special class $C^{\infty}_{6/5,\,3/2}$ of solenoidal vector fields vanishing at infinity. Here, outer forces are trivial. Then, the class $C^{\infty}_{6/5,\,3/2}$ is invariant (Theorem 2). This is a new result.

The second invariant is a special parameter λ (see (68)) which is connected with a velocity changing of E^2, where E is a kinetic energy of a fluid flow. If $\lambda \geq 1$ or kinetic energy at a special moment is not less any mean depending on λ for $\lambda < 1$ (i.e., changing of E^2 at moment $t = 0$ is negligible), then an ideal, global and smooth motion is determined. In other words, a global regular solution exists (Theorem 7). This is an essential and qualitative improvement of the classical result together with a new a priori estimate given by Theorems 8–10. These theorems are new results in principle.

Finally, the other parameters $\varepsilon, 0 < \varepsilon < 1$, (see formula (87)) and $\mu, 1 < \mu < \lambda^{-4}$, or $\mu = \infty$ may also be very useful (see formula (69), Lemma 50). The first of them is a dissipation coefficient of kinetic energy . The last parameter holds a time interval of a solution regularity. These three numerical characteristics $\lambda, \varepsilon, \mu$ are invariant with respect to the scaling procedure.

By the way, the first attempts to estimate invariant norms were implicitly undertaken in [12,16].

An introduction of a special invariant class of vector fields and invariant parameters gives the main idea for the proof of basic results. The first step is connected with a change of the construction offered in [19]. These changes concern solution approximations. The special kind of them gives many uniform a priori estimates. Approximations of a velocity function are built on a fundamental system with a condition for Laplacians of approximative solutions. They must be a finite part of the Fourier series. Simultaneously, approximations of a pressure function are being built. Jointly with a hydrodynamical potential, these approximations give the following facts and properties of local solutions:

(1) solutions are bounded with respect to a uniform norm and therefore it belongs to any class $L_{p,\,q}$;
(2) there is a universal time interval $[0, T_0)$ where bounded solutions exist;
(3) more exact necessary conditions of a hypothetical turbulence phenomenon if it is;
(4) a lower estimate of the kinetic energy which influences an existence of a global smooth solution.

The last two items are very important. If dissipation of kinetic energy is large (close to the unit), then blow up is probable.

To the structure of the paper. In the first part (Section 2), there are considered solutions' properties of the Cauchy problem in a local form if initial data is smooth enough. Here, there is given a modification of classical results with some supplements (see Theorem 1).The rest of this part contains technical lemmas which are proved by application of hydrodynamic potentials and multiplicative inequalities from Appendix (Appendix A). In the second part (Sections 3 and 4), there are existence conditions of global solutions studied in this problem, conditions for local solutions' extensions if the kinetic energy is small and close to the minimum. A more precise hypothetic blow up time interval is found. Here, three basic parameters λ, μ, ε are very useful.

The third part (Sections 5 and 6) contains the proof of main statements (Theorems 7–9), which are based on properties of invariant parameters λ, μ, ε.

I think, in this way, it is convenient to remove any restrictions on a smoothness in some contrast to the traditional way. The main idea is connected with an invariant form of an a priori estimate for gradient norms of a velocity. In addition, other norms are estimated in class L_6 and, after that, it is done in class L_2. In particular, it is shown that there is a bad solution of a class L_6 with some good properties. As the corollary, this solution has many uniformly bounded norms with respect to time argument. Only after that, by routine calculations, we prove the bad solution from above belongs to a class L_2. Precisely, this step distinguishes from classical way for the second time (see [7]).

In the considered problem, a boundedness of solutions depends on a smoothness of initial data. At least, initial data from the Sobolev class W_2^3 gives the same in principle.

The offered construction doesn't permit diminishing the index of smoothness.

In the final (Section 7), we explain the principal difference between the Navier–Stokes equations in space and plane.

A part of local results in modification (Section 2) and invariants as tools (Section 4) were announced by author in [20–22].

NOTATION. Now, let us consider the Cauchy problem ($n = 3$):

$$D_t u_k + \sum_{i=1}^{n} u_i u_{k,\,i} = \nu \Delta u_k - P_{,\,k},\ k = 1, 2, \ldots, n, \tag{1}$$

$$div u = 0,\ u(0, x) = \varphi(x), \tag{2}$$

where u is a velocity of flow, P is a pressure function, symbols

$$D_t u = \frac{\partial u}{\partial t}, u_{k,\,i} = \frac{\partial u_k}{\partial x_i},\ u_{k,\,ij} = \frac{\partial^2 u_k}{\partial x_i \partial x_j}, ..., P_{,\,k} = \frac{\partial P}{\partial x_k}$$

indicate a partial differentiation or differentiation in distributions, $\triangle$ is the Laplace operator, and ν is a positive constant (viscosity coefficient). A mapping φ has all derivatives and satisfies conditions of averaged growth: $\varphi \in L_{6/5}(R^3)$, $\varphi_{,\,i} \in L_{3/2}(R^3)$. The other derivatives belong to classes $L_r(R^3)$ for any $r > 1$. Furthermore, this class is denoted by symbol $C^{\infty}_{6/5,\,3/2}$. A class $C_0^{\infty}(R^n)$ is the class of infinitely smooth mappings with a compact support. A norm in a space $L_p(\Omega)$ is defined by formula:

$$\|v\|_p = \Big(\int_{\Omega} |v(x)|^p dx\Big)^{1/p}.$$

A mixed norm is defined by equality:

$$\|u\|_{p,\,q} = \Big(\int_0^T \Big(\int_{\Omega} |u(t,x)|^p dx\Big)^{\frac{q}{p}} dt\Big)^{\frac{1}{q}}.$$

A symbol $D^{\alpha} v$ denotes a partial differentiation or distributions with respect to a multi–index α. An order of the derivative is indicated by $|\alpha|$. Jacobi matrix of a mapping v with respect to spatial variables is denoted by ∇v . Its modulus is

$$|\nabla v| = \Big(\sum_{i,\,j} v^2_{i,\,j}\Big)^{\frac{1}{2}}. \tag{3}$$

Functions' properties from the Sobolev classes $W^l_p(\Omega)$ are given, for example, in [23–25]. A norm in this functional space is defined by

$$\|v\|_{W^l_p(\Omega)} = \sum_{|\alpha| \le l} \|D^{\alpha} v\|_p.$$

Let v be a mapping that is determined on the whole space. For the Riesz potential, we apply notation:

$$I_{\alpha}(v)(x) = \frac{1}{\gamma(\alpha)} \int_{R^n} \frac{v(y)dy}{|x-y|^{n-\alpha}}, \tag{4}$$

where $\gamma(\alpha) = \pi^{\frac{n}{2}} \Gamma(\frac{\alpha}{2}) / \Gamma(\frac{n-\alpha}{2})$ and Γ is the Euler gamma–function. The properties of these potentials can be found in [24].

The agreement about summation. Everywhere in this article, the repeated indices give a summation if it is not done reservation specially. For example,

$$u_i u_{j,\,i} = \sum_{i=1}^n u_i u_{j,\,i},\ u_{i,\,j} u_{j,\,i} = \sum_{i,j=1}^n u_{i,\,j} u_{j,\,i},\ u_i u_{j,\,i} \triangle u_j = \sum_{i,j=1}^n u_i u_{j,\,i} \triangle u_j,$$

etc.

Furthermore, $S_T = [0, T] \times R^3$. A number T_0 we define by formula:

$$T_0 = \Big(\frac{9}{4}\Big)^4 \frac{\nu^3}{\|\nabla \varphi\|_2^4}. \tag{5}$$

We apply the definition of a weak solution given in [7] everywhere.

2. Preliminaries. Boundedness and Smoothness Properties of Local Solutions in the Cauchy Problem

Here, with some compliments, a local result described by Theorem 1 is basic in this section. The rest contains only technical statements.

Theorem 1. *Let T_0 be a number from formula (5) and a mapping $\varphi \in C^{\infty}_{6/5,\,3/2}$. Then, on the set S_{T_0}, there exist weak solutions u and P of problems (1) and (2) with the following properties:*

(1) *mappings u and P uniformly continuous and bounded on the set S_T for every number T, $0 < T < T_0$;*

(2) *the solution u belongs to Sobolev classes $W^2_2(S_T)$ and $W^1_6(S_T)$ for every number T, $0 < T < T_0$, moreover, all norms*

$$\|u\|_p,\ \|\nabla u\|_p,\ \|D_t u\|_p,\ \|u_{,\,ij}\|_p,\ \|\nabla D_t u\|_2$$

are uniformly bounded in spaces $L_p(R^3)$, $2 \le p \le 6$, by a constant $C = C(\nu, \varphi, T)$ depending on ν, φ and T only, in addition $\|u\|_2 \le \|\varphi\|_2$;

(3) *gradients ∇u_i, $i = 1,2,3$, ∇P are bounded on the set S_T for every number T, $0 < T < T_0$;*

(4) *the solution P satisfies uniform estimates:*

$$\|\nabla P\|_q \le C,\ \frac{3}{2} < q < \infty,\ \|\nabla D_t P\|_q \le C,\ \|P_{,\,ij}\|_q \le C,$$

for all numbers q, $\frac{3}{2} < q \le 3$, and $t \in [0, T]$, $T < T_0$, with constants C depending on ν, φ, T and q only;

(5) *solutions u and P are classical solutions that is for any $T < T_0$ they belong to the class $C^{\infty}((0, T_0) \times R^3) \bigcap C(S_T)$.*

The proof of the theorem is given to the end of this section. We note items 1, 3, 4 compliment well-known Ladyzhenskaya's results (see [7]). Item (2) contains new uniform estimate for norms of derivatives. Hence, it follows a boundedness of weak solutions and a finiteness of its mixed norms. Moreover, we have an existence of weak solution with required properties on the interval $[0, T_0)$ with the finite length. To the studying of the smoothness property for weak solutions, the mixed norms were applied by O. Ladyzhenskaya in [26] (see, also [7]). They were applied by other authors (see, for example, [8,10,14]). Item (5) is a particular case from [27]. However, from this theorem, a deeper result follows (see Theorem 7).

2.1. A Priori Estimates of Gradients' Norms

Lemma 1. *Suppose that a mapping $w : S_{T_0} \to R^3$ belongs to a class C^2 and $w(0,x) = \varphi(x)$. If, for every $t \in [0, T_0)$, Laplacian supports are subsets of some ball with a fixed radius and $w \in L_6(R^3)$, $\nabla w \in L_2(R^3) \bigcap L_6(R^3)$, then for all mappings w satisfying condition:*

$$\frac{1}{2}\frac{d}{dt}\|\nabla w\|_2^2 + \nu\|\triangle w\|_2^2 = \int_{R^3} w_i w_{k,\,i} \triangle w_k dy, \tag{6}$$

the following estimate holds:

$$\|\nabla w\|_2 \le \frac{\|\nabla\varphi\|_2}{\left(1 - t/T_0\right)^{1/4}}$$

for all $t \in [0, T_0)$, where T_0 from formula (5).

Proof. We take from Corollary A4 the second inequality. Then, from (6), we obtain:

$$\frac{1}{2}\frac{d}{dt}\|\nabla w\|_2^2 + \nu\|\triangle w\|_2^2 \le a_1\|\nabla w\|_2^{3/2}\|\triangle w\|_2^{3/2}. \tag{7}$$

Let $y = ||\triangle w||_2/||\nabla w||_2^3$. Then, (7) can be rewritten in the form:

$$\frac{1}{2||\nabla w||_2^6}\frac{d}{dt}||\nabla w||_2^2 \leq a_1 y^{3/2} - \nu y^2.$$

The maximal mean on the right-hand side is $\frac{27a_1^4}{256\nu^3}$. Therefore, integrating the inequality

$$\frac{1}{2||\nabla w||_2^6}\frac{d}{dt}||\nabla w||_2^2 \leq \frac{27a_1^4}{256\nu^3}$$

over the interval $[0, t]$, we get:

$$\frac{1}{||\nabla \varphi||_2^4} - \frac{1}{||\nabla w||_2^4} \leq \frac{27a_1^4}{64\nu^3}t.$$

Furthermore, we take a number a_1 from Corollary A4 and obtain the required estimate. □

Lemma 2. *Let T_0 be a constant from Lemma 1. Assume a mapping $w : S_{T_0} \to R^3$ belongs to a class C^3 and $w(0,x) = \varphi(x)$, $D_t w(0,x) = \psi(x)$. Suppose that, for every t, there are fulfilled conditions:*

(1) Laplacian supports $\triangle w$, $\triangle D_t w$ are subsets of a ball with a fixed radius;

(2) mappings

$$w,\ D_t w \in L_6(R^3),\ \nabla w,\ \nabla D_t w \in L_2(R^3) \bigcap L_6(R^3);$$

(3) with constants k_1, l the inequalities hold:

$$||\nabla w||_2 \leq k_1||\nabla \varphi||_2,\ \int_0^t ||\triangle w||_2^2 dt \leq l;$$

(4) the equality

$$\frac{1}{2}\frac{d}{dt}||\nabla D_t w||_2^2 + \nu||\triangle D_t w||_2^2 = \int_{R^3}\left(D_t w_i \cdot w_{k,\,i} + w_i D_t w_{k,\,i}\right)\triangle D_t w_k dx \tag{8}$$

is true. Then, for every segment, $[0, T]$ where $T < T_0$ the estimate $||\nabla D_t w||_2 \leq k_2||\nabla\psi||_2$ holds with a constant k_2 which depends on $\nu, T, k_1, l, ||\nabla \varphi||_2$ only.

Proof. The integral on the right-hand side in formula (8) we rewrite with two integrals J_1 and J_2. Applying Corollary A4, we make estimates for every integral. In integral J_1, a triple of mappings u, v, w is the triple $D_t w,\ w,\ D_t w$. In integral J_2, a required triple is the triple $w,\ D_t w,\ D_t w$. Therefore, condition (3) yields estimates:

$$J_1 \leq a||\nabla D_t w||_2||\nabla w||_2^{1/2}||\triangle w||_2^{1/2}||\triangle D_t w||_2$$

$$\leq a\sqrt{k_1}||\nabla \varphi||_2^{1/2}||\nabla D_t w||_2||\triangle w||_2^{1/2}||\triangle D_t w||_2,$$

$$J_2 \leq a||\nabla w||_2||\nabla w||_2^{1/2}||\triangle w||_2^{1/2}|| \leq ak_1||\nabla \varphi||_2||\nabla D_t w||_2^{1/2}||\triangle w||_2^{3/2}.$$

Hence, from (8), we get:

$$\frac{1}{2}\frac{d}{dt}||\nabla D_t w||_2^2 + \nu||\triangle D_t w||_2^2 \leq \tag{9}$$

$$a\sqrt{k_1}||\nabla \varphi||_2^{1/2}||\nabla D_t w||_2^{1/2}||\triangle D_t w||_2\left(||\triangle w||_2^{1/2} + \sqrt{k_1}||\nabla \varphi||_2^{1/2}||\triangle D_t w||_2^{1/2}\right).$$

Let $g(t) = ||\nabla D_t w||_2$, $h(t) = ||\triangle D_t w||_2/g(t)$. Then, formula (9) can be transformed to the formula:

$$\frac{1}{2}\frac{d}{dt}\ln g(t) + \nu h^2(t) \leq a\sqrt{k_1}||\nabla \varphi||_2^{1/2}||\triangle w||_2^{1/2}h(t) + ak_1||\nabla \varphi||_2 h^{3/2}(t).$$

Let us integrate over segment $[0, t]$ this inequality. For the next step, we apply to each term the Hölder inequality for three and two factors, respectively getting quantities $h^2(t)$ and $||\triangle w||_2^2$. Hence, from condition (3), we obtain:

$$\frac{1}{2}\ln\frac{g(t)}{g(0)} + \nu\int_0^t h^2(t)dt \leq a\sqrt{k_1||\nabla\varphi||_2\sqrt[4]{t}}\Big(\int_0^t ||\triangle w||_2^2 dt\Big)^{1/4}\Big(\int_0^t h^2(t)dt\Big)^{1/2} +$$

$$+ak_1||\nabla\varphi||_2\sqrt[4]{t}\Big(\int_0^t h^2(t)dt\Big)^{3/4} \leq a\sqrt{k_1||\nabla\varphi||_2\sqrt[4]{lt}}\sqrt{y} + ak_1||\nabla\varphi||_2\sqrt[4]{t}y^{3/4},$$

where $y = \int_0^t h^2(t)dt$. Let M be a maximal mean of the function

$$F(y) = a_1\sqrt{y} + a_2 y^{3/4} - \nu y,$$

where $a_1 = a\sqrt{k_1||\nabla\varphi||_2\sqrt[4]{lt}}$, $a_2 = ak_1||\nabla\varphi||_2\sqrt[4]{t}$. Then, the last estimates give $g(t) \leq e^{2M}g(0)$. From the definition of function g, we have $g(0) = ||\nabla\psi||_2$. □

2.2. A Priori Estimates of Laplacian Norms

Lemma 3. *Let w, T_0 be a mapping and a number from Lemma 1. Then, for every number T, $0 < T < T_0$, there exists a constant $l = l(\nu, \varphi, T)$ such that*

$$\int_0^t ||\triangle w||_2^2 dt \leq l$$

for all $t \in [0, T]$.

Proof. We transform inequality (7) applying the estimate from Lemma 1. Then,

$$\frac{1}{2}\frac{d}{dt}||\nabla w||_2^2 + \nu||\triangle w||_2^2 \leq a\frac{||\nabla\varphi||_2^{3/2}}{(1-t/T_0)^{3/8}}||\triangle w||_2^{3/2}.$$

This inequality we integrate over the segment $[0, t]$. Then, we estimate the right-hand side applying the Hölder inequality and underlining the integral with the term $||\triangle w||_2^2$. If $\beta(t) = \int_0^t ||\triangle w||_2^2 dt$, then we get

$$\frac{1}{2}||\nabla w||_2^2\Big|_0^t + \nu\beta(t) \leq a||\nabla\varphi||_2^{3/2}\beta^{3/4}(t)\Big(\int_0^t (1-t/T_0)^{-3/2}dt\Big)^{1/4}.$$

The direct calculations of the integral on the right-hand side and the estimate

$$\frac{1}{\sqrt{1-b}} - 1 \leq \frac{b}{\sqrt{1-b}}$$

give the inequality:

$$\frac{1}{2}||\nabla w||_2^2 + \nu\beta(t) \leq a||\nabla\varphi||_2^{3/2}\beta^{3/4}(t)\frac{\sqrt[4]{2t}}{(1-t/T_0)^{1/8}} + \frac{1}{2}||\nabla\varphi||_2^2.$$

Take out the first term on the left hand. Then, the required estimate for function $\beta(t)$ will be obvious. If $\beta(t) \leq ||\nabla\varphi||_2$, then the estimate is acceptable. If $\beta(t) \geq ||\nabla\varphi||_2$, then we have:

$$\nu\beta(t) \leq a||\nabla\varphi||_2^{3/2}\beta^{3/4}(t)\frac{\sqrt[4]{2t}}{(1-t/T_0)^{1/8}} + \frac{1}{2}||\nabla\varphi||_2^{5/4}\beta^{3/4}(t).$$

Hence, it follows the lemma. □

Lemma 4. *Let w be a mapping from Lemma 2 and a number T_0 from Lemma 1. Then, for every number T, $0 < T < T_0$, there exists a constant $l_1 = l_1(\nu, \varphi, T)$ such that*

$$\int_0^t \|\triangle D_t w\|_2^2 dt \leq l_1$$

for all $t \in [0, T]$.

Proof. For the mapping w, inequality (9) is fulfilled. Its right-hand side we estimate relying on Lemma 2. Then,

$$\frac{1}{2}\frac{d}{dt}\|\nabla D_t w\|_2^2 + \nu\|\triangle D_t w\|_2^2 \leq \tag{10}$$

$$a\sqrt{k_1}k_2\|\nabla\varphi\|_2^{1/2}\|\nabla\psi\|_2\|\triangle w\|_2^{1/2}\|\triangle D_t w\|_2 + ak_1\sqrt{k_2}\|\nabla\varphi\|_2^{1/2}\|\nabla\psi\|_2\|\triangle D_t w\|_2^{3/2}.$$

Let C be a maximal coefficient of factors

$$\|\triangle w\|_2^{1/2}\|\triangle D_t w\|_2,\ \|\triangle D_t w\|_2^{3/2}.$$

Therefore, from formula (10), we have inequality:

$$\frac{1}{2}\frac{d}{dt}\|\nabla D_t w\|_2^2 + \nu\|\triangle D_t w\|_2^2 \leq C\|\triangle w\|_2^{1/2}\|\triangle D_t w\|_2 + C\|\triangle D_t w\|_2^{3/2}.$$

This inequality we integrate over segment$[0, t]$ and its right-hand side we estimate applying the Hölder inequality and underlining terms with norms $\|\triangle D_t w\|_2$. If

$$\beta_1(t) = \int_0^t \|\triangle, D_t w\|_2^2 dt$$

then we have the estimate:

$$\frac{1}{2}\|\nabla D_t w\|_2^2\Big|_0^t + \nu\beta_1(t) \leq C\sqrt[4]{t}\Big(\int_0^t \|\triangle w\|_2^2 dt\Big)^{1/4}\beta_1^{1/2}(t) + C\sqrt[4]{t}\beta_1^{3/4}(t).$$

We increase the right side using Lemma 3 and deduce the left side taking out the first positive term. Then, we obtain:

$$\nu\beta_1(t) \leq C\sqrt[4]{lt}\beta_1^{1/2}(t) + C\sqrt[4]{t}\beta_1^{3/4}(t) + \frac{1}{2}\|\nabla\psi\|_2^2.$$

Hence, we get the lemma in the same way as Lemma 3. If $\beta_1(t) > \|\nabla\psi\|_2$, then, from

$$\|\nabla\psi\|_2^2 < \beta_1^{1/2}(t)\|\nabla\psi\|_2^{3/2},$$

we obtain the lemma inequality. If $\beta_1(t) \leq \|\nabla\psi\|_2$, then the estimate is acceptable. □

2.3. Basic Space of Solenoidal Vector Fields and Orthogonal Systems

Let us consider solenoidal vector fields $\varphi : R^3 \to R^3$ from class C^∞ with a compact support of $\triangle\varphi$. A closure of this class is defined by the norm:

$$\|\varphi\| = \|\varphi\|_6 + |\nabla\varphi\|_2 + \sum_{i,j}\|\varphi,_{ij}\|_2. \tag{11}$$

We denote its by $J_0^2(R^3)$. From Lemmas A1 and A2, it follows that elements $u \in J_0^2(R^3)$ are represented by the Riesz potentials; moreover, u, $\nabla u \in L_6(R^3)$. Otherwise, each element is defined uniquely by its Laplacian. The class $J_0^2(R^3)$ is a separable space as a subspace of the Sobolev classes

$W^l_p(R^3), 1 < p < \infty$. Therefore, there exists a countable system $(\psi^n)_{n=1,...}$ of infinite smooth vector fields satisfying conditions:

(1) $div\ \psi^n = 0$;
(2) supports of $\triangle\psi^n$ are compact sets;
(3) the closure of a linear span in norm (11) coincides with the space $J^2_0(R^3)$.

Now, we apply the Sonin–Shmidt orthogonalization to the fundamental system $(\psi^n)_{n=1,...}$ and construct a countable system of mappings $(b^n)_{n=1,...}$, which would be with the orthogonality property of Laplacians in the space $L_2(R^3)$. That is, the scalar product

$$(\triangle b^n, \triangle b^m) = \int_{R^3} \triangle b^n_i \triangle b^m_i dx = \delta_{ij}, \tag{12}$$

where δ_{ij} is Kronecker's symbol. Then, every mapping b^n is a finite linear combination of mappings (ψ^k). Therefore, a support of $\triangle b^n$ is a compact set. Let

$$\triangle b^n = a^n. \tag{13}$$

The system (a^n) is complete for the space $J^2_0(R^3)$; that is, the following proposition is true.

Lemma 5. *If in the space $L_2(R^3)$ for a some vector field $u \in J^2_0(R^3)$ the scalar product $(u, a^n) = 0$ for every $n = 1, 2, \ldots$ then $u = 0$.*

Proof. From chosen mappings a^n, the equality $(u, \triangle\psi^n) = 0$ for each element of the fundamental system $(\psi^n)_{n=1,...}$ follows. The Stokes theorem gives

$$\int_{|x|\leq r} u_k \triangle\psi^n_k dx = -\int_{|x|\leq r} u_{k,\,i}\psi^n_{k,\,i} dx + \int_{|x|=r} u_k \psi^n_{k,\,i}\frac{x_i}{r} dS.$$

The integral over the sphere vanishes as $r \to \infty$. Actually, from Corollary A2 of Lemma A2, we have:

$$\Big|\int_{|x|=r} u_k \psi^n_{k,\,i}\frac{x_i}{r} dS\Big| \leq \frac{C_1}{r^2}\int_{|x|=r} |u(x)| dS.$$

Furthermore, we apply Lemma A4 ($\alpha = 2,\ p = 6$) taking into consideration a continuity of u. The passage to the limit yields the equality $(u, \triangle\psi^n) = -(\nabla u, \nabla\psi^n)$ or $(\nabla u, \nabla\psi^n) = 0$. We take a sequence of finite and smooth mappings

$$(\eta^n)_{n=1,...},\ \eta^n \in J^2_0(R^3),\ \eta^n = \sum_i^n \beta_i \psi^i,$$

which converges to the vector field u in the space $J^2_0(R^3)$. Hence, $(\nabla u, \nabla u) = 0$. The summability of u in the space $L_6(R^3)$ proves lemma equality. □

Remark 1. *To the fundamental system of mappings (ψ^n) we can adjoin any solenoidal vector field $\varphi \in C^\infty_0(R^3), \varphi \neq 0$ or any vector field from the class $J^2_0(R^3)$ as the first element of this system.*

2.4. Successive Approximations of Solutions and Its Estimates: Velocity

Let $(a^n_{1,...})$ be an orthonormal system of mappings in the space $L_2(R^3)$ constructed above with the completeness property in $J^2_0(R^3)$ and conditions (12) and (13). Moreover, $a^n \in C^\infty_0(R^3)$ for all n and $a^1 = \frac{\triangle\varphi}{\|\triangle\varphi\|_2}$ where a vector field $\varphi \in C^\infty_0(R^3)$ is initial data in problems (1) and (2).

For successive approximations v^n, we define changing Ladyzhenskaya's construction in ([7], p. 197). Set

$$\triangle v^n(t,x) = \sum_{q=1}^{n} c_{qn}(t)a^q(x). \tag{14}$$

Then, an approximative solution v^n is built as a hydrodynamical potential

$$v^n(t,x) = -\frac{1}{4\pi}\sum_{q=1}^{n} c_{qn}(t)\int_{R^3}\frac{a^q(y)dy}{|x-y|}. \tag{15}$$

Functions c_{qn} are solutions of a system of differential equations:

$$\left(D_t v^n, a^q\right) - \nu\left(\triangle v^n, a^q\right) + \int_{R^3} v_i^n v_{k,\,i}^n a_k^q dx = 0,\ q = 1,2\ldots,n, \tag{16}$$

with initial data: $c_{qn}(0) = ||\triangle\varphi||_2\delta_{q1}$, where δ_{qr} is Kronecker's delta. Hence,

$$\triangle v^n(0,x) = \triangle\varphi(x),\ v^n(0,x) = \varphi(x). \tag{17}$$

Now, we find an existence interval of a smooth solution in system (16). For every equation from (16), we multiply by functions c_{qn} and sum them. As a result, we have:

$$\left(D_t v^n, \triangle v^n\right) - \nu|\triangle v^n\|_2^2 + \int_{R^3} v_i^n v_{k,\,i}^n \triangle v_k^n dx = 0.$$

From Corollary A3, we get:

$$\left(D_t \nabla v^n, \nabla v^n\right) + \nu||\triangle v^n||_2^2 = \int_{R^3} v_i^n v_{k,\,i}^n \triangle v_k^n dx. \tag{18}$$

From Lemmas A1 and A2 vector fields $v^n \in L_6(R^3)$, $\nabla v^n \in L_6(R^3)\bigcap L_2(R^3)$. Equalities (17) and (18) are conditions of Lemma 1 for mappings v^n. Therefore, in system (16), an existence of smooth solutions on some interval $[0,t_0)$ is guaranteed by well-known theorems for ordinary differential equations. By Lemma 1 (see estimates), these solutions can be extended on the interval $[0,T_0)$ where T_0 is the constant in Lemma 1 (see also (5)). Thus, we proved the following statement.

Lemma 6. *Let $[0,T_0)$ be an interval from Lemma 1. Then, for every $t \in [0,T_0)$, approximations v^n constructed by formulas (14) and (16) satisfy conditions:*

(1) $||\nabla v^n||_2 \le ||\nabla\varphi||_2\left(1 - t/T_0\right)^{-1/4}$,
(2) $||\nabla v^n||_6 \le A||\triangle v^n||_2$,

where a constant A from Lemma A1.

Proof. Item (1) follows from Lemma 1. Item (2) is the corollary of the second representation in (A1), Lemma A1 and arguments in the proof of Corollary A4. □

Lemma 7. *Let $[0,T_0)$ be a constant of Lemma 1. Then, for every segment $[0,T]$, $T < T_0$, approximations v^n, which are constructed by formulae (14)–(16), satisfy inequalities:*

(1) $||\nabla D_t v^n||_2 \le k_2||\nabla(\nu\triangle\varphi - \varphi_i\varphi_{,\,i})||_2$, *where a number $k_2 = k_2(\nu,\varphi,T)$ depends on ν,φ,T only;*
(2) $||\nabla D_t v^n||_6 \le A||\triangle D_t v^n||_2$ *with the constant A from Lemma A1.*

Proof. The item (2) can be proved in the same way as the estimate (2) from Lemma 6. Let us prove item (1). We differentiate equalities (16) with respect to t. Then, from each, we multiply by the derivative $c'_{qn}(t)$ and add together in final. As a result, we have

$$\left(D_{tt}v^n, \triangle D_t v^n\right) - \nu\|\triangle D_t v^n\|_2^2 + \int_{R^3}\left(D_t v_i^n v_{k,\,i}^n + v_i^n D_t v_{k,\,i}^n\right)\triangle D_t v_k^n dx = 0.$$

A support of $\triangle D_t v^n$ is a compact set. The Stokes theorem and Corollary A3 give:

$$\left(D_{tt}\nabla v^n, \nabla D_t v^n\right) + \nu\|\triangle D_t v^n\|_2^2 = \int_{R^3}\left(D_t v_i^n v_{k,\,i}^n + v_i^n D_t v_{k,\,i}^n\right)\triangle D_t v_k^n dx. \tag{19}$$

From Lemmas A1 and A2, we have

$$v^n, D_t v^n \in L_6(R^3),\ \nabla v^n,\ D_t\nabla v^n \in L_6(R^3)\bigcap L_2(R^3).$$

By Lemma 3, the vector field v^n satisfies the inequality:

$$\int_0^t \|\triangle v^n\|_2^2 dt \leq l(\nu, \varphi, T).$$

Then, mappings v^n satisfy Lemma 2. This implies:

$$\|\nabla D_t v^n\|_2 \leq k_2\|\nabla D_t v^n(0,x)\|_2 \tag{20}$$

with some constant $k_2 = k_2(\nu, \varphi, T)$.

Let us estimate the right-hand side of (20). In (16), we take $t = 0$. Then, we multiply them by numbers $c'_{qn}(0)$ respectively and add them together. As a result, formula (17) gives

$$\left(D_t v^n(0,x), \triangle D_t v^n(0,x)\right) - \left(\triangle D_t v^n(0,x), T^1(x)\right) = 0, \tag{21}$$

where

$$T^1 = \nu\triangle\varphi - \varphi_i\varphi_{,\,i}. \tag{22}$$

We move derivatives with the factor $\triangle D_t v^n$ in (21) using Corollary A3 and a finiteness of mapping φ. Then, we obtain

$$\|\nabla D_t v^n(0,x)\|_2^2 = \left(\nabla D_t v^n(0,x), \nabla T^1(x)\right).$$

Apply Cauchy–Bunyakovskii's inequality. Hence, we get the required estimate

$$\|\nabla D_t v^n(0,x)\|_2 \leq \|\nabla T^1(x)\|_2.$$

Thus, from (20), a lemma follows. □

Lemma 8. *Let T_0 be a constant of Lemma 1. Then, on every segment $[0,T]$, $0 < T < T_0$, approximations v^n from formulae (14)–(16) satisfy conditions:*

(1) $\|\triangle v^n\|_2 \leq C = C(\nu, \varphi, T)$;

(2) $\int_0^t \|\triangle D_t v^n\|_2^2 dt \leq l_1$;

where constants C and l_1 depend on ν, φ, T only.

Proof. Condition (1) follows from (18). We apply the Cauchy–Bunyakovskii inequality and estimates (1) of Lemmas 6 and 7 to the scalar product $(D_t\nabla v^n, \nabla v^n)$. Then, $|(D_t\nabla v^n, \nabla v^n)| \leq C_1(\nu, \varphi, T) = C_1$. The right-hand side from (18) is estimated by applying Corollary A4, where we take the triple v_n, v_n, v_n. From (18), we have

$$\nu\|\triangle v^n\|_2^2 \leq C_1 + a\|\nabla v^n\|_2^{3/2}\|\triangle v^n\|_2^{3/2}.$$

Apply again estimate (1) of Lemma 6. Then,

$$\nu\|\triangle v^n\|_2^2 \leq C_1 + C_2\|\triangle v^n\|_2^{3/2}.$$

This implies condition (1). Vector fields v^n satisfy Lemma 2 (see the proof of Lemma 6). Then, Lemma 4 gives estimate (2). □

Lemma 9. *Let T_0 be a constant of Lemma 1. Then, approximations v^n from (14)–(16) are bounded by a constant C on the set S_T for every $T < T_0$ where a constant C depends on ν, φ, T only.*

Proof. For approximation v^n, we use integral representation (A2). One should replace integration over whole space by integrations over ball $|y-x| \leq 1$ and its complement. Then, $v^n(t,x) = \frac{1}{4\pi}(J_1 + J_2)$. Every term is estimated by application of Hölder's inequality. We have

$$|J_1| \leq \|\nabla v^n\|_6\Big(\int_{|y-x|\leq 1}\frac{dy}{|x-y|^{2,4}}\Big)^{5/6}, |J_2| \leq \|\nabla v^n\|_2\Big(\int_{|y-x|\geq 1}\frac{dy}{|x-y|^4}\Big)^{1/2}.$$

Hence,

$$|J_1| \leq C_1|\nabla v^n\|_6,\ |J_2| \leq C_2|\nabla v^n\|_2,$$

where C_1, C_2 are universal constants. The norm $\|\nabla v^n\|_6$ is estimated in two steps. In the first step, we apply inequality 2) from Lemma 6. After that, we use inequality (1) from lemma 8. To estimate another norm $\|\nabla v^n\|_2$, we can apply inequality (1) from Lemma 6. Hence, we get a boundedness of all vector fields v^n by a general constant. □

Lemma 10. *Let T_0 be a constant of Lemma 1. Then, for every exponent $p \in [3/2, 6]$ and every segment $[0,T]$, $T < T_0$, approximations v^n from formulae (14)–(16) satisfy the inequality $\|v_i^n v^n_{,i}\|_p \leq C$ with a constant C depending on ν, φ, T, p only.*

Proof. If $p = 6$, then the statement follows from Lemma 9 and estimates by item (2) of Lemma 6 and item (1) of Lemma 8. If $p = 3/2$, then we apply Hölder's inequality. Hence, we have $\|v_i^n v^n_{,i}\|_{3/2} \leq \|v^n\|_6\|\nabla v^n\|_2$. Estimates of Lemma 6 and Lemma 8 prove the lemma for this exponent. An intermediate exponents is verified by Lemma A5. □

Lemma 11. *Let T_0 be a constant of Lemma 1. Then, for all $t \in [0, T_0)$, approximations v^n from formulae (18)–(20) satisfy inequalities*

$$\|v^n_{,ij}\|_2 \leq M|\triangle v^n\|_2,\ \|D_t v^n_{,ij}\|_2 \leq M|\triangle D_t v^n\|_2,$$

$i,j = 1,2,3$, with a universal constant M.

Proof. The statement of lemma is the corollary well-known results about integral differentiation with a weak singularity (see [28]). From the second representation of Lemma A2, we obtain two equalities: $v^n_{,ij} = k_{ij}\triangle v^n + T_{ij}(\triangle v^n)$, $D_t v^n_{,ij} = k_{ij}\triangle D_t v^n + T_{ij}(\triangle D_t v^n)$, where k_{ij} are some constants, T_{ij} are singular integral operators. Its boundedness in the space L_2 gives the required estimates. □

Lemma 12. *Let T_0 be a constant of Lemma 1. Then, for every segment $[0,T]$, $T < T_0$, for all exponents $p \in [1, 3/2]$ and each triple $i,j,k = 1,2,3$ approximations v^n from (14)–(16) satisfy inequalities:*

(1) $\|v^n_{i,j} v^n_{j,ik}\|_p \leq C$;
(2) $\|v^n_{i,j} D_t v^n_{j,ik}\|_p \leq C\|\triangle D_t v^n\|_2$;
(3) $\|D_t v^n_{i,j} v^n_{j,ik}\|_p \leq C\|\triangle D_t v^n\|_2^{3/p-2}$;
(4) $\|v^n_{i,j} D_t v^n_{j,i}\|_p \leq C$;

where constants C depend on ν, φ, T, p only.

Proof. Apply Hölder's inequality. Then,

$$\int |h_{i,\,j} g_{j,\,ik}|^p dx \le \Big(\int |h_{i,\,j}|^{2p/(2-p)} dx\Big)^{1-p/2} \Big(\int |g_{j,\,ik}|^2 dx\Big)^{p/2}. \tag{23}$$

Denote $h = v^n$, $g = v^n$. An exponent $2p/(2-p) \in [2,6]$. Then, the first factor in (23) is estimated by Lemma 1 with an assumption $r = 2$, $s = 6$. Uniform estimate (1) follows from Lemma 6 and Lemma 8. In the same way taking a pair $h = v^n$, $g = D_t v^n$ we get estimate (2). Now, denote $h = D_t v^n$, $g = v^n$. To the first factor from the right-hand side of (23) we apply Lemma A5 relying on $r = 2$, $s = 6$, $t = 3 - 3/p$. The norm $\|\nabla D_t v^n\|_2$ has a uniform estimate with respect to t and n by Lemma 7. Apply the both estimates of this lemma and obtain estimate (3). The other estimates (4) and (1) we prove in the same way. □

2.5. Successive Approximations of Solutions and Its Estimates: Pressure

Let v^n be an approximation from formulae (14)–(16). Fix T, $T < T_0$ where T_0 is the constant from Lemma 1. Consider a hydrodynamical potential

$$P^n(t,x) = \frac{1}{4\pi} \int_{R^3} \frac{v^n_{i,\,j}(t,y) v^n_{j,\,i}(t,y) dy}{|x-y|}. \tag{24}$$

A product $v^n_{i,\,j} v^n_{j,\,i} \in L_1(R^3) \bigcap L_3(R^3)$. This follows from estimates of Lemma 6, Lemma 8 and Hölder's inequality. By Lemma A4 on every segment $[0,T]$, we have:

$$\|v^n_{i,\,j} v^n_{j,\,i}\|_p \le C(\nu, \varphi, T, p) = C, \ 1 \le p \le 3. \tag{25}$$

Lemma A1 implies a uniform estimate with respect to t and n:

$$\|P^n\|_q \le A(p,q) C(\nu, \varphi, T, p) \tag{26}$$

for any exponent $q > 3$, where $\frac{1}{q} = \frac{1}{p} - \frac{2}{3}$.

Let us decompose integral in (24) by two integrals J_1 and J_2: over ball $|y-x| < 1$ and over its exterior. Every integral we estimate by Hölder's inequality or a simple estimation. Then,

$$4\pi J_1 \le \|v^n_{i,\,j} v^n_{j,\,i}\|_3 \Big(\int_{|y-x|<1} \frac{dy}{|x-y|^{1,5}}\Big)^{2/3} \le C_1, \ 4\pi J_2 \le \|v^n_{i,\,j} v^n_{j,\,i}\|_1 \le C_2.$$

Thus, with some constant $C = C(\nu, \varphi, T)$ on the set S_T for all n, we obtain:

$$|P^n(t,x)| \le C. \tag{27}$$

Function P^n has derivatives in distributions:

$$P^n_{,\,i}, \ D_t P^n, \ D_t P^n_{,\,i}, \ P^n_{,\,ij}, \ D_t P^n_{,\,ij}.$$

The differentiation of the integral from (24), the summation and a simple estimation give:

$$|\nabla P^n(t,x)| \le \frac{1}{2\pi} \int \frac{|\nabla v^n(t,y)|^2 dy}{|x-y|^2}, \ |\nabla D_t P^n(t,x)| \le \frac{1}{2\pi} \int \frac{|v^n_{i,\,j}(t,y)| D_t v^n_{j,\,i}(t,y)| dy}{|x-y|^2}. \tag{28}$$

By Lemma A1 for exponents $p \in (1,3]$ and $q > 3/2$ where $\frac{1}{q} = \frac{1}{p} - \frac{1}{3}$, we have:

$$\|\nabla P^n(t,x)\|_q \le 2A(p,q) \| |\nabla v^n|^2 \|_p. \tag{29}$$

The right-hand side of (29) is bounded upper by a constant $C = C(\nu, \varphi, T, p)$. Here, we apply inequalities from Lemma 6, Lemma 8 and Lemma A5. Therefore,

$$\|\nabla P^n(t,x)\|_q \leq C. \tag{30}$$

Derivatives

$$v^n_{i,\,j},\ D_t v^n_{i,\,j} \in L_6(R^3) \bigcap L_2(R^3).$$

Thus, $D_t \nabla P^n \in L_q(R^3)$ for any exponent $q > 3/2$. By Lemma A1, we obtain:

$$\|\nabla D_t P^n\|_q \leq 2A(p,q)\|v^n_{i,\,j} D_t v^n_{j,\,i}\|_p. \tag{31}$$

Consider two cases: $1 < p \leq 3/2$ and $3/2 < p \leq 3$.

Let $1 < p \leq 3/2$. Then, the right-hand side of (31) is bounded by a constant $C = C(\nu, \varphi, T, p)$. This follows from estimate 4 of Lemma 12.

Let $3/2 < p \leq 3$. Then, the exponent $6p/(6-p) \in [2,6]$. Applying Hölder's inequality, we get

$$\|v^n_{i,\,j} D_t v^n_{j,\,i}\|_p \leq \|v^n_{i,\,j}\|_{6p/(6-p)} \|D_t v^n_{j,\,i}\|_6.$$

The first factor is estimated uniformly by a some constant $C = C(\nu, \varphi, T, p)$. This is proved by application Lemma A5, Lemmas 6 and 8. The second factor is estimated by inequality (2) from Lemma 7. Hence, for an exponent q, $q > 3$, $\frac{1}{q} = \frac{1}{p} - \frac{1}{3}$, we get:

$$\|v^n_{i,\,j} D_t v^n_{j,\,i}\|_p \leq C\|\triangle D_t v^n\|_2.$$

Applying the integral representation for derivative $D_t P^n$ in the same way we prove another uniform estimate $\|D_t P^n\|_q \leq C$ for every exponent q, $q > 3$.

As the final result from (26), (27), (30), (31), we obtain the following statement.

Lemma 13. *Let T_0 be a constant from Lemma 1. Let P^n be a function defined by (24). Then, on every segment $[0,T]$, $T < T_0$, with some constants $C_1 = C(\nu, \varphi, T)$, $C_2 = C(\nu, \varphi, T, q)$, there are fulfilled uniform estimates with respect to $t \in [0,T]$ and n:*

(1) $|P^n(t,x)| \leq C_1$ *for all* $x \in R^3$;
(2) $\|\nabla P^n\|_q \leq C_2$ *for every* $q > 3/2$;
(3) $\|\nabla D_t P^n\|_q \leq C_2$ *for every* $q \in (3/2, 3]$;
(4) $\|\nabla D_t P^n\|_q \leq C_2\|\triangle D_t P^n\|_2$, $\|P^n\|_q \leq C_2$, $\|D_t P^n\|_q \leq C_2$ *for every* $q > 3$.

Lemma 14. *Suppose that T_0 is the constant from Lemma 1. Let P^n be a function defined by (24). Then, on every segment $[0,T]$, $T < T_0$, with some constants $C_2 = C(\nu, \varphi, T, q)$, there are fulfilled uniform estimates with respect to $t \in [0,T]$ and n:*

(1) $\|P^n_{,\,km}\|_q \leq C_2$;
(2) $\|D_t P^n_{,\,km}\|_q \leq C_2 \max(1, \|\triangle D_t v^n\|_2)$;

for every $q \in (3/2, 3]$ and every pair of numbers $k, m = 1, 2, 3$.

Proof. These estimates follow from Lemma A1, Lemma 12 and integral representations for derivatives extracting from (24). Apply Lemma A1 and item (1) of Lemma 12. Then, we obtain the first inequality. In the same way, we get the second inequality with an application of estimates (2) and (3) from Lemma 12. □

Lemma 15. *Let T_0 be a constant of Lemma 1. Let P^n be a function defined by (24). Then, on every segment $[0,T]$, $T < T_0$, with some constant $C = C(\nu, \varphi, T, q)$, there are fulfilled uniform estimates with respect to $t \in [0,T]$ and n: $\|P^n_{,klm}\|_q \leq C$ for every $q \in (1, 3/2]$, $k,l,m = 1,2,3$.*

Proof. It is sufficient to repeat the proof of Lemma 11 with the application of formula (24). □

Lemma 16. *Let T_0 be a constant from Lemma 1. Supposing that P^n is the function defined by (24), then*

$$\triangle P^n = -v^n_{i,\,j} v^n_{j,\,i}.$$

Proof. This follows from proposition A3. □

2.6. Estimates of Uniform Continuity of Approximations in Spaces $L_2(R^3)$ and $C(S_T)$

Now, we estimate the integral continuity modulus of gradients and Laplacians for approximations following [7]. Let T_0 be a constant from Lemma 1. Let T, T_1 be arbitrary numbers such that $T < T_1 < T_0$. Assume $t \in [0,T]$, $t+h \in [0,T_1]$. Equations (16) we write by the following form:

$$\Big(D_t v^n(t+h,\cdot), a^q\Big) - \nu\Big(\triangle v^n(t+h,\cdot), a^q\Big) + \int v^n_i(t+h,x) v^n_{k,\,i}(t+h,x) a^q_k dx = 0, \tag{32}$$

$$q = 1, \ldots, n.$$

Every equality we multiply by difference $c_{qn}(t+h) - c_{qn}(t)$ respectively and add together them. Setting $z = v^n(t+h,x) - v^n(t,x)$, we have

$$\Big(\frac{\partial z}{\partial h}, \triangle z\Big) = \nu(\triangle v^n(t+h,\cdot), \triangle z) - \int v^n_i(t+h,x) v^n_{k,\,i}(t+h,x) \triangle z_k dx.$$

To the scalar product on the right-hand side, we apply Cauchy–Bunyakovskii's inequality. The integral (J is its mean) we estimate by Corollary A4 for the triple v^n, v^n, z. Then,

$$|(\triangle v^n(t+h,\cdot), \triangle z)| \leq \|\triangle v^n\|_2 \|\triangle z\|_2,$$

$$|J| \leq a\|\nabla v^n\|_2^{3/2} \|\triangle v^n\|_2^{1/2} \|\triangle z\|_2.$$

Every factor from the right-hand side of these inequalities is bounded by a constant $C = C(\nu, \varphi, T_1)$ uniformly with respect to t, n, h. This follows from estimates of Lemmas 6–8, definition of z and the choice of means h, T_1. Since

$$\Big(\frac{\partial z}{\partial h}, \triangle z\Big) = -\frac{1}{2}\frac{d}{dh}\|\nabla z\|_2^2,$$

then we get inequalities:

$$-C \leq \frac{1}{2}\frac{d}{dh}\|\nabla z\|_2^2 \leq C.$$

Integrating it over segments $[0,h]$ if $h > 0$ and $[h,0]$ if $h < 0$ in any case we have: $\|\nabla z\|_2^2 \leq 2C|h|$. Thus, the following statement is proved.

Lemma 17. *Let T_0 be a constant from Lemma 1. Let T, T_1, $T < T_1 < T_0$ be arbitrary but fixed numbers. Then, there exists a constant $C = C(\nu, \varphi, T_1)$ such that, for all approximations v^n, there is a fulfilled inequality:*

$$\|\nabla v^n(t+h,\cdot) - \nabla v^n(t,\cdot)\|_2 \leq C\sqrt{|h|},$$

whenever $t \in [0,T]$, $t+h \in [0,T_1]$.

Lemma 18. *Let T_0 be a constant from 1. Let $T, T_1,\ T < T_1 < T_0$ be arbitrary but fixed numbers. Then, there exists a constant $C = C(\nu, \varphi, T_1)$ such that for all approximations v^n there is fulfilled inequality:*

$$\|\triangle v^n(t+h,\cdot) - \triangle v^n(t,\cdot)\|_2 \leq C\sqrt[8]{|h|^3}$$

whenever $t \in [0,T],\ t+h \in [0,T_1]$.

Proof. Formulae (16) and (32) yield equalities:

$$\Big(D_t z, a^q\Big) - \nu\Big(\triangle z, a^q\Big) + \int (z_i v^n_{k,\,i}(t+h,x) + v^n_i(t,x) z_{k,\,i}) a^q_k dx = 0,\ q = 1,\ldots,n,$$

where $z = v^n(t+h,x) - v^n(t,x)$. Every equality we multiply by factor $c\prime_{qn}(t+h)$ respectively and add together them. Furthermore, in the second term, we replace differentiation on variable t by differentiation on variable h. Hence, we obtain:

$$\Big(D_t z, \triangle D_t v^n(t+h,\cdot)\Big) - \nu\Big(\triangle z, \frac{\partial}{\partial h}\triangle z\Big) =$$

$$-\int (z_i v^n_{k,\,i}(t+h,x) + v^n_i(t,x) z_{k,\,i}) \triangle D_t v^n_k(t+h,x) dx = -L_1 - L_2.$$

Here, L_1, L_2 are integrals from the first and the second products sums, respectively. Hence,

$$\Big(\nabla D_t z, \nabla D_t v^n(t+h,\cdot)\Big) + \frac{\nu}{2}\frac{\partial}{\partial h}\|\triangle z\|_2^2 = L_1 + L_2. \tag{33}$$

The scalar products on the left-hand side of (33) are bounded uniformly. This follows from estimates of Lemmas 6–8. Therefore, we have:

$$-C - L_1 - L_2 \leq \frac{\nu}{2}\frac{\partial}{\partial h}\|\triangle z\|_2^2 = C + L_1 + L_2 \tag{34}$$

with some constant $C = C(\nu, \varphi, T_1)$. A uniform boundedness of integrals L_1, L_2 follows from Corollary A4. For the verification, we take mappings triples $z,\ v^n(t+h,\cdot),\ D_t v^n(t+h,\cdot)$ and $v^n(t,\cdot),\ z,\ D_t v^n(t+h,\cdot)$, respectively. Finally, applying estimates from Lemma 6, Lemma 8 and Lemma 17, we obtain:

$$|L_1| \leq a\|\nabla z\|_2 \|\nabla v^n(t+h,\cdot)\|_2^{1/2} \|\triangle D_t v^n(t+h,\cdot)\|_2^{1/2} \|\triangle z\|_2 \leq \tag{35}$$

$$C_1\sqrt{|h|}\|\triangle D_t v^n(t+h,\cdot)\|_2^{1/2},$$

$$|L_2| \leq a\|\nabla v^n\|_2 \|\nabla z\|_2^{1/2} \|\triangle z\|_2^{1/2} \|\triangle D_t v^n(t+h,\cdot)\|_2 \leq C_2\sqrt[4]{|h|}\|\triangle D_t v^n(t+h,\cdot)\|_2, \tag{36}$$

where constants $C_m = C_m(\nu, \varphi, T_1),\ m = 1,2$ depend on ν, φ, T_1 only.
We integrate (34) over segments $[0,h]$ if $h > 0$ and $[h,0]$ if $h < 0$. Assume $h > 0$ without restriction of the generality. Then, from (35) after Hölder's inequality application and inequality (2) of Lemma 8, we get:

$$\int_0^h |L_1| dh \leq C_1 h^{5/4}\Big(\int_0^h \|\triangle D_t v^n(t+h,\cdot)\|_2^2 dh\Big)^{1/4} \leq C_1\sqrt[4]{I_1} h^{5/4},$$

where C_1 is a new constant. From (36) in the same way, we obtain another estimate:

$$\int_0^h |L_2| dh \leq C_2\sqrt{I_1} h^{3/4}.$$

Integrating (34) and, gathering last estimates, we get lemma inequality. □

Lemma 19. *Let T_0 be a number from Lemma 1. Let $T,\ T_1,\ T < T_1 < T_0$ be arbitrary but fixed numbers. Then, there exists a constant $C = C(\nu, \varphi, T_1)$ such that for all approximations v^n and P^n there are fulfilled inequalities:*

$$|v^n(t+h,x) - v^n(t,z)| \leq C(|h|^{0,375} + |x-z|^{0,5}),$$

$$|P^n(t+h,x) - P^n(t,z)| \leq C(|h|^{0,375} + |x-z|^{0,5})$$

whenever $t \in [0,T],\ t+h \in [0,T_1],\ |h| \leq 1,\ x,z \in R^3$.

Proof. We have $|f(t+h,x) - f(t,z)| \leq |f(t+h,x) - f(t,x)| + |f(t,x) - f(t,z)|$. Therefore, one should find uniform estimates for every modulus on the right-hand side considering mappings $v^n,\ P^n$. From representation (A2), it follows: $|v^n(t+h,x) - v^n(t,x)| \leq \frac{1}{4\pi}(J_1 + J_2)$, where

$$J_1 = \int_{|y-x|\leq 1} \frac{|\nabla v^n(t+h,\ y) - \nabla v^n(t,\ y)|dy}{|x-y|^2},$$

$$J_2 = \int_{|y-x|\geq 1} \frac{|\nabla v^n(t+h,\ y) - \nabla v^n(t,\ y)|dy}{|x-y|^2}.$$

To every integral, we apply again Hölder's inequality. Then,

$$J_1 \leq \|\nabla v^n(t+h,\cdot) - \nabla v^n(t,\cdot)\|_6 \Big(\int_{|y-x|\leq 1} |x-y|^{-12/5} dy\Big)^{5/6},$$

$$J_2 \leq \|\nabla v^n(t+h,\cdot) - \nabla v^n(t,\cdot)\|_2 \Big(\int_{|y-x|\geq 1} |x-y|^{-4} dy\Big)^{1/2}.$$

The second representation in (A2) and Lemma A1 yield estimate:

$$\|\nabla v^n(t+h,\cdot) - \nabla v^n(t,\cdot)\|_6 \leq A\|\Delta v^n(t+h,\cdot) - \Delta v^n(t,\cdot)\|_2.$$

Therefore, previous inequalities and estimates from Lemma 17 and Lemma 18 give formula:

$$|v^n(t+h,x) - v^n(t,x)| \leq C|h|^{0,375}, \tag{37}$$

where C is a constant depending on $\nu,\ \varphi, T_1$ only.
Let us estimate the second modulus applying Poisson's formula (see (A1)). Then,

$$|v^n(t,x) - v^n(t,z)| \leq \frac{|x-z|}{4\pi} \int \frac{|\Delta v^n(t,y)|dy}{|x-y||z-y|} = \frac{|x-z|}{4\pi} J_3.$$

From the inequality,

$$J_3 \leq \|\Delta v^n(t,\cdot)\|_2 \Big(\int |x-y|^{-2}|z-y|^{-2} dy\Big)^{1/2},$$

with some constant C_1, we obtain:

$$J_3 \leq C_1 \|\Delta v^n(t,\cdot)\|_2 |x-z|^{-1/2}.$$

Previous estimates and Lemma 8 (estimate (1)) yield:

$$|v^n(t,x) - v^n(t,z)| \leq C|x-z|^{0,5}, \tag{38}$$

where a constant C depends on ν, φ, T_1 only. Thus, the first estimate follows from (37) and (38).

In the same way, we prove an inequality of the kind (38) for the function P^n (formula (24)). The norm $\|\nabla v^n\|_4$, which appears after applying Holder's inequality, we must estimate by Lemma A5. Then,

$$\|\nabla v^n\|_4 \leq \|\nabla v^n\|_2^{1/2}\|\nabla v^n\|_6^{1/6}.$$

Furthermore, Lemma 6 (estimates (1), (2)) and lemma 8 (estimate (1)) yield the inequality $\|\nabla v^n\|_4 \leq C$, where $C = C(\nu, \varphi, T)$ is some universal constant. Then, it follows:

$$|P^n(t,x) - P^n(t,z)| \leq C_1|x-z|^{0,5}. \tag{39}$$

A difference $L = P^n(t+h,x) - P^n(t,x)$ is represented in the following form:

$$L = \frac{1}{4\pi}\int \frac{(v^n_{i,j}(t+h,y) - v^n_{i,j}(t,y))(v^n_{j,i}(t+h,y) - v^n_{j,i}(t,y))dy}{|x-y|}.$$

To obtain this formula, we change summation index for a separate terms (use (24)) and apply Hölder's inequality for three factors and two factors. We make estimates separately on a ball $|y-x| \leq 1$ and its exterior. Let $m = \left(\int_{|y-x|\leq 1}|x-y|^{-2}dy\right)^{1/2}$. Then,

$$\left|\int_{|y-x|\leq 1}(\cdot)dy\right| \leq m\|\nabla v^n(t+h,\cdot) - \nabla v^n(t,\cdot)\|_6(\|\nabla v^n(t+h,\cdot)\|_3 + \|\nabla v^n(t,\cdot)\|_3),$$

$$\left|\int_{|y-x|\geq 1}(\cdot)dy\right| \leq \|\nabla v^n(t+h,\cdot) - \nabla v^n(t,\cdot)\|_2(\|\nabla v^n(t+h,\cdot)\|_2 + \|\nabla v^n(t,\cdot)\|_2).$$

In the last case, as the first step, we make a simple estimate, thereupon, we apply Hölder's inequality. The analogous arguments that are used above for the proof of the first estimate in lemma and formula (39) yield the inequality:

$$|P^n(t+h,x) - P^n(t,x)| \leq C|h|^{0,375}, \tag{40}$$

where $C = C(\nu, \varphi, T_1)$ is some constant depending on ν, φ, T_1 only. Uniform estimates (39) and (40) prove the second inequality of lemma. □

2.7. Weak Limits Properties of Approximation Sequences

Lemma 20. *Let T_0 be a number from Lemma 1 and $T < T_0$ be a positive number. Then, the sequence of mappings $(v^n)_{n=1,\dots}$ defined by (14)–(16) is bounded in the space $W^1_6(S_T)$ and the sequence $(P^n)_{n=1,\dots}$ constructed by formula (24) is bounded in spaces $W^1_q(S_T)$, $q > 3$.*

Proof. Estimate (2) from Lemma 6 and estimate (1) from Lemma 8 yield inequality $\|\nabla v^n\|_6 \leq C$. It is fulfilled with some constant C whenever n and $t \in [0,T]$. For all mappings v^n, $D_t v^n$ integral representation (A2) is true. Then, by Lemma A1, we obtain:

$$\|v^n\|_6 \leq A\|\nabla v^n\|_2, \; \|D_t v^n\|_6 \leq A\|\nabla D_t v^n\|_2.$$

From inequalities (1) of Lemmas 6 and 8, we conclude that there exist constants C_1, C_2, such that $\|v^n\|_6 \leq C_1$, $\|D_t v^n\|_6 \leq C_2$. All norms are uniformly bounded with respect to t. Hence, the sequence $(v^n)_{n=1,\dots}$ is bounded in $W^1_6(S_T)$.
Uniform boundedness of these norms $\|P^n\|_q$, $\|\nabla P^n\|_q$, $\|D_t P^n\|_q$, $q > 3$, with respect to t and n follows from Lemma 13. Therefore, the sequence $(P^n)_{n=1,\dots}$ is bounded in spaces $W^1_q(S_T)$. □

Remark 2. *The spaces $W^1_6(S_T)$, $W^1_q(S_T)$ are reflexive. Hence, every bounded set from it is a weakly compact set (see [?]). Then, by Lemma 20, sequences $(v^n)_{...}$, $(P^n)_{...}$ are bounded in these spaces. It is possible to extract a weakly converging subsequences from its. Let*

$$u(t,x) = \lim_{k\to\infty} v^{n_k}(t,x), \ P(t,x) = \lim_{k\to\infty} P^{n_k}(t,x) \tag{41}$$

be weak limits of these subsequences. Without restriction of generality, we assume that these subsequences converge to the own weak limits on every compact set of S_T. This follows from Arzela's theorem and Lemma 19.

Lemma 21. *Let u and P be weak limits from (41). Then,*

(1) *mappings u and P are uniformly continuous on a set S_T, $T < T_0$, moreover, $u(0,x) = \varphi(x)$;*
(2) *mappings u and P are bounded on a set S_T;*
(3) *the mapping $u \in W^1_6(S_T)$ and there exists a constant $C = C(\nu, \varphi, T)$ such that following inequalities are true: $\|u\|_6 \le C$, $\|\nabla u\|_6 \le C$, $\|D_t u\|_6 \le C$ whenever $t \in [0,T]$;*
(4) *$\|\nabla u\|_2 \le C\|\nabla\varphi\|_2$, $\|\nabla D_t u\|_2 \le C\|\nabla T^1\|_2$ whenever $t \in [0,T]$, where vector field T^1 from (22), a constant $C = C(\nu, \varphi, T)$;*
(5) *u has distributions of the second and third orders: $u_{,\,ij}$, $D_t u_{,\,ij}$, in addition, for all $t \in [0,T]$, there are fulfilled inequalities: $\|\triangle u\|_2 \le C$, $\int_0^t \|\triangle D^t u\|_2^2 dt \le l_1$ where constants C, l_1 from Lemma 8;*
(6) *the function $P \in W^1_q(S_T)$ for every $q > 3$, in this case, there exists a constant $C = C(\nu, \varphi, T, q)$ such that, for all $t \in [0,T]$ estimates $\|P\|_q \le C$, $\|D_t P\|_q \le C$ are true;*
(7) *there exist constants $C_i = C_i(\nu, \varphi, T, q)$ such that $\|\nabla P\|_q \le C_1$ for every $q > 3/2$ and $\|\nabla D_t P\|_q \le C_2$ for every $q \in (3/2, 3]$;*
(8) *the function P has distributions of the second and third orders: $P_{,\,km}$, $P_{,\,kmj}$, $D_t P_{,\,i}$, in addition, there exists a number $C = C(\nu, \varphi, T, q)$ such that, for all $t \in [0,T]$, the following inequalities hold:*

$$\|P_{,\,km}\|_q \le C, \ \|D_t P_{,\,i}\|_q \le C \text{ for every } q \in (3/2, 3] \text{ and } \|P_{,\,kmj}\|_q \le C, \text{for every } q \in (1, 3/2].$$

Proof. Property (1) follows from Remark 2. A uniform continuity follows from Lemma 19 and a uniform convergence of subsequences $(v^{n_k})_{k=1,...}$ and $(P^{n_k})_{k=1,...}$ on compact subsets of S_T.

Property (2) follows from a uniform convergence on compact sets, Lemma 9 and Lemma 13 (item (1)).

Property (3) follows from norm semicontinuity of a weak limit in reflexive spaces.

Property (4) follows from Lemma 11. A uniform boundedness of norms $\|v^n_{,\,ij}\|_2$ (see Lemma 8 and Lemma 11) and norms boundedness $\|D_t v^n_{,\,ij}\|_2$ in the space $W^1_2(S_T)$ (see Lemma 8 and Lemma 11) guarantee an existence of distributions $u_{,\,ij}$, $D_t u_{,\,ij}$. Estimates of its norms follow from a semicontinuity of a weak limit norm.

Properties (5)–(8) are proved in the same way. For the verification, we apply Lemmas 13–15. □

Lemma 22. *Weak limits from (41) satisfy equalities:*

$$P_{,\,k} = -\frac{1}{4\pi}\int \frac{u_{i,\,j}(t,y)u_{j,\,i}(t,y)(x_k - y_k)dy}{|x-y|^3}, \ u_{,\,j} = \frac{1}{4\pi}\int \frac{\Delta u(t,y)(x_j - y_j)dy}{|x-y|^3}.$$

Proof. The first equality is fulfilled for mappings v^n and P^n. The sequence $(\nabla v^n)_{n=1,...}$ is bounded in the space $W^1_2(S_T)$. In addition, estimates of norms $\|\nabla v^n\|_2$, $\|v^n_{,\,ij}\|_2$, $\|D_t v^n_{,\,i}\|_2$ are uniform with respect to t and n (see Lemmas 6–8 and 11). Apply Sobolev–Kondrashov's embedding theorem (see [23], pp. 83–94) to the sequence $(\nabla v^n)_{n=1,...}$. As a bounded set, it is embedded in the space $L_q([0,T]\times\Omega)$ for every ball $\Omega \subset R^3$. An exponent q satisfies condition

$$\frac{1}{q} - \frac{1}{2} + \frac{1}{m} > 0, \ q < 4.$$

In this case, a dimension of spatial domain $[0,T] \times \Omega)$ $m = 4$. Thus, we can assume that a subsequence $(\nabla v^{n_k})_{k=1,\ldots}$ converges strongly to a mapping ∇u in the space $L_q([0,T] \times \Omega)$, $q < 4$, for every ball $\Omega \subset R^3$. Denote the integral from the first equality of the lemma by $Q_k(t,x)$. Let $d^n = P^n_{,m} - Q_m$. From equality

$$v^n_{i,j} v^n_{j,i} - u_{i,j} u_{j,i} = (v^n_{i,j} - u_{i,j})(v^n_{j,i} + u_{j,i}),$$

we deduce:

$$|d^n(t,x)| \leq \frac{1}{4\pi} \int \frac{|\nabla v^n(t,y) - \nabla u(t,y)||\nabla v^n(t,y) + \nabla u(t,y)|dy}{|x-y|^2}.$$

Multiply this inequality by $|\eta|$ where $\eta \in C_0(S_T)$ an arbitrary test–function. Thereupon, integrate over the set S_T and change integration order. Then,

$$\Big|\int_{S_T} d^n \eta dx\Big| \leq \int_0^T \int_{R^3} I_2(|\eta|)|\nabla v^n - \nabla u||\nabla v^n + \nabla u|dydt = \int_0^T (K_1 + K_2)dt,$$

where I_2 is the Riesz potential, K_1 is the interior integral calculating over ball $|y| < r$, and K_2 is the interior integral calculating over exterior of this ball. Estimate every integral applying Hölder's inequality. Thus, we have

$$K_2 \leq \Big(\int_{|y|\geq r} I_2(|\eta|)dy\Big)^{1/2} \|\nabla v^n - \nabla u\|_3 \|\nabla v^n + \nabla u\|_6.$$

The second and the third factors on the right-hand side we estimate by constants independent of t and n (see Lemmas 6, 8, 21 with conditions (3)–(4) and Lemma A5). A radius r is fixed so that the first factor is less an arbitrary positive number ε. Then, $K_2 \leq C\varepsilon$. Integral K_1 we estimate on a subsequence. Then,

$$K_1 \leq \Big(\int_{|y|\leq r} |\nabla v^{n_k} - \nabla u|^3 dy\Big)^{1/3} \|I_2(|\eta|)\|_2 \|\nabla v^n + \nabla u\|_6.$$

The second and the third factors are uniformly bounded by a some constant C. Therefore, the inequality:

$$\int_0^T K_1 dt \leq C\Big(\int_0^T \int_{|y|<r} |\nabla v^{n_k} - \nabla u|^3 dydt\Big)^{1/3} \sqrt[3]{T^2}$$

is fulfilled. The middle factor is not greater ε if a number k is large enough. This follows from condition of a strong convergence on a bounded set. Combining all estimates above, we obtain the inequality

$$|\int_{S_T} d^{n_k} \eta dx| \leq C\varepsilon T + C\varepsilon \sqrt[3]{T^2}.$$

This means that $d^{n_k} \to 0$ weakly because a function η is an arbitrary. The first equality is proved. The second equality is proved in the same way. Consider the difference $d^n = v^n_{,j} - R_j$ where R_j is the integral of the second equality. In the integral $d^n(t,x)$, we replace the variable by $y = x + z$. Thereupon, we multiply the equality by a test–function $\eta \in C_0(S_T)$ and integrate its over set S_T. Change integration order and carry over Laplace operator to function η. Then,

$$\Big|\int_{S_T} d^n \eta dxdt\Big| \leq \frac{1}{4\pi} \int_0^T \int_{R^3} \frac{1}{|z|^2} \int_{R^3} |v^n(t,x+z) - u(t,x+z)||\triangle \eta(t,x)|dxdzdt.$$

Replace variables in the interior integral by $x = y - z$ and change integration order. Hence, we get:

$$\Big|\int_{S_T} d^n \eta dxdt\Big| \leq \int_0^T \int_{R^3} |v^n(t,y) - u(t,y)||I_1(\triangle \eta)(y)|dydt.$$

Integration with respect to y we make separately over ball $|y| < r$ and its exterior. The furthest arguments are conducted in the same way as above. A distinction in the following. In this time, we use an uniform convergence of a subsequence $(v^{n_k})_{k=1,...}$ on compact sets (see Remark 2). □

2.8. Weak Solutions and Gradients Boundedness

Lemma 23. *Let u and P be weak limits from (41). Then, for every solenoidal vector field $\psi \in C_0^\infty(R^3)$ and almost everywhere $t \in [0, T]$, there is fulfilled integral identity:*

$$(D_t u, \triangle\psi) - \nu(\triangle u, \triangle\psi) + \int u_i u_{j,\,i} \triangle\psi_j dx + (\nabla P, \triangle\psi) = 0.$$

Proof. Equalities (16) multiply by a test–function $\eta \in C_0^\infty([0, T])$ and integrate its over segment $[0, T]$. If a subsequence $(v^{n_k})_{k=1,...}$ converges weakly, then, for all $q = 1, \ldots, n_k$, we have

$$\int_0^T (D_t v^{n_k}, \triangle a^q)\eta(t)dt - \nu \int_0^T (\triangle v^{n_k}, \triangle a^q)\eta(t)dt + \int_0^T \int v_i^{n_k} v_{j,\,i}^{n_k} \triangle a_j^q \eta(t)dxdt = 0.$$

Fix a some number q. Then, the passage to the limit gives the equality

$$\int_0^T (D_t u, \triangle a^q)\eta(t)dt - \nu \int_0^T (\triangle u, \triangle a^q)\eta(t)dt + \int_0^T \int u_i u_{j,\,i} \triangle a_j^q \eta(t)dxdt = 0. \quad (42)$$

This is explained by a weak convergence of a sequence $(v_i^{n_k} v_{,\,i}^{n_k})_{k=1,...}$ to the mapping $u_i u_{,\,i}$. It is given by support compactness of a vector field a^q, by uniform boundedness with respect to t and n of norms $\|\nabla v^n\|_p$, $2 \le p \le 6$, and a uniform convergence of subsequence $(v^{n_k})_{k=1,...}$ on compact subsets of S_T. A function η is an arbitrary. Therefore, from (42), we obtain

$$(D_t u, \triangle a^q) - \nu(\triangle u, \triangle a^q) + \int u_i u_{j,\,i} \triangle a_j^q dx = 0.$$

It is already fulfilled for every natural number q. The construction of vector fields a^q permits this integral identity to extend on elements of the fundamental system $(\psi^n_{n=1,...})$ (see (12) and (13)), i.e.,

$$(D_t u, \triangle\psi^n) - \nu(\triangle u, \ \triangle\psi^n) + \int u_i u_{j,\,i} \triangle\psi_j^n dx = 0. \quad (43)$$

We show that identity (43) is true for every solenoidal vector field $\psi \in C_0^\infty(R^3)$. Let $(\xi^m)_{m=1,...}$ be a sequence of a finite linear combinations of mappings ψ^n, which converges to a vector field $\psi \in C_0^\infty(R^3)$ in the space $J_0^2(R^3)$. Then,

$$\|\nabla\xi^m - \nabla\psi\|_2 \to 0, \ \|\xi^m_{,\,ij} - \psi_{,\,ij}\|_2 \to 0$$

and equality (43) for mappings ξ^m is true. Mappings $\triangle u$, $u_i u_{,\,i}$ belong to the space $L_2(R^3)$ for a.e. t. Then,

$$(\triangle u, \triangle\xi^m) \to \triangle u, \triangle\psi), \int u_i u_{j,\,i} \triangle\xi_j^m dx \to \int u_i u_{j,\,i} \triangle\psi_j dx$$

a.e. as $m \to \infty$. Let us show

$$(D_t u, \triangle\xi^m) \to (D_t u, \triangle\psi)$$

as the same condition is. Consider the equality of scalar products

$$-(D_t u, \triangle\xi^m) = (D_t u_{,\,j}, \xi^m_{,\,j})$$

and note that the right side tends to $(D_t u_{,\,j}, \psi_{,\,j})$ (see Lemma 21 item (4)). On the other side, $-(D_t u_{,\,j}, \psi_{,\,j}) = (D_t u, \triangle\psi)$. Condition (43) is true for an arbitrary $\psi \in C_0^\infty(R^3)$. From $(\nabla P, \triangle\psi) = 0$, we have the lemma. □

Lemma 24. *(see ([7], pp. 41–44), see also [29].) Let $B \subset R^3$ be an arbitrary ball. Then, a space $L_2(B)$ of any vector fields has a decomposition by a direct sum $L_2(B) = G(B) \oplus J_0(B)$ of orthogonal subspaces. A subspace $G(B)$ is the space of gradients ∇g where $g : B \to R$ is locally square–integrable function with a finite norm $\|\nabla g\|_2$. A space $J_0(B)$ is the closure with respect to the norm $L_2(B)$ of all solenoidal vector fields from the class $C_0^\infty(B)$.*

Lemma 25. *If u and P are weak limits (41), then there are fulfilled equalities:*

$$D_t u_k - \nu \triangle u_k + u_i u_{k,\,i} + P_{,\,k} = 0,\ k = 1,2,3,$$

a.e. on a set S_T for any $T \in [0, T_0)$.

Proof. Let

$$H_k = D_t u_k - \nu \triangle u_k + u_i u_{k,\,i} + P_{,\,k}.$$

Denote $h_2 = -\nu \triangle u$, $h_3 = u_i u_{,\,i}$, $h_6 = D_t u + \nabla P$. Every vector field $h_p,\ p = 2,3,6$, belongs to the space $L_p(R^3)$ (see Lemma 21). Mappings norms h_p are bounded by constants independent of $t \in [0,T]$. From the first equality of Lemma 22, we gather $(H, \nabla g) = 0$, where $g \in C_0^\infty(R^3)$ is an arbitrary. We assume the mapping H and its generators h_p belong to the class $C^\infty(R^3)$. Otherwise, we take averages with a kernel from $C_0^\infty(R^3)$ for them. For averages, the equality $(H, \nabla g) = 0$ and the equality of Lemma 23 are kept. This follows from behind an arbitrary choice of a smooth function g and a field $\psi \in C_0^\infty(R^3)$. Then, $div\ H = 0$. Moreover, a smoothness H and the equality of Lemma 23 imply $(\triangle H,\ \psi) = 0$. From Lemma 24 on every ball $B \subset R^3$, we have $\triangle H = \nabla h$. A function h is infinitely smooth. This is given by smoothness $\triangle H$. Then, $div\ \triangle H = \triangle h$. On the other hand, $div\ \triangle H = \triangle div\ H = 0$. Therefore, the function h is a harmonic function. Hence, and from above, there is $\triangle^2 H = 0$. By Lemma A7, we have $H = 0$. Making an average parameter tending to zero, we obtain this equality in the general case. □

Lemma 26. *Let u and P be weak limits from (41). Then, there exists a number $C = C(\nu, \varphi, T)$ such that, for almost everywhere, $t \in [0,T]$ following conditions are fulfilled:*

(1) $\|\triangle u\|_6 \le C$;

(2) $|\nabla u_k(t,x)| \le C,\ |\nabla P(t,x)| \le C,\ k = 1,2,3.$

Proof. From Lemma 25, we conclude that Laplacian $\triangle u$ is the linear combination of three vector fields ∇P, $D_t u$, $u_i u_{,\,i}$. Coordinates u_i are bounded on the set S_T by Lemma 21 item (2). Then, from Lemma 21 (see estimates (3) and (6)), it follows the first part of the lemma.
Gradients boundness ∇u_i we obtain from the second integral representation of Lemma 22 and estimate $\|\triangle u\|_6 \le C$. In the next step, we repeat the proof of Lemma 9.

Gradients boundedness ∇P we get from the first integral representation of Lemma 22 and gradients boundedness ∇u_i with repeating of the proof from Lemma 9. □

2.9. Weak Solutions, Integral Equations and Energetic Inequality

Let $\Gamma(x,t) = (4\pi\nu t)^{-n/2} e^{-|x|^2/4\nu t}$ be a Weierstrass kernel. Furthermore, we consider mixed norms for mappings defined on the set $S_T = [0,T] \times R^n$.

Lemma 27. *(See [13], Theorem 2.1.) A vector field $u : S_T \to R^n$ with a finite mixed norm $\|u\|_{p,\,q}$ is a weak solution of problems (1) and (2) if and only if when u is a solution of integral equation*

$$u + B(u,u) = f, \tag{44}$$

where B is a some nonlinear integral operator, $f(t,x) = \int \Gamma(x-y,t)\varphi(y)dy$.

Lemma 28. *(See [13], Theorem 3.4.) Let u be a solution of integral Equation (44) with a finite mixed norm $\|u\|_{p,\,q}$ where $p, q \geq 2,\ \frac{3}{p} + \frac{2}{q} \leq 1$. Let k be a positive integer such that $k+1 < p,\ q < \infty$. If mixed norms of derivatives*

$$D^{\alpha}\frac{\partial^j f}{\partial t^j}$$

with exponents $p_1 = \frac{p}{|\alpha|+2j+1},\ q_1 = \frac{q}{|\alpha|+2j+1}$ are finite whenever $|\alpha| + 2j \leq k$, then also mixed norms of

$$D^{\alpha}\frac{\partial^j u}{\partial t^j}$$

are finite for the same means α, j, p_1, q_1.

Remark 3. *The proof of this result relies on Calderon–Zygmund's theorem and a boundedness of singular integral operators of parabolic type (see [30]).*

Remark 4. *Norms $D^{\alpha}\frac{\partial^j u}{\partial t^j}$ are bounded by a constant that depends on exponents p, q, derivative order and the mixed norm $\|u\|_{p,\,q}$. It follows directly from the proof of the theorem in [13].*

Lemma 29. *If u is a weak limit from (41), then there exists a number $C = C(\nu, \varphi, T, p, q)$ such that $\|u\|_{p,\,q} \leq C$ whenever $p,\ q \geq 2,\ \frac{3}{p} + \frac{2}{q} \leq 1$.*

Proof. Let $T < T_0$ be a positive arbitrary number. Integrate the equality of Lemma 25 over segment $[0, t]$ where $t < T$. Then, continuity and absolute continuity on lines of mapping u give:

$$u(t, x) - \varphi(x) = \int_0^t (\nu\triangle u(\tau, x) - u_i(\tau, x)u_{,\,i}(\tau, x) - \nabla P(\tau, x))d\tau.$$

Every integrable term has finite norms

$$\|\triangle u\|_2,\ \|\nabla P\|_2,\ \|u_i u_{,\,i}\|_2.$$

In addition, every norm is bounded by a constant $C = C(\nu, \varphi, T)$ depending on ν, φ, T only. It follows from Lemma 21 (see estimates (5) and (7)) for the first and the second norms. A boundedness of the third norm follows from mapping boundedness u (see Lemma 21 item (2)) and the estimate from item (4) (see Lemma 21) . Therefore, $\|u\|_2 \leq C$. A boundedness of vector field u (see Lemma 21 item (2)) gives a uniform estimate $\|u\|_p \leq C$ whenever $p \geq 2$. Then, any mixed norm $\|u\|_{p,\,q}$ is finite whenever $p,\ q$ from lemma condition. □

Lemma 30. *If u is a weak limit from (41), then a mixed norm $\|\triangle u\|_{6/5,\,4} < \infty$.*

Proof. Let initial data $\varphi \in C^{\infty}_{6/5,\,3/2}$. Function f from Lemma 27 is represented by integral

$$f(t, x) = \frac{1}{\pi^{3/2}}\int e^{-|z|^2}\varphi(x + \sqrt{4\nu t}z)dz.$$

For $\varphi \in C^{\infty}_{6/5,\,3/2}$, there is true Lemma 34. Therefore, the mapping f and any of its derivatives have a finite mixed norm $\|\cdot\|_{p,\,q}$. By Lemma 25 and Lemma 29, the vector field u is a weak solution of problems (1) and (2) with a finite mixed norm $\|u\|_{p,\,q}$ whenever $p, q \geq 2$. Then, from Lemma 27, we conclude that u is a solution of integral Equation (44). From Lemma 28, we obtain a finiteness of mixed norms for the second derivatives $\|D^{\alpha}u\|_{p_1,\,q_1}$, where $p_1 = p/3,\ q_1 = q/3,\ |\alpha| = 2, j = 0$. Let $p = 18/5, q = 12$. Then, we have the statement of the lemma. □

Lemma 31. *(Energetic condition.) Let u and P be weak limits from* (41). *Then,*

$$\|u\|_2^2 + 2\nu \int_0^t \|\nabla u(\tau, x)\|_2^2 d\tau = \|\varphi\|_2^2$$

for every $t \in [0, T_0)$ *where* T_0 *from Lemma* 1.

Proof. Note that weak solutions satisfy conditions:

$$J_1 = \int P_{,k} u_k dx = 0, \; J_2 = \int u_i u_{k,\,i} u_k dx = 0. \tag{45}$$

From the first equality of Lemma 22, we have:

$$J_1 = \frac{1}{4\pi} \int_{R^3} u_{i,\,j}(t, y) u_{j,\,i}(t, y) \int_{R^3} \frac{u_k(t, x)(x_k - y_k)}{|x - y|^3} dx dy. \tag{46}$$

Integrals commutation is possible since the integral over R^6 is a finite. It follows from

$$\int_{R^6} |\cdot| dx dy \leq 4\pi \int_{R^3} |\nabla u(\tau, y)|^2 I_1(|u|) dy,$$

Tonnelli's theorem, boundedness and summability of $u, \; \nabla u$ with any exponent not less than two and Lemma A1. Here, I_1 is the Riesz potential. The interior integral in (46) is equal to zero since

$$\int_{|x-y|<r} \frac{u_k(t, x)(x_k - y_k)}{|x - y|^3} dx = - \int_{|x-y|=r} \frac{u_k(t, x)(x_k - y_k)}{r^2} dS =$$
$$-\frac{1}{r} \int_{|x-y|<r} div \; u dx = 0$$

for any radius r.

Let us prove the second equality from (45). The second equality of Lemma 22 implies:

$$J_2 = \frac{1}{4\pi} \int_{R^3} \triangle u_k(t, y) \int_{R^3} \frac{u_i(t, x) u_k(t, x)(x_i - y_i)}{|x - y|^3} dx dy. \tag{47}$$

Integrals commutation we prove in the same way. There is inequality:

$$\int_{R^6} |\cdot| dx dy \leq 4\pi \int_{R^3} |\triangle u(\tau, y)|^2 I_1(|u|^2) dy.$$

The right-hand side is a finite because $\triangle u \in L_p, \; 2 \leq p \leq 6$ (see Lemma 21 item (5), Lemma 26 item (1), Lemma A5). In addition, $I_1(|u|^2) \in L_p, \; p > 3/2$ by Lemma A1. To interior integral in (47) we apply the Stokes formula. Then,

$$\int_{|x-y|<r} \frac{u_i(t, x) u_k(t, x)(x_i - y_i)}{|x - y|^3} dx =$$
$$\int_{|x-y|<r} \frac{u_i(t, x) u_{k,\,i}(t, x)}{|x - y|} dx - \frac{1}{r} \int_{|x-y|=r} u_i u_k \frac{x_i - y_i}{r} dS.$$

A product $u_i u_k$ belongs to the space $W_p^1(R^3)$ whenever $p > 1$. Then, the integral over surface tends to zero as $r \to \infty$ (to apply Lemma A4 with exponent $\alpha = 1$ and a mean p, close to unit). Hence, and from (47) we have:

$$J_2 = \frac{1}{4\pi} \int_{R^3} \triangle u_k(t, y) \int_{R^3} \frac{u_i(t, x) u_{k,\,i}(t, x)}{|x - y|} dx dy. \tag{48}$$

In the iterated integral

$$\int_{|y|<r} \triangle u_k(t,y) \int_{R^3} \frac{u_i(t,x)u_{k,\,i}(t,x)}{|x-y|} dxdy,$$

we change integration order because the double integral is finite (see above). Hence, we get:

$$\int_{|y|<r} \triangle u_k(t,y) \int_{R^3} \frac{u_i(t,x)u_{k,\,i}(t,x)}{|x-y|} dxdy = \tag{49}$$

$$\int_{R^3} u_i(t,x)u_{k,\,i}(t,x) \int_{|y|<r} \frac{\triangle u_k(t,y)}{|x-y|} dydx.$$

The interior integral in the right-hand side of (49) is uniformly bounded with respect to $r > 1$. This follows from a boundedness of the Riesz potential $I_2(|\triangle u|)$. It is proved in the same way as Lemma 9 with applications Lemma 26 item (1) and Lemma 21 item (5). Furthermore, we use Lebesgue's theorem. Then, (48) and (49) give the equality of iterated integrals:

$$J_2 = \int_{R^3} u_i(t,x)u_{k,\,i}(t,x)I_2(\triangle u_k)(x)dx. \tag{50}$$

The mapping $u \in \mathring{J}^2_0(R^3)$ (norm defined by (15)). Lemma A1 shows that Poisson's formula is true for elements of the space $\mathring{J}^2_0(R^3)$. Then, $I_2(\triangle u) = -u$. Therefore, we have $J_2 = -J_2$ from (50). The second equality from (45) is proved.

Let us show that vector field u satisfies the equality

$$\int_{R^3} u_k \triangle u_k dx = -\|\nabla u\|_2^2 \tag{51}$$

a.e. on $[0,T]$. We have the equality of iterated integrals:

$$\int_{R^3} \triangle u_k(t,x) \int_{|y|<r} \frac{u_{k,\,j}(t,x)(x_j-y_j)}{|x-y|^3} dydx = \tag{52}$$

$$\int_{|y|<r} u_{k,\,j}(t,y) \int_{R^3} \frac{\triangle u_k(t,x)(x_j-y_j)}{|x-y|^3} dydx.$$

A finiteness of double integral follows from a boundedness ∇u (see Lemma 26, item (2)) and properties of the Riesz potential $I_1(|\triangle u|)$. Let $r \to \infty$. The interior integral on the left-hand side of (52) tends to $4\pi u_k(t,x)$ in the space $L_6(R^3)$ for almost every t. (See Lemma A1 and equality (A2), which is true for elements of the space $\mathring{J}^2_0(R^3)$). The norm $\|\triangle u\|_{6/5}$ is finite a.e. by Lemma 30. In (52), we make the passage to the limit. The interior integral on the right-hand side of (52) is replaced by application of Lemma 22. Then, we get (51). To finish the proof, we are helped with the following steps. Every equality from Lemma 25 we multiply by function u_k. Thereupon, we add together them and integrate over space R^3. From (45) and (49), we have

$$(D_t u, u) + \nu\|\nabla u\|_2^2 = 0.$$

Hence, we get the required equality. □

2.10. Proof of Theorem 1

Observe that all estimates in proved lemmas above depend on norms $\|\nabla\varphi\|_2$, $\|\nabla T^1\|_2$ (see (22)), $\|\triangle\varphi\|_2$ or $\|\varphi\|_2$ only and don't depend on a diameter of Laplacian support $\triangle\varphi$.

If $\varphi \in C^\infty_{6/5,\,3/2}$, then, by Lemma A4 integrals,

$$\frac{1}{r^2}\int_{|y-x|=r} |\varphi(y|dS, \frac{1}{r}\int_{|y-x|=r} |\nabla\varphi(y|dS$$

tend to zero as $r \to \infty$. Therefore, equalities from Lemma A2 are true for mappings of the class $C^\infty_{6/5,\,3/2}$. In addition, we have summability φ with any exponent $p > 6/5$ and $\nabla\varphi$ with any exponent $p > 3/2$ (see Lemma 32).

1. Assume that initial data $\varphi \in C^\infty_{6/5,\,3/2}$ and its Laplacian support is a compact set. Let T_0 be a constant from Lemma 1. Then, item (3) follows from Lemma 26, and items (1) and (4) we get from Lemma 21.

Let us prove estimates of item (2). A uniform boundedness with respect to t of norms

$$\|u\|_6,\ \|\nabla u\|_6,\ \|D_t u\|_6$$

we obtain from Lemma 21 (item (3)). An uniform boundedness of norms

$$\|\nabla u\|_2,\ \|\nabla D_t u\|_2,\ \|\triangle u\|_2, \|\nabla P\|_2$$

follows from Lemma 21 (see items (4), (5), (7)). The estimate of norm $\|u\|_2$ follows from Lemma 31. A uniform boundedness of norms $\|u_{,\,ij}\|_6$ we get by Lemma 26. A uniform boundedness for norm $\|D_t u\|_2$ is the corollary of Lemma 25 because $D_t u$ is the finite linear combination of terms with uniform bounded norms in the space $L_2(R^3)$. Uniform estimates of norms in spaces $L_p(R^3)$, $2 < p < 6$ we take from Lemma A5. The occurrence of vector field u in spaces $W^1_2(S_T)$ and $W^1_6(S_T)$ we get from the uniform estimates proved above. By Lemma 25 and Lemma 21 (see items (5) and (7)), we obtain $D^2_{tt}u = \nu\triangle D_t u - u_i D_t u_{,\,i} - (D_t u_i)u_{,\,i} - D_t\nabla P$. Hence, it follows a finiteness of norm $\|D^2_{tt}u\|_2$ since u and ∇u are bounded. Therefore, $u \in W^2_2(S_T)$.

Let us prove item (5) using mixed norms (see [8,26,27]). Weak solutions u and P belong to class $C(S_T)$ (see item 1) of this theorem). In Lemma 28, we put $p = q$ assuming it is very large. Now, we fix an order of derivatives: $m > 1$. Then, by Lemma 27 and Lemma 28, derivative norms $\|D^\alpha D^j_t u\|$, $|\alpha| \le m$ are bounded in the space $L_r(S_T)$ where an exponent $r \ge 6$ is an arbitrary but fixed. A boundedness of weak solution u and its summability in $L_2(S_T)$ imply the belonging $u \in L_r(S_T)$, $r \ge 2$. Exponents means r, $p = q$ we choose by large numbers so that the next conditions are fulfilled:

(1) for any ball lying in S_T, all conditions of Sobolev's embedding theorem in a space of continuous functions are certainly valid ([23], p. 64);
(2) at least, all derivatives of the order up to $m - 1$ satisfy also all conditions Sobolev's theorem from above.

Since an integer number m is an arbitrary, then a weak solution u belongs to the class $C^\infty((0, T_0) \times R^3)$. A smoothness of function P we obtain from Lemma 25 and the smoothness of vector field u. The continuity is proved in item 1.

2. Let initial data $\varphi \in C^\infty_{6/5,\,3/2}$. We take a test-function $\eta \in C^\infty(R^3)$ such that $\eta(x) = 1$ if $|x| \le 1$ and $\eta(x) = 0$ if $|x| \ge 2$. Consider a solenoidal vector field

$$\Phi^r(x) = \eta(x/r)\varphi(x) - \nabla Q(x).$$

Then, $\triangle Q(x) = \frac{1}{r}\eta_{,\,i}(x/r)\varphi_i(x)$. A function Q is Poisson's integral

$$Q(x) = -r^{-1}I_2(\eta_{,\,i}(\cdot/r)\varphi_i)(x).$$

Hence, we have:

$$|\nabla Q(x)| \le r^{-1}I(|\nabla\eta(\cdot/r)| \cdot |\varphi|)(x) \le Mr^{-1}I_1(|\varphi|)(x)$$

where I_1 is the Riesz potential, M is the maximal mean of $|\nabla\eta|$. From Lemma A1, we obtain

$$\|\nabla Q\|_2 \le AMr^{-1}\|\varphi\|_{6/5} = O(r^{-1}).$$

Direct calculations yield:

$$\|\Phi^r\|_2^2 = \int \Big(r^{-2}|\nabla\eta(x/r)|^2 + \eta^2(x/r)|\nabla\varphi(x)|^2 + 2r^{-1}\eta\eta,_k(x/r)\varphi_i(x)\varphi_{i,k}(x)\Big)dx +$$

$$\|\nabla Q\|_2^2 - 2r^{-1}\int\Big(\eta,_k(x/r)\varphi_j(x)Q,_{jk}(x) + \eta(x/r)\varphi_{j,k}(x)Q,_{jk}(x)\Big)dx.$$

Without the second term in the first integral, the rest of the integrals of all terms in the right-hand tend to zero as $r \to \infty$. This is guaranteed by a test-function η and a boundedness of the second derivatives $Q,_{jk}$. The last follows from representation of function Q by Poisson's integral and definition of the class $C^\infty_{6/5,\,3/2}$. In this case, we have two equalities:

$$Q,_{kj}(x) = \frac{1}{4\pi r}\int_{R^3}(\eta,_i(y/r)\varphi_i(y)),_k\frac{x_j - y_j}{|x-y|^3}dy, \tag{53}$$

$$Q,_{kj}(x) = c_{kj}r^{-1}\eta,_i(x/r)\varphi_i(x) + r^{-1}T_{kj}(\eta,_i(\cdot/r)\varphi_i)(x), \tag{54}$$

where c_{kj} are universal constants, and T_{kj} are singular integral operators. Therefore, as $r \to \infty$, then

$$\|\nabla\Phi^r\|_2 \to |\nabla\varphi\|_2. \tag{55}$$

A vector field $\Phi^r \in C^\infty_{6/5,\,3/2}$. A summability of the vector field and its derivatives follows from (53) and (54), the equality

$$Q,_j(x) = c_{ij}\eta(x/r)\varphi_i(x) + T_{ij}(\eta(\cdot/r)\varphi_i)(x)$$

and Lemma A1. In addition, $\Phi^r \to \varphi$ in the space $J_0^2(R^3)$, $D^\alpha\Phi^r \to D^\alpha\varphi$ in the space $L_2(R^3)$. Laplacians supports $\triangle\Phi^r$ are compact sets. Therefore, there exist solutions u^r and P^r with an initial data $u^r(0,x) = \Phi^r(x)$ satisfying theorem with the number

$$T_0(r) = \left(\frac{9}{4}\right)^4\frac{\nu^3}{\|\Phi^r\|_2^4}.$$

From (55), we have $T_0(r) \to T_0$ as $r \to \infty$. Fix a number $T < T_0$. From the remark at the beginning of the proof, we conclude all estimates of the theorem for solutions u^r, P^r. They are uniform with respect to r for $r > r_0$. Hence, sets of mappings $(u^r)_{r>r_0}$, $(P^r)_{r>r_0}$ are bounded in spaces $W_2^1(S_T)$ and $W_6^1(S_T)$. Extract subsequences $(u^{r_k})_{k=1,\ldots}$, $(P^{r_k})_{k=1,\ldots}$, which converge weakly. Let u and P be its weak limits, respectively. These limits satisfy the next properties:

(1) Lemma 21 is true for them (this is verified in the same way as the proof of Lemma 21 for subsequences);
(2) Lemma 25 is true for them;
(3) Lemma 26 is fulfilled for them. Thus, u and P are weak solutions of problems (1) and (2). Lemma 27 and Lemma 29 are true for vector field u. Conditions of growth for a mapping $\varphi \in C^\infty_{6/5,\,3/2}$ show correctness of Lemma 28 for weak solutions from above.

Furthermore, we realize the proof from the first part (see item (1) above). Therefore, the theorem is true also in this case. Theorem 1 is proved.

3. Homotopic Property of Cauchy Problem Solutions in Class $C^\infty_{6/5,\,3/2}$

If initial data $\varphi \in C^\infty_{6/5,\,3/2}$, then the Cauchy problem solutions from Theorem 1 have the next homotopic property.

Theorem 2. *Let u and P be solutions of problems* (1) *and* (2) *from Theorem* 1*. Then, for every fixed mean* $t \in (0, T_0)$ *(see* (5)*) mappings* $u,\ P,\ D_t u \in C^{\infty}_{6/5,\,3/2}$*. Moreover, all norms*

$$\|u\|_{6/5}, \|\nabla u\|_{3/2}, \ \|D^{\alpha}u\|_r, \ \|D^{\beta}P\|_r, \ \|D^{\beta}D_t u\|_r$$

if $r > 1$, $|\alpha| \geq 2$, $|\beta| \geq 0$, *are uniformly bounded on every segment* $[0, T]$ *where* $T < T_0$.

Proof of Theorem 2 (it is given below) is relied on for the next simple properties of mappings $v \in C^{\infty}_{6/5,\,3/2}$. For every vector field v and its derivatives of the first order, these are true for both (A1) and representation (Riesz's formula):

$$v(x) = \frac{1}{4\pi}\int_{R^3} \frac{v_{,j}(y)(x_j - y_j)dy}{|x-y|^3}, \ v_{,j}(x) = \frac{1}{4\pi}\int_{R^3} \frac{v_{,jk}(y)(x_j - y_j)dy}{|x-y|^3}. \tag{56}$$

The second equality we obtain by application of the Stokes theorem to the integral from (56) calculating over a spherical layer $\varepsilon \leq |y - x| \leq r$. From Lemma A4,

$$\int_{|y-x|=r} \frac{|v_{,}(y)|dS}{r^2} \to 0$$

as $r \to \infty$ since $\nabla v \in W^1_{3/2}(R^3)$. Then, the passage to limit as $r \to \infty$, $\varepsilon \to 0$ implies the second equality (56). The first equality is proved in the same way.

We have

$$|v_{,j}(x)| \leq \frac{\pi}{2} I_1(|\nabla v_{,j}|)(x),$$

where I_1 is the Riesz potential from (4). Hardy–Littlewood–Sobolev's inequality (see Lemma A1) implies

$$\|I_1(\nabla v_{,j}|)\|_q \leq A\|\nabla v_{,j}\|_p,$$

where $\frac{1}{q} = \frac{1}{p} - \frac{1}{3}$, $1 < p < q$. Consider only $p \in (1, 3)$. Two last estimates yield $\nabla v \in L_q(R^3)$ for every $q > 3/2$. Analogously with the above, we show for the mapping v and a number $q \in [3/2, 3)$ the belonging $v \in L_r(R^3)$ whenever $r \geq 3$. The logarithmic convex of norm $\|v\|_p$ and Lemma A5 yield norm finiteness $\|v\|_p$ for $p \geq 6/5$. Thus, we proved the next statement.

Lemma 32. *Let* $v \in C^{\infty}_{6/5,\,3/2}$*. Then,* $v \in L_p(R^3)$, $\nabla v \in L_q(R^3)$ *whenever* $p \geq 6/5, q \geq 3/2$.

Remark 5. *Write Poisson's formula (the representation by Riesz's integral* I_2*) for mappings* v, ∇v *and* $D^{\alpha}v$*. Then, we have a boundedness of every vector field* $v \in C^{\infty}_{6/5,\,3/2}$ *and its derivatives.*

Let

$$P(x) = \frac{1}{4\pi}\int_{R^3} \frac{v_{i,\,j}(y)v_{j,\,i}(y)dy}{|x-y|} \tag{57}$$

(the repeated index gives summation).

Lemma 33. *Let* $v \in C^{\infty}_{6/5,\,3/2}$ *and div* $v = 0$*. Then, the function P and all its derivatives belong to the space* $L_r(R^3)$ *whenever* $r > 1$.

Proof. The integral from (57) we integrate by parts twice over a spherical layer $\varepsilon \leq |y - x| \leq r$. Lemma A4 and the passage to the limit as $r \to \infty$, $\varepsilon \to 0$ imply:

$$P(x) = \frac{|v(x)|^2}{3} - T_{ij}(v_i v_j)(x), \tag{58}$$

where T_{ij} is a singular integral operator with a kernel

$$k_{ij} = \frac{\partial^2}{\partial y_i \partial y_j} \frac{1}{4\pi|x-y|}.$$

Lemma A1 and well-known Calderon–Zygmund's theorem give a summation of function P for any finite exponent $r > 1$. Since

$$P_{,k}(x) = -\frac{1}{4\pi} \int_{R^3} \frac{v_{i,\,j}(y) v_{j,\,i}(y)(x_k - y_k)dy}{|x-y|^3},$$

then, analogously with the above, we get:

$$P_{,k}(x) = -\frac{1}{3} v_j(x) v_{k,\,j}(x) + T_{ik}(v_j v_{i,\,j})(x). \tag{59}$$

Hence, we obtain a summability of ∇P whenever finite $p > 1$. A summability of the other derivatives follows from equalities:

$$D^\beta P_{,k}(x) = -\frac{1}{3} D^\beta (v_j v_{k,\,j})(x) + T_{ik}(D^\beta (v_j v_{i,\,j}))(x). \tag{60}$$

□

Lemma 34. *If $v \in C^\infty_{6/5,\,3/2}$, then Poisson's and Riesz's formulae are true:*

$$v(x) = -\frac{1}{4\pi} \int_{R^3} \frac{\triangle v(y) dy}{|x-y|}, \ v(x) = \frac{1}{4\pi} \int_{R^3} \frac{v_{,\,j}(y)(x_j - y_j)dy}{|x-y|^3}.$$

Lemma 35. *Suppose a function P and all its derivatives are summaable in space R^3 whenever $r > 1$. Let $v, w \in C^\infty_{6/5,\,3/2}$. Then,*

$$\int_{R^3} v_k \triangle w_k dx = -\int_{R^3} v_{k,\,j} w_{k,\,j} dx, \ \int_{R^3} w_k P_{,\,k} dx = -\int_{R^3} P div\ w dx.$$

Proof. Apply the second representation from Lemma 34 and make the commutation of integrals. Then,

$$\int_{R^3} v_k \triangle w_k dy = \frac{1}{4\pi} \int_{R^3} v_{k,\,j}(y) \int_{R^3} \frac{\triangle w_k(y)(x_j - y_j)dx}{|x-y|^3} dy = -\int_{R^3} v_{k,\,j} w_{k,\,j} dy$$

(see the first equality of Lemma A4). Changing of integration order is possible because the integral

$$J = \int_{R^6} \frac{|\nabla v(y)||\triangle w(x)| dx dy}{|x-y|^2}$$

is a finite. Really, we have

$$J = \gamma(1) \int_{R^3} |\triangle w| I_1(|\nabla v|) dx.$$

Then, a finiteness follows from a summability of the Riesz potential $I_1(|\nabla v|)$ with exponent 3 (see A1) and the summability of $\triangle w$ with exponent $3/2$. The first equality is proved. To prove the second formula, we observe a finiteness of integrals

$$J_1 = \int_{R^6} \frac{|w(x)||\triangle P(y)| dx dy}{|x-y|^2},$$

$$J_2 = \int_{R^6} \frac{|div\, w(x)||\triangle P(y)|dxdy}{|x-y|} = 4\pi \int_{R^3} |divw| I_2(|\triangle P|)dx.$$

Thereupon, we have:

$$\int_{R^3} w_k P_{,k} dx = \frac{1}{4\pi} \int_{R^3} w_k(x) \int_{R^3} \frac{\triangle P(y)(x_k - y_k)dy}{|x-y|^3} dx =$$

$$\frac{1}{4\pi} \int_{R^3} \triangle P(y) \int_{R^3} \frac{\triangle w_k (x_k - y_k)dx}{|x-y|^3} dy = \frac{1}{4\pi} \int_{R^3} \triangle P(y) \int_{R^3} \frac{div\, w(x)dx}{|x-y|}.$$

□

Proof of Theorem 2. Items (1), (2) and (3) from Theorem 1, Lemma 27 and Lemma 28 yield a finiteness of mixed norms $\|D^\alpha u\|_{p_1, q_1}$ where $p_1 = \frac{p}{|\alpha|+1}$, $q_1 = \frac{q}{|\alpha|+1}$, whenever $p,\ q \geq 2$, $\frac{3}{p} + \frac{2}{q} \leq 1$. For derivatives of the second order, in particular, we have a finiteness of norm $\|\triangle u\|_{6/5,\,4}$ (see 30). Integrate (1) over segment $[0,t]$ where $t \leq T < T_0$. The solution P is represented by (57). Then, from (59), we get

$$u_k(t,x) - \varphi_k(t,x) = \tag{61}$$

$$\int_0^t (\nu \triangle u_k(\tau,\ x) + T_{ik}(u_j u_{j,\,i})(\tau,x) - \frac{2}{3} u_j(\tau,\ x) u_{j,\,i}(\tau,x)) d\tau.$$

Estimate norms in $L_{6/5}$ of every term in (61) in the usual way. We apply Hölder's inequality to interior and exterior integrals. Then,

$$\int_{R^3} \Big| \int_0^t u_i u_{k,\,i} d\tau \Big|^{6/5} dx \leq t^{1/5} \int_{S_t} u_i u_{k,\,i} d\tau dx \leq \tag{62}$$

$$T^{1/5} \Big(\int_{S_t} |u|^2 d\tau dx \Big)^{3/5} \Big(\int_{S_t} |\nabla u_k|^3 d\tau dx \Big)^{2/5},$$

$$\int_{R^3} \Big| \int_0^t \triangle u_k d\tau \Big|^{6/5} dx \leq t^{1/5} \int_0^t \|\triangle u_k\|_{6/5}^{6/5} d\tau \leq \tag{63}$$

$$T^{9/10} \Big(\int_0^T \|\triangle u_k\|_{6/5}^4 d\tau \Big)^{3/10} < \infty.$$

The singular integral operator T_{ik} is bounded. Hence, from (61)–(63) and item (2) of Theorem 1, we obtain a uniform estimate of norm $\|u\|_{6/5}$ with respect to $t \in [0,T]$.

In the same way, we prove a summability of gradient ∇u with any exponent $p \geq 3/2$. From Lemma 27 and Lemma 28, whenever $p,q \geq 2$, $\frac{3}{p} + \frac{2}{q} \leq 1$, we get a finiteness of mixed norms $\|D^\alpha u\|_{p_1, q_1}$ for derivatives of the third order where $p_1 = \frac{p}{4}$, $q_1 = \frac{q}{4}$ since $\alpha = 3,\ j = 0$. In particular, we have a finiteness of norm $\|\triangle \nabla u\|_{3/2,\,3/2}$.

Let us differentiate (1) with respect to x_m. Thereupon, we integrate its over $[0,t]$ where $t \leq T < T_0$. Formulae (57) and (60) yield

$$u_{k,\,m} - \varphi_{k,\,m} = \int_0^t (\nu \triangle u_{k,\,m} + T_{ik}((u_j u_{j,\,i}),_m) - \frac{2}{3}(u_j u_{j,\,i}),_m) d\tau. \tag{64}$$

Hence, for exponent $p = 3/2$, we obtain estimates, which are similar estimates (62) and (63). A boundedness u, uniform estimates of norm $\|\nabla u\|_3$ (see item (2) from Theorem 1) on segment $[0,T]$, a finiteness of mixed norm $\|\triangle \nabla u\|_{3/2,\,3/2}$ give a uniform boundedness of norms $\|\nabla u\|_{3/2}$.

Let derivative order $|\alpha| \geq 2$. Then, (62) takes the form:

$$D^\alpha u_k - D^\alpha \varphi_k = \int_0^t (\nu \triangle D^\alpha u_k + T_{ik}(D^\alpha(u_i u_{j,\,i})) - \frac{2}{3} D^\alpha(u_j u_{j,\,i})) d\tau. \tag{65}$$

Fix an exponent $r > 1$. Choose numbers $p, q = (|\alpha| + 3)r$. Then, we have a finiteness of the mixed norm $\|D^\alpha u\|_{r,\,r}$. It follows

$$\Big|\int_{R^3}\int_0^t \triangle D^\alpha u d\tau dx\Big|^r \le t^{r-1}\int_{S_t}|\triangle D^\alpha u|^r d\tau \le T^{r-1}\int_{S_T}|D^\alpha \triangle u|^r d\tau dx < \infty. \tag{66}$$

Terms in derivative $D^\alpha(u_i u_{k,\,i})$ without coefficients have a form: $D^\beta u_i D^\gamma u_{k,\,i}$, where $|\beta| + |\gamma| = |\alpha|$. Then,

$$\Big|\int_{R^3}\int_0^t D^\beta u_i D^\gamma u_k d\tau dx\Big|^r \le T^{r-1}\int_{S_T}|D^\beta u|^r |D^\gamma \nabla u|^r d\tau dx = T^{r-1} J. \tag{67}$$

To the right-hand side, we apply Hölder's inequality with exponents $\frac{|\alpha|+3}{|\beta|+1}$ and $\frac{|\alpha|+3}{|\gamma|+2}$. Therefore,

$$J \le \|D^\beta u\|^{r/p_1}_{p_1,\,p_2}\|D^\gamma \nabla u\|^{r/p_2}_{p_1,\,p_2},$$

where $p_1 = \frac{p}{|\beta|+1}$, $p_2 = \frac{p}{|\gamma|+2}$. All these mixed norms are bounded. This follows from Lemma 27 and Lemma 28. Hence, formulae (65)–(67) and a boundedness of a singular integral operator give uniform boundedness of all norms with respect to $t \in [0, T]$.

The solution P is represented by (57) with replacing v by u. A summability follows from Lemma 33 whenever $r > 1$. Equalities (58) and (59) and a uniform boundedness derivatives norms of vector field u prove a uniform boundedness of norms $\|D^\beta P\|_r$ where $r > 1$. From (1) and proved uniform estimates from above, we have necessary statement for derivative $D_t u$. Theorem 2 is proved. □

4. Basic Parameters and Extension of the Cauchy Problem Solutions

Now, we define two from three basic parameters. They have a key part for an extension of the Cauchy problem solutions as solutions with initial data from the class $C^\infty_{6/5,\,3/2}$. A functional $l(\varphi)$ and the first parameter λ we define by

$$l(\varphi) = \|\varphi\|_2 \cdot \|\nabla\varphi\|_2,\ \lambda = \Big(\frac{4\sqrt[4]{3}}{3a_1}\Big)^2 \frac{\nu^2}{l(\varphi)} = \frac{81\nu^2}{8l(\varphi)}, \tag{68}$$

where the constant a_1 from Corollary A4. By Theorem 2, the solution of the Cauchy problem with condition $\varphi \in C^\infty_{6/5,\,3/2}$ can be extended as the solution in any time t. Moreover, extended solutions keep uniform estimates of all norms from Theorem 2 on extended segments $[0, T] \subset [0, T_*)$. In other words, the class $C^\infty_{6/5,\,3/2}$ is kept. If $[0, T_*)$ is the maximal interval of solution existence, then the second parameter is defined by:

$$\mu = \frac{T_*}{T_0}, \tag{69}$$

where T_0 from (5).

The third parameter ε is defined below by (87).

4.1. *Solutions Extension in Global with Condition* $l(\varphi) < \frac{81\nu^2}{8}$

Lemma 36. *Let u be a solution of problems (1) and (2) from Theorem 2. Then, functions*

$$\eta_1(t) = \|\nabla u\|_2,\ \eta_2(t) = \|\triangle u\|_2,\ \eta_3(t) = \int_{R^3} u_i u_{k,\,i}\triangle u_k dx,\ \eta_4(t) = \|u\|_2$$

are continuous functions on the interval $[0, T_*)$.

Proof. Let $s \in (0, T_*)$. Fix $t \in (s, \tau_1(s))$ where the function τ_1 from Lemma 45. Choose a segment $[T, T_1] \subset (s, \tau_1(s))$ assuming $t,\ t+h \in [T, T_1]$. Denote $z = z(t, h, x) = u(t+h, x) - u(t, x)$. Take equalities (1) with time argument $t + h$. Thereupon, we multiply them by $\triangle z$ getting the scalar product and integrate over R^3. The derivative $D_t u \in C^{\infty}_{6/5,\,3/2}$ for all $t \in [0, T_*)$. It follows from Theorem 2. Then, by Lemma 35 (the scalar product in L_2 we write as (f, g)), we have:

$$\frac{1}{2}\frac{d}{dh}\|\nabla z\|_2^2 = -\nu(\triangle u(t+h,\cdot), \triangle z) + \int_{R^3} u_i(t+h,x)u_{k,\,i}(t+h,x)\triangle z_k dx.$$

Here, the right-hand side is bounded uniformly (see Theorem 2). Then, $\|\nabla z\|_2^2 = O(h)$ as $h \to 0$. Triangle inequality implies the continuity of function η_1.

We write equality (1) for time arguments t, $t+h$ and subtract it. Thereupon, the difference we multiply by $\triangle D_t u(t+h,x)$ getting the scalar product and integrating over the whole space. As a result, we have

$$(D_t z, \triangle D_t u(t+h,\cdot)) - \nu(\triangle z, \triangle D_h z) +$$

$$\int_{R^3} (z_i u_{k,\,i}(t+h,x) + u_i(t,x) z_{k,\,i}) \triangle D_t u_k(t+h,x) dx = 0.$$

Uniform estimates from Theorem 2 and an integrability for any exponent a.e. imply the equality:

$$\frac{\nu}{2}\frac{d}{dh}\|\triangle z\|_2 = O(1).$$

Then, we have the continuity of function η_2.

Function continuity of η_3 follows also from uniform estimates of Theorem 2. Difference $\eta_3(t) - \eta_3(t_0)$ is considered as the sum of three integrals with combinations:

$$u_i(t,\ x) - u_i(t_0,x),\ u_{k,\,i}(t,\ x) - u_{k,\,i}(t_0,x),\ \triangle u_k(t,\ x) - \triangle u_k(t_0,x).$$

Every integral we estimate by Hölder's inequality so that there appear norms:

$$\|u_i(t,\cdot) - u_i(t_0,\cdot)\|_6,\ \|u_{k,\,i}(t,\cdot) - u_{k,\,i}(t_0,\cdot))\|_2,$$

$$\|\triangle u_k(t,\cdot) - \triangle u_k(t_0,\cdot))\|_2.$$

The first of these norms is estimated through the second norm by the inequality from Lemma A1 with application of the second representation in Lemma 34. Therefore, on every segment $[0, T]$ with some constants C_1, C_2, we have:

$$|\eta_3(t) - \eta_3(t_0)| \leq C_1|\eta_1(t) - \eta_1(t_0)| + C_2|\eta_2(t) - \eta_2(t_0)|.$$

Hence, the first statement follows. Let us prove function continuity of η_4. The estimate

$$\|u(t,\cdot) - u(t_0,\cdot))\|_6 \leq A\|\nabla u(t,\cdot) - \nabla u(t_0,\cdot))\|_2$$

was called above. The logarithmic convex inequality $\|v\|_2 \leq \|v\|_{6/5}^{1-\theta}\|v\|_6^{\theta}$ where $\frac{1}{2} = \frac{5}{6}(1-\theta) + \frac{\theta}{2}$ and Theorem 2 (see item (2)) about uniform boundedness of norms) give the statement of the lemma. Here, it is enough to take $v = u(t,\cdot) - u(t_0,\cdot)$. □

Lemma 37. *Let $\varphi \in C^{\infty}_{6/5,\,3/2}$ and $l(\varphi) < \left(\frac{v}{a_1}\right)^2$, where $l(\varphi)$ is defined by (68), the number a_1 from Corollary A4. Then, solution u of problems (1) and (2) from Theorem 1 satisfies inequality: $\|\nabla u\|_2 \leq \|\nabla\varphi\|_2$.*

Proof. Equality (1) we multiply by $\triangle u$ getting the scalar product and integrating over the whole space. Then, from Theorem 2 and Lemma 35, we have:

$$\frac{1}{2}\frac{d}{dt}\|\nabla u\|_2^2 = \eta_3(t) - \nu\eta_2^2(t),$$

where $\eta_i,\ i = 2,3$, from Lemma 32. Now, we show that the function $\eta(t) = \eta_3(t) - \nu\eta_2^2(t)$ is negative. Note $\eta(0) < 0$. Suppose the opposite. Then,

$$\nu\|\triangle\varphi\|_2^2 \leq \int \varphi_i\varphi_{k,\,i}\triangle\varphi_k dx.$$

From Corollary A4 (it is extended on the class $C^\infty_{6/5,\,3/2}$ by Lemma 34), we have estimate:

$$\nu\|\triangle\varphi\|_2^2 \leq a_1\|\nabla\varphi\|_2^{3/2}\|\triangle\varphi\|_2^{3/2}.$$

Since

$$\|\nabla\varphi\|_2 \leq \|\varphi\|_2^{1/2}\|\triangle\varphi\|_2^{1/2}$$

(it follows from Lemma 35), then the last two inequalities imply $\nu^2 \leq a_1^2 l(\varphi)$. We have a contradiction. Let $[0, t_0)$ be a maximal interval where function $\eta < 0$. Suppose $t_0 < T_0$. Continuity condition (see Lemma 32 and Theorem 2) gives $\eta(t_0) = 0$.
Repeating arguments from above, we obtain estimate:

$$\nu^2 \leq a_1^2 l(u(t_0,\cdot)). \tag{70}$$

With the other hand, function η_1 from Lemma 32 is a decreasing function on interval $[0, t_0)$. Therefore, $\|\nabla u(t_0,\cdot)\|_2 < \|\nabla\varphi\|_2$. Since $\|u\|_2 \leq \|\varphi\|_2$, then $l(u(t_0,\cdot)) < l(\varphi)$. Compare this inequality with (70). Then, we have a contradiction. □

Lemma 38. *Let $\varphi \in C^\infty_{6/5,\,3/2}$ and*

$$q^{\alpha_{m-1}}\left(\frac{\nu}{a_1}\right)^2 \leq l(\varphi) < q^{\alpha_m}\left(\frac{\nu}{a_1}\right)^2,$$

where $l(\varphi)$ is defined by (68), numbers a_1 from Corollary A4,

$$q = \frac{4}{3}\sqrt[4]{3},\ \alpha_0 = 0,\ \alpha_m = 2 - \frac{1}{2^{m-1}},\ m = 1,2,\ldots.$$

Then, for solution u of problems (1) and (2) from Theorem 1, there exists a number $t_0 \in (0, T_0)$ such that

$$l(u(t_0,\cdot)) < q^{\alpha_{m-1}}\left(\frac{\nu}{a_1}\right)^2.$$

Proof. Suppose the opposite. Then, on interval $[0, T_0)$, the inequality holds:

$$q^{2\alpha_{m-1}}\left(\frac{\nu}{a_1}\right)^4 \leq l^2(u).$$

Integrate it over this interval. Since

$$2\nu\|\nabla u\|_2^2 = -\frac{d}{dt}\|u\|_2^2,$$

then

$$q^{2\alpha_{m-1}}\left(\frac{\nu}{a_1}\right)^4 T_0 \leq -\frac{1}{4\nu}\|u\|_2^4\Big|_0^{T_0}.$$

Take out a nonpositive term on the right-hand side and input the mean T_0 from (5). Then,

$$q^{\alpha_m}\left(\frac{\nu}{a_1}\right)^2 \leq l(\varphi).$$

We have a contradiction with the condition. □

Theorem 3. *Let $\varphi \in C^{\infty}_{6/5,\,3/2}$ and $l(\varphi) < \left(\frac{\nu}{a_1}\right)^2$ where $l(\varphi)$ is defined by (68) and the number a_1 from Corollary A4. Then, problems (1) and (2) have unique solution u and solution P such that are defined on the set $[0,\infty) \times R^3$. In addition, these solutions have properties (1)–(5) from Theorem 1 on every fixed segment $[0,T]$ and satisfy Theorem 2. Moreover, the norm $\|\nabla u\|_2$, as a function of argument t, is a decreasing function on the set $[0,\infty)$.*

Proof. If $T < T_0$, then the statement of theorem follows from Theorems 1, 2 and Lemma 37. A finiteness of mixed norms $\|u\|_{p,\,q}$ we get from a boundedness of the vector field u and estimates $\|u\|_2 \le \|\varphi\|_2$. Solution uniqueness in the class $L_{p,\,q}$, $\frac{3}{p} + \frac{2}{q} \le 1$ is proved in [7,8,26] (see also [13]).
Norm monotonicity $\|\nabla u\|_2$ as a function on time argument t follows from condition $\eta < 0$ (see proof of Lemma 37).
Let $[0, T^*)$ be an interval of the maximal length such that there exist solutions with the estimates of Theorem 2.
Suppose $T^* < \infty$. Let $t_0 < T^*$ and $T^* - t_0 < 0,5T^*$. By Theorem 2 mapping, $u(t_{0,}\cdot)$ belongs to class $C^{\infty}_{6/5,\,3/2}$. Therefore, by Theorem 1 with this initial data, there is the unique solution w of the Cauchy problem that can be built that can be considered as the extension of solution u (see Lemma 36 and Theorem 1). Extension of u is the unique solution of problems (1) and (2) that satisfies the theorem, at least, on the interval $[0, t_0 + T_2)$, where

$$T_2 = \left(\frac{9}{4}\right)^4 \frac{\nu^3}{\|\nabla u(t_0,\cdot)\|_2^4}.$$

We have $T_2 \ge T_0$ from condition $\|\nabla u\|_2 \le |\nabla\varphi\|_2$, $w(t,x) = u(t_0 + t, x)$ for means $t < T^* - t_0$. Hence, the solution u is extended with the half-interval $[0, T^*)$ on an interval of more length $[0, T^* + 0,5T_0)$. We have a contradiction. □

Theorem 4. *Let $\varphi \in C^{\infty}_{6/5,\,3/2}$ and*

$$\left(\frac{\nu}{a_1}\right)^2 \le l(\varphi) < \left(\frac{4\sqrt[4]{3}\nu}{3a_1}\right)^2 = \frac{81\nu^2}{8},$$

where $l(\varphi)$ is defined by (68), number a_1 from Corollary A4. Then, problems (1) and (2) have a unique solution u and a solution P that are defined on the set $[0,\infty) \times R^3$. In addition, these solutions have properties (1)–(5) from Theorem 1 on every fix segment $[0,T]$ and satisfy Theorem 2. The norm $\|\nabla u\|_2$, as a function of t, is not decreasing function on the set $[T_0,\infty)$, where constant T_0 from (5).

Proof. Let u and P be solutions of problems (1) and (2) from Theorem 2. The proof proceeds from induction with respect to number m from Lemma 38. Let $m = 1$. By Lemma 38, there exists a number $t_0 \in (0, T_0)$ such that

$$l(u(t_0,\cdot)) < \left(\frac{\nu}{a_1}\right)^2.$$

By Theorems 1–3, there exists a global solution w of problems (1) and (2) with changed initial data $w(0,x) = u(t_0,x)$. This is the unique smooth extension of solution u that satisfies the proving theorem. Assume the theorem is true for a some natural number m. That is, every solution u has a global extension with properties of the theorem if, for this u, there exists a number $t_0 \in (0, T_0)$ such that

$$l(u(t_0,\cdot)) < q^{\alpha_m}\left(\frac{\nu}{a_1}\right)^2.$$

Now, we take initial data φ such that

$$q^{\alpha_m}\left(\frac{\nu}{a_1}\right)^2 \leq l(\varphi) < q^{\alpha_{m+1}}\left(\frac{\nu}{a_1}\right)^2.$$

By Lemma 38, there exists $t_0 \in (0, T_0)$ satisfying

$$l(u(t_0, \cdot)) < q^{\alpha_m}\left(\frac{\nu}{a_1}\right)^2.$$

By Theorem 2 and the induction hypothesis, there exists a global solution w of problems (1) and (2) with a new initial data $w(0, x) = u(t_0, x)$. By a uniqueness theorem, it is the unique smooth extension of solution u that satisfies the proving theorem. By the induction principle, the theorem is proved because

$$q^{\alpha_m} a_1^{-2} \to \frac{16\sqrt{3}}{9} a_1^{-2}$$

as $m \to \infty$. □

4.2. Critical λ Parameter Mean and the First Hypothetical Turbulent Solution

Furthermore, it is important in principle an invariant form of a priori estimate for the Cauchy problem solution. An invariance follows from Lemmas 1, 6, 20 and 25, Remark 2, norm semicontinuity of $\|\nabla u\|_2$ and Theorem 1.

Lemma 39. *The solution u of problems (1) and (2) from Theorem 1 satisfies estimate:*

$$\|\nabla u\|_2^2 \leq \|\nabla \varphi\|_2^2 (1 - t/T_0)^{-1/2}. \quad (71)$$

Lemma 40. *Let $\varphi \in C^\infty_{6/5,\,3/2}$ and $\lambda \geq 1$ i.e.,*

$$l(\varphi) \geq \left(\frac{4\sqrt[4]{3}\nu}{3a_1}\right)^2.$$

Let u be a solution of problems (1) and (2) from Theorem 1 If

$$l(u(t_0, \cdot)) < \left(\frac{4\sqrt[4]{3}\nu}{3a_1}\right)^2$$

for a some number $t_0 \in [0, T_0)$, then solution u can be extended by a global solution with properties (1)–(5) from Theorem 1 and estimates from Theorem 2.

Proof. We construct the extension in the same way as in the proof of Theorem 4. □

Lemma 41. *Let $\varphi \in C^\infty_{6/5,\,3/2}$ and parameter $\lambda \leq 1$ (see (68)). If u is the solution of problems (1) and (2) from Theorem 1, then on interval $[0, T_0)$, the inequality holds:*

$$\|\varphi\|_2^2\left(1 - \lambda^2 + \lambda^2\sqrt{1 - t/T_0}\right) \leq \|u\|_2^2.$$

Proof. We integrate the inequality of Lemma 39 over the segment $[0, t]$. Since

$$\frac{1}{2}\frac{d}{dt}\|u\|_2^2 + \nu\|\nabla u\|_2^2 = 0, \quad (72)$$

then, applying Newton–Leibnitz's formula, we obtain the statement. □

Lemma 42. *Let $\varphi \in C^{\infty}_{6/5,\,3/2}$ and parameter $\lambda = 1$ (see (68)). Let u be a solution of problems (1) and (2) from Theorem 1. Suppose, on the interval $[0, T_0)$, there is fulfilled estimate:*

$$l(u(t,\cdot)) \geq \Big(\frac{4\sqrt[4]{3}\nu}{3a_1}\Big)^2.$$

Then,

$$\|u\|_2^2 = \|\varphi\|_2^2\Big(1 - t/T_0\Big)^{1/2}, \|\nabla u\|_2^2 = \|\nabla\varphi\|_2^2\Big(1 - t/T_0\Big)^{-1/2}. \tag{73}$$

If $\lim_{t\uparrow T_0} \|u\|_2 > 0$, then there exists a number $t_0 \in (0, T_0)$ such that

$$l(u(t_0,\cdot)) < \Big(\frac{4\sqrt[4]{3}\nu}{3a_1}\Big)^2.$$

Proof. Both parts we raise to the second power and integrate over the interval $[0, T_0)$. From (72), we get:

$$\Big(\frac{4\sqrt[4]{3}\nu}{3a_1}\Big)^4 T_0 \leq \int_0^{T_0} \|u\|_2^2\|\nabla u\|_2^2 dt = \frac{1}{4\nu}\Big(\|\varphi\|_2^4 - \lim_{t\to T_0} \|u\|_2^4\Big), \tag{74}$$

where the number T_0 from Theorem 1. Since $\lambda = 1$, then

$$\Big(\frac{4\sqrt[4]{3}\nu}{3a_1}\Big)^4 T_0 = \frac{\|\varphi\|_2^4}{4\nu}.$$

Therefore, the limit in (74) is equal to zero because, in (74), it must be equalities. This is possible only if, on the interval $[0, T_0)$ (see Lemma 36), it is fulfilled:

$$l^2(u(t,\cdot)) = \Big(\frac{4\sqrt[4]{3}\nu}{3a_1}\Big)^4 T_0. \tag{75}$$

Integrate (72) over the interval $[0, T_0)$. As the result, we have:

$$2\nu \int_0^{T_0} \|\nabla u\|_2^2 dt = \|\varphi\|_2^2$$

(we take into consideration in formula (74) the limit vanishes). Apply the estimate of Lemma 39. Then,

$$4\nu T_0\|\nabla\varphi\|_2^2 \geq \|\varphi\|_2^2.$$

Hence, we have the inequality $\lambda \geq 1$. Since $\lambda = 1$, then the inequality from Lemma (71) must be as the equality. The second formula of lemma is proved. The first follows from (75) and condition $\lambda = 1$. The last statement of lemma we prove from the opposite in the same way. □

Lemma 43. *Let initial data $\varphi \in C^{\infty}_{6/5,\,3/2}$, $\varphi \neq 0$, and parameter $\lambda = 1$. There doesn't exist solution u of problems (1) and (2) satisfying (73). It is always true inequality $\lim_{t\uparrow T_0} \|u\|_2 > 0$.*

Proof. If such solution exists, then, from (73), we obtain

$$\frac{1}{2}\frac{d}{dt}\|\nabla u\|_2^2 = \Big(\frac{8}{81}\Big)^2 \frac{\|\nabla u\|_2^6}{\nu^3}.$$

Here, the identical equality is impossible because, for any solution u, the inequality (see (7)) is fulfilled:

$$\frac{1}{2}\frac{d}{dt}\|\nabla u\|_2^2 + \nu\|\triangle u\|_2^2 \leq a_1\|\nabla u\|_2^{3/2}\|\triangle u\|_2^{3/2}.$$

Apply estimates from the proof of Lemma 1. Then, we obtain:

$$\frac{1}{2}\frac{d}{dt}\|\nabla u\|_2^2 \leq \left(\frac{8}{81}\right)^2 \frac{\|\nabla u\|_2^6}{\nu^3}.$$

Compare this inequality with the identity above. Therefore, we must have the equalities for intermediate estimates of Corollary A4 and Lemma 1. Since we used Cauchy–Bunyakovskii's inequality in the Hilbert space $L_2(R^3)$, then there exists a constant c such that

$$u_{i,\,j} = c\Big(u_{k,\,i}u_{k,\,j} - \frac{\delta_{ij}}{3}|\nabla u|_2^2\Big)$$

for any $i, j = 1, 2, 3$. Hence, we have $u_{i,\,j} = u_{j,\,i}$ for each pair $i,\ j$ and $\triangle u \equiv 0$, respectively. From Lemma A6, it follows $u \equiv 0$—a contradiction. The lemma is proved. □

Lemma 44. *Let initial data $\varphi \in C^\infty_{6/5,\,3/2}$, $\varphi \neq 0$, and parameter $\lambda < 1$. If u is the solution of problems (1) and (2), then*

$$\lim_{t\to T_0} \|u\|_2^2 > \|\varphi\|_2^2(1-\lambda^2).$$

Proof. Suppose the opposite. Then, we have the equality in Lemma 41. It implies the second equality from (73). Repeating the proof of Lemma 43, we obtain a contradiction. □

Lemma 45. *Let u be a solution of problems (1) and (2) with initial data $\varphi \in C^\infty_{6/5,\,3/2}$, $\varphi \neq 0$. Then, a function*

$$\tau_1(t) = t + \left(\frac{9}{4}\right)^4 \frac{\nu^3}{\|\nabla u\|_2^4}$$

and a function

$$\lambda(t) = \left(\frac{4\sqrt[4]{3}\nu}{3a_1}\right)^2 \Big/ \|u\|_2\|\nabla u\|_2$$

with condition $\lambda(0) = \lambda \geq 1$ are not decreasing functions on the interval $[0,\ T_*)$ where constant a_1 from Corollary A4.

Proof. From Theorem 2, we have safety of class $C^\infty_{6/5,\,3/2}$ for every $t \in [0,\ T_*)$ if u satisfies lemma conditions. The both functions are continuous (see Lemma 36 and Theorem 2). Inequality (84) (see below) is true for any mean $\lambda(0) = \lambda$. Rewrite its in another form:

$$\frac{1}{\|\nabla u\|_2^4}\frac{d}{dt}\|\nabla u\|_2^2 \leq \frac{27a_1^4}{128\nu^3}\|\nabla u\|_2^2 \tag{76}$$

and integrate its over the segment $[t, s]$. Simple transformations give:

$$\|u(t,\cdot)\|_2^2(\lambda^2(t)-1) \leq \|u(s,\cdot)\|_2^2(\lambda^2(s)-1). \tag{77}$$

Hence, and from lemma condition, it follows $\lambda(s) \geq 1$ for all $s \in [0, T_*)$. Furthermore, we use inequality $\|u(s,\cdot)\|_2 \leq \|u(t,\cdot)\|_2$ and get the monotonicity of the second function. For the monotonicity, the first function follows from inequality (84) because, in this case, $\tau_1' \geq 0$. □

4.3. Solutions Extension in Global with Condition $l(\varphi) \geq \frac{81\nu^2}{8}$: Necessary Conditions For Hypothetical Turbulence Solutions

Lemma 46. *Let $\varphi \in C^{\infty}_{6/5,\,3/2}$. Suppose that $l(\varphi) \geq \frac{81\nu^2}{8}$ and parameter λ from (68). If $\lambda = 1$ or $\lambda < 1$ and the solution u of problems (1) and (2) from Theorem 2 satisfies*

$$\lim_{t\to T_0} \|u\|_2^2 \geq \|\varphi\|_2^2\sqrt{1-\lambda^4}.$$

Then, there exists a number $t_0 \in (0, T_0)$ such that inequality is fulfilled:

$$l(u(t_0,\cdot)) < \Big(\frac{4\sqrt[4]{3}\nu}{3a_1}\Big)^2,$$

where constant a_1 from Corollary A4.

Proof. Assume $\lambda = 1$. Then, the statement follows from Lemmas 42 and 43. Let $\lambda < 1$. Suppose the opposite. Then, we have:

$$\Big(\frac{4\sqrt[4]{3}\nu}{3a_1}\Big)^2 \leq l(u(t,\cdot)).$$

Hence, and from (72), we get

$$\Big(\frac{4\sqrt[4]{3}\nu}{3a_1}\Big)^4 \leq \|u\|_2^2\|\nabla u\|_2^2 = -\frac{1}{4\nu}\frac{d}{dt}\|u\|_2^4. \tag{78}$$

Let

$$\alpha = \Big(\frac{4\sqrt[4]{3}}{3a_1}\Big)^4\frac{\nu^5}{\|\varphi\|_2^4}. \tag{79}$$

Integrate (78) over segment $[0, t]$. Then, we obtain:

$$\frac{\|u\|_2}{\sqrt[4]{1-4\alpha t}} \leq \|\varphi\|_2. \tag{80}$$

Make the passage to the limit in (80) as $t \uparrow T_0$ and compare the new estimate with the inequality from lemma condition. Taking (5), (68) and (80), we conclude $\lim_{t\to T_0}\|u\|_2^2 = \|\varphi\|_2^2\sqrt{1-\lambda^4}$. Consider a function $\beta(t) = \|u\|_2^2 - \sqrt{1-4\alpha t}\|\varphi\|_2^2$. It vanishes at boundary points of $[0, T_0]$; moreover, $\beta \leq 0$ (see (80)). Let $I \subset (0, T_0)$ be an interval, where the function β vanishes at boundary points and $\beta < 0$ on its interior. Then, there exists a point $t_0 \in I$, where $\beta'(t_0) = 0$. Hence, from (72), we get:

$$\nu\|\nabla u(t_0,\cdot)\|_2^2 = \frac{\alpha\|\varphi\|_2^2}{\sqrt{1-4\alpha t_0}}.$$

From (80), we have:

$$\nu\|u(t_0,\cdot)\|_2^2\|\nabla u(t_{0,}\cdot\|_2^2 = \frac{\alpha\|\varphi\|_2^2\|u(t_0,\cdot)\|_2^2\|}{\sqrt{1-4\alpha t_0}} \leq \alpha\|\varphi\|_2^4. \tag{81}$$

Compare the left and right sides of this formula and, after we apply (79). Then,

$$l(u(t_0,\cdot)) \leq \Big(\frac{4\sqrt[4]{3}\nu}{3a_1}\Big)^2.$$

The hypothesis from proof beginning gives the equality:

$$l(u(t_0,\cdot)) = \left(\frac{4\sqrt[4]{3}\nu}{3a_1}\right)^2.$$

Therefore, in (81), the inequality must be by the equality. Hence, we get $\beta(t_0) = 0$. This goes to a contradiction with the choice of the interval I. It implies $\beta = 0$. Hence, $\|u\|_2^2 = \|\varphi\|_2^2\sqrt{1-4\alpha t}$. Respectively, from (72), we have

$$\nu\|\nabla u(t,\cdot)\|_2^2 = \frac{\alpha\|\varphi\|_2^2}{\sqrt{1-4\alpha t}}.$$

Multiply these equalities. From (78), we obtain:

$$l(u(t,\cdot)) = \left(\frac{4\sqrt[4]{3}\nu}{3a_1}\right)^2.$$

In particular, by Lemma 36, $l(\varphi) \geq \frac{81\nu^2}{8}$. This is impossible with the considering lemma condition. This contradiction proves the lemma. □

Now, we shall study properties of unextended solutions of problems (1) and (2) if such solutions exist. Let $[0, T_*)$ be an interval of the maximal length, where solutions u and P of problems (1) and (2) have properties from Theorem 2. Then, $T_* \geq T_0$ and $T_* = \mu T_0$ (see (69)). Hence, $\mu \geq 1$. Therefore, J. Leray's estimate from [3] can be given in invariant form in the following statement.

Lemma 47. *Let $\varphi \in C^\infty_{6/5,\,3/2}$ and $l(\varphi) > \frac{81\nu^2}{8}$. Suppose that $[0, T_*)$ is the maximal interval where solutions u and P of problems (1) and (2) have solution properties from Theorem 2. If this interval is a finite, then the following estimate holds:*

$$\|\nabla u\|_2^2 \geq \sqrt{\frac{1}{\mu}}\|\nabla\varphi\|_2^2\Big(1 - t/T_*\Big)^{-1/2}. \tag{82}$$

Proof. A function $\eta_1(t) = \|\nabla u\|_2$ is unbounded in some left neighborhood of the point T_*. Suppose the opposite. Then, for every point t_0, there exists solution v of problems (1) and (2) with initial data $v(0,\ x) = u(t_0,\ x)$ which satisfy Theorems 1 and 2.

This solution gives the unique extension u on the interval $[t_0, t_0 + l)$, where $l \geq 9^4\nu^3/(4M)^4$ and M is the supremum of η_1. A point t_0 is an arbitrary, therefore, solutions u and P can be extended on the interval $[0, T_* + l)$. In addition, they have solutions' properties from Theorem 2 on this interval. This contradicts the choice of interval with the maximal length.

Now, we prove estimate (82). For solution u, we have:

$$\frac{1}{2}\frac{d}{dt}\|\nabla u\|_2^2 + \nu\|\triangle u\|_2^2 = \int u_i u_{k,\,i}\triangle u_k dx, \tag{83}$$

which follows from (1). It is true for every mean $t \in [0, T_*)$ by Theorem 2 and Lemma 35 because the solution $u \in C^\infty_{6/5,\,3/2}$. The integral representation from Lemma 34 permits to apply Corollary A4 and estimate of the right-hand side in the last equality. Repeating the proof of Lemma 1, we obtain

$$\frac{1}{\|\nabla u\|_2^6}\frac{d}{dt}\|\nabla u\|_2^2 \leq \frac{27a_1^4}{128\nu^3}. \tag{84}$$

Integrate (84) over segment $[t, s]$. Then,

$$\frac{1}{\|\nabla u(t,\cdot)\|_2^4} - \frac{1}{\|\nabla u(s,\cdot)\|_2^4} \leq \frac{27a_1^4}{128\nu^3}(s-t). \tag{85}$$

Replace in (84) s on s_m, where $s_m \uparrow T_*$ and $\eta_1(s_m) \to \infty$ as $m \to \infty$. The passage to the limit in (85) gives estimate (82). □

Lemma 48. *Let $\varphi \in C^{\infty}_{6/5,\,3/2}$ and $l(\varphi) > \frac{81\nu^2}{8}$. For finite interval $[0, T_*)$ of the maximal length, the parameter μ is not greater than the number λ^{-4}, where λ is defined by (68).*

Proof. In the inequality

$$2\nu \int_0^{T_*} \|\nabla u\|_2^2 dt \le \|\nabla \varphi,\|_2^2$$

we apply Lemma (79). From (5) and (68), we get the statement. □

Lemma 49. *Let $\varphi \in C^{\infty}_{6/5,\,3/2}$, $l(\varphi) > \left(\frac{4\sqrt[4]{3}\nu}{3a_1}\right)^2 = \frac{81\nu^2}{8}$ and λ be a parameter from (68). If the interval $[0, T_*)$ has the maximal finite length and solutions u, P of problems (1) and (2) have properties from Theorem 2 on this interval, then unextended solutions satisfy conditions:*

$$\|\varphi\|_2^2 \lambda^2 \sqrt{\mu - t/T_0} + \|u(T_*, \cdot)\|_2^2 \le \|u\|_2^2 \le \qquad (86)$$
$$\|\varphi\|_2^2 \Big(1 - \sqrt{\mu}\lambda^2 + \lambda^2 \sqrt{\mu - t/T_0}\Big).$$

Proof. Consider a function

$$\omega(t) = \|u\|_2^2 - \|\varphi\|_2^2 \lambda^2 \sqrt{\mu - t/T_0}.$$

From (5) and (68), (72), we have:

$$\omega'(t) = 2\nu\Big(-\|\nabla u\|_2^2 + \|\nabla \varphi\|_2^2 \big(\mu - t/T_0\big)^{-1/2}\Big), \ \omega'(t) \le 0.$$

Therefore, $\omega(0) \ge \omega(t) \ge \omega(T_* - 0)$. Hence, it follows the first inequality from (86).

Integrate over $[0, t]$ the inequality of Lemma (79). From (72), we obtain:

$$\|\varphi\|_2^2 - \|u\|_2^2 \ge 4T_*\nu \frac{\|\nabla \varphi\|_2^2}{\sqrt{\mu}} \Big(1 - \sqrt{1 - t/T_*}\Big).$$

Applying (5), (68) and (69), we get:

$$1 - \frac{\|u\|_2^2}{\|\varphi\|_2^2} \ge \sqrt{\mu}\lambda^2 \Big(1 - \sqrt{1 - t/T_*}\Big).$$

Therefore, we have the second inequality in (86). □

Theorem 5. *Set $\varphi \in C^{\infty}_{6/5,\,3/2}$ and $l(\varphi) > \left(\frac{4\sqrt[4]{3}\nu}{3a_1}\right)^2$ with a constant a_1 from A4. Let u be a solution of problems (1) and (2) from Theorem 2. If u satisfies condition*

$$\lim_{t \to T_0} \|u(t, \cdot)\|_2^2 \ge \|\varphi\|_2^2 \sqrt{1 - \lambda^4},$$

then problems (1) and (2) have global solutions u and P. Moreover, they have properties (1)–(5) from Theorem 1 on every segment $[0, T]$, $T > 0$ and satisfy conditions of Theorem 2 there. As a function of argument t, the product $\|u\|_2 \|\nabla u\|_2$ is a decreasing function on the set $[T_0, \infty)$, where constant T_0 from (5).

Proof. Let t_0 a number from Lemma 46. Without norm monotonicity, the statement of theorem follows from Theorem 2 and Lemma 40. The product $\|u\|_2 \|\nabla u\|_2$ is a decreasing function on the set $[t_0, \infty)$. It follows from Theorem 4. Therefore, the theorem is proved. □

Now, we give one result that is connected with a local solutions' extension. If $\lambda < 1$, then we introduce the third parameter ε, which gives a dissipation quantity of a kinetic energy. It is defined by formula:

$$\lim_{t \to T_0} \|u(t,\cdot)\|_2^2 = \|\varphi\|_2^2(1 - \varepsilon\lambda^2). \tag{87}$$

We observe from Lemmas 43 and 44 that the parameter ε satisfies strong inequalities: $0 < \varepsilon < 1$. This is very important for the furthest. The usefulness of this parameter is explained by the following result.

Theorem 6. *Suppose initial data $\varphi \in C^\infty_{6/5,\,3/2}$ and*

$$l(\varphi) > \left(\frac{4\sqrt[4]{3}\nu}{3a_1}\right)^2 = \frac{81\nu^2}{8},$$

where $l(\varphi)$ is defined by (68). If solution u of problems (1) and (2) from Theorem 2 satisfies (87), then this solution has an extension on the set S_{T_3} where

$$T_3 = \frac{T_0}{4}\left(\varepsilon + \frac{1}{\varepsilon}\right)^2.$$

This extension has properties (1)–(5) from Theorem 1 on every segment $[0,T] \subset [0,T_3)$.

Proof. Now, we consider only that solutions which don't have any global and smooth extension. Take $t = T_0$. From theorem condition and the second inequality of Lemma 49, we obtain: $-\varepsilon \le -\sqrt{\mu} + \sqrt{\mu - 1}$. Hence, we get: $\mu \ge \frac{1}{4}\left(\varepsilon + \frac{1}{\varepsilon}\right)^2$. Then, the statement of the theorem follows from the definition of parameter μ. The theorem is proved. □

Lemma 50. *Suppose $\lambda < 1$. A finite mean of parameter μ satisfies inequalities:*

$$\frac{1}{4}\left(\varepsilon + \frac{1}{\varepsilon}\right)^2 < \mu \le \lambda^{-4}.$$

Proof. In the first inequality of Lemma 49, we take $t = T_*$. Then, we get the necessary upper estimate. The strong lower estimate doesn't follow from (71) yet. Let

$$\tau(\varepsilon) = \frac{1}{2}\left(\varepsilon + \frac{1}{\varepsilon}\right).$$

Consider a function

$$\varrho(t) = \|u\|_2^2 - \|\varphi\|_2^2\left(1 - \tau(\varepsilon)\lambda^2 + \lambda^2\sqrt{\tau^2(\varepsilon) - t/T_0}\right).$$

We observe $\varrho(0) = \varrho(T_0) = 0$ (see formula (87)). Hence, there exists a number $\xi \in (0, T_0)$ such that $\varrho'(\xi) = 0$. Then,

$$\|\nabla u(\xi,\cdot)\|_2^2 = \frac{\|\nabla\varphi\|_2^2}{\sqrt{\tau^2(\varepsilon) - \frac{\xi}{T_0}}}$$

or

$$T_0\tau^2(\varepsilon) = \xi + \left(\frac{9}{4}\right)^4 \frac{\nu^3}{\|\nabla u(\xi,\cdot)\|_2^4} = \tau_1(\xi),$$

where function τ_1 from Lemma 45. Since this function does not decrease then for every t, $\xi < t < T_0\tau^2(\varepsilon)$, we have $T_0\tau^2(\varepsilon) \le \tau_1(t)$. Therefore,

$$\|\nabla u\|_2^2 \le \|\nabla\varphi\|_2^2\left(\tau^2(\varepsilon) - t/T_0\right)^{-1/2},$$

which holds for every $t, \xi < t < T_0\tau^2(\varepsilon)$. Integrating this inequality over interval $[T_0, \tau^2(\varepsilon)T_0)$ from formula (87), we gather:

$$\|u(T_0\tau^2(\varepsilon)), \cdot)\|_2^2 \geq \|\varphi\|_2^2(1 - \tau(\varepsilon)\lambda^2). \tag{88}$$

If $\mu = \tau^2(\varepsilon)$, then, in formula (86), we must have identical equalities (see (88)). It implies the identity

$$\|\nabla u\|_2^2 = \sqrt{\frac{1}{\mu}\|\nabla\varphi\|_2^2}\Big(1 - t/T_*\Big)^{-1/2}.$$

Take $t = 0$. Hence, we get $\mu = 1$. This contradicts Theorem 6, from which we obtain $\mu \geq \tau^2(\varepsilon) > 1$ because $0 < \varepsilon < 1$. The last proves the strong lower estimate for μ. □

5. Main Results, Existence of Global Regular Solutions, and Sufficient Conditions

Now, we prove the basic result which is described by Theorem 7.

Theorem 7. *Let $\varphi \in C^\infty_{6/5,\,3/2}$ be initial data, the parameter λ from (68) and the number T_0 from (5), the vector field u from Theorem 1. If parameter $\lambda \geq 1$ or in opposite case*

$$\lim_{t\uparrow T_0} \|u(t,\cdot)\|_2^2 \geq \|\varphi\|_2^2\sqrt{1-\lambda^4}, \tag{89}$$

then the Cauchy problems (1) and (2) have global solutions $u(t,x) = (u_1(t,x), u_2(t,x), u_3(t,x))$ and $P = P(t,x)$ with the following properties:

(1) *mappings u and P are uniformly continuous and bounded on a set S_T for every number $T,\ T > 0$;*
(2) *for every numbers $T > 0,\ p \geq 6/5,\ q \geq 3/2,\ r > 1$ and multi-indices $|\alpha| \geq 2,\ |\beta| \geq 0$ all norms*

$$\|u\|_p,\ \|\nabla u\|_q,\ \|D^\alpha u\|_p,\ \|D^\beta P\|_2,\ \|D^\beta D_t u\|_r$$

are uniformly bounded on the segment $[0,T]$, moreover $\|u\|_2 \leq \|\varphi\|_2$;
(3) *gradients $\nabla u_i,\ i = 1,2,3,\ \nabla P$ are bounded on the set S_T for every $T > 0$;*
(4) *solution u has a finite mixed norm $\|u\|_{p,\,q}$ on the set S_T for every $T > 0$ and every pair of exponents $p, q \geq 2$;*
(5) *solutions u, P belong to class $C^\infty((0,T) \times R^3) \bigcap C(S_T)$ i.e., these solutions are classical.*

If parameter $\lambda \geq 1$, then the function $l(t) = \|u\|_2\|\nabla u\|_2$ is a decreasing function on the interval $[0,\infty)$. If $\lambda < 1$ and condition (89) is fulfilled then the function $l = l(t)$ is a decreasing function on the interval $[T_0,\infty)$.

Proof. Let $\lambda > 1$. Then, the statement follows from Theorem 4.
Let $\lambda = 1$. In this case, the theorem arises from Lemmas 42, Lemma 43 and Theorem 4. The monotonicity of the function l follows from Lemma 45.
Let $\lambda < 1$ and condition (89) is fulfilled. Then, the statement of the theorem arises from Lemmas 45 and 46, Theorem 5. The theorem is proved. □

Theorem 8. *Let $\varphi \in C^\infty_{6/5,\,3/2}$ be initial data, the parameter $\lambda < 1$ (see (68)) a vector field u is a weak solution from the Cauchy problems (1) and (2). If on an interval $[0,T)$ an inequality*

$$\|u(t,\cdot)\|_2^2 \geq \|\varphi\|_2^2\Big(1 - \lambda^2\sqrt{\frac{t}{T_0}}\Big) \tag{90}$$

is fulfilled, then the weak solutions $u(t,x) = (u_1(t,x), u_2(t,x), u_3(t,x))$ and $P = P(t,x)$ are regular on interval $[0,T)$ and satisfy Theorem 5.

Proof. Let be $[0, T_*) \subseteq [0, T)$ a maximal interval where the weak solution u is regular. Suppose $T_* < T$. Then, from (86) and (90), we have

$$\|u(t,\cdot)\|_2^2 = \|\varphi\|_2^2\Big(1 - \sqrt{\mu}\lambda^2 + \lambda^2\sqrt{\mu - \frac{t}{T_0}}\Big).$$

Since solution u is regular on the interval $[0, T_0)$, then, by differentiation of the previous identity at point $t = 0$, we obtain $\mu = 1$. From Lemma 50, we have a contradiction. Therefore, $T_* = T$. □

Does there exist weak solution u satisfying opposite inequality (90) if $t > T_0$? It is unknown.

6. The Cauchy Problem with Less Smoothness of Initial Data

In addition, the invariant class $C^\infty_{6/5,\,3/2}$ Sobolev space $\overset{\circ}{W}{}^3_2(R^3)$ as the closure of infinitely smooth vector fields is another important invariant class, which satisfy existence condition of global solutions. Different exceptions for solenoidal vector fields from Sobolev classes $\overset{\circ}{W}{}^3_2(R^3)$ and $W^3_2(R^3)$ were shown in [31]. Therefore, we consider the first space from them.

Let $\varphi \in \overset{\circ}{W}{}^3_2(R^3)$ be a solenoidal vector field. Set $(\varphi^m)_{m=1,\ldots}$ a sequence of of finite, solenoidal and infinitely smooth vector fields, which converges to the field φ in the space $\overset{\circ}{W}{}^3_2(R^3)$. We observe that $\varphi^m \in C^\infty_{6/5,\,3/2}$. Let $(u^m)_{m=1,\ldots}$, $(P^m)_{m=1,\ldots}$ be sequences of solutions in the Cauchy problem for Navier–Stokes equations with the initial dates φ^m. Then, all pairs u^m, P^m satisfy all uniform estimates of Lemma 21 on any compact set of the interval $[0, T_0^m)$ where $T_0^m = 9^4\nu^3/4^4\|\nabla\varphi^m\|_2^4$ since upper bounds in these inequalities depend on a set and ν, $\|\nabla\varphi^m\|_2$, $\|\varphi^m\|_2$. Therefore, on every fixed segment $[0, T] \subset [0, T_0]$, we can take these constants as common for all u^m, P^m because $\varphi^m \to \varphi$ in the space $\overset{\circ}{W}{}^3_2(R^3)$. Then, without loss of generality, we assume that the sequence $(u^m)_{m=1,\ldots}$ converges weakly in the space $W^1_6(S_T)$ to a field u_0. In addition, we suppose that $(\triangle u^m)_{m=1,\ldots}$, $(\nabla D_t u^m)_{m=1,\ldots}$ and $(\nabla P^m)_{m=1,\ldots}$ converge weakly in $L^1_2(S_T)$ to $\triangle u^0$, $\nabla D_t u^0$ and ∇P^0. More generally, weak limits u_0, P^0 satisfy all conclusions of Lemma 21 and they are weak solutions of problems (1) and (2). From the equality (1) for couple u_0, P^0 and items (2), (4), (5), (8) differentiating (1), we obtain that distributions $\triangle u^0_{,j}$, $j = 1, 2, 3$, belong to the space $L_2(R^3)$ for almost everywhere t. Thus, the class $\overset{\circ}{W}{}^3_2(R^3)$ is invariant similar to the class $C^\infty_{6/5,\,3/2}$. For this case, in the same way, we can define the basic parameters λ, μ, ε. After that, one should note that the statement of Lemma 36 will be true when initial data $\varphi \in \overset{\circ}{W}{}^3_2(R^3)$. Repeating the proof of Theorem 7, we obtain the following result.

Theorem 9. *Let $\varphi \in \overset{\circ}{W}{}^3_2(R^3)$ be initial data, the parameter λ from (68) and the number T_0 from (5). Let be a vector field u is a weak solution of the Cauchy problems (1) and (2). If parameter $\lambda \geq 1$ or in opposite case*

$$\varliminf_{t \uparrow T_0} \|u(t,\cdot)\|_2^2 \geq \|\varphi\|_2^2\sqrt{1-\lambda^4},$$

then the Cauchy problems (1) and (2) have global solutions $u(t,x) = (u_1(t,x), u_2(t,x), u_3(t,x))$ and $P = P(t,x)$ with the following properties:

(1) *mappings u and P are uniformly continuous and bounded on a set S_T for every number T, $T > 0$;*
(2) *for every numbers $T > 0$, $p \geq 2$, $q \geq 2$, all norms*

$$\|u\|_2,\ \|\nabla u\|_2,\ \|\triangle u\|_2,\ \|\nabla D_t u\|_2,\ \|\nabla P\|_2$$

are uniformly bounded and mixed norms $\|u\|_{p,\,q}$, $\|D_t\triangle u\|_{2,\,2}$ are finite on segment $[0, T]$;

(3) *solutions u, P belong to class* $C^\infty((0,T) \times R^3) \bigcap C(S_T)$ *i.e., these solutions are classical.*
If parameter $\lambda \geq 1$, *then the function* $l(t) = \|u\|_2 \|\nabla u\|_2$ *is a decreasing function on the interval* $[0, \infty)$.
If $\lambda < 1$ *and*

$$\lim_{t \uparrow T_0} \|u(t,\cdot)\|_2^2 \geq \|\varphi\|_2^2 \sqrt{1-\lambda^4},$$

then the function $l = l(t)$ *is a decreasing function on the interval* $[T_0, \infty)$.

Theorem 10. *Let* $\varphi \in W_2^3(R^3)$ *be initial data in problems (1) and (2). If parameter* $\lambda > 1$, *then the solution u from Theorem 9 satisfies:*

(1) *a power of norm* $\|u\|_2^4$ *is a convex function;*
(2) *there is fulfilled:*

$$\|\nabla u\|_2^2 \leq \|\nabla \varphi\|_2^2 \frac{\lambda^2}{\lambda^2 - 1}.$$

Proof. It follows from Lemma 45 because this lemma is true for solution u from Theorem 9. □

7. Integral Identities for Solenoidal Vector Fields: Dimensions Comparison

Some review and results about integral identities for solenoidal vector fields are given by authors in [32,33]. Here, we reduce one from these identities, which shows the essential distinction for the Navier–Stokes equations between space and plane.

Let $u, v, w : R^n \to R^n$ be any triple of solenoidal vector fields from the class $C_0^2(R^n)$. Denote

$$c_{ki}(u) = u_{k,\,i} - u_{k,\,i}, \quad k, i = 1, 2, \ldots, n.$$

Lemma 51. *(see [32]) For every triple* $u,\ v,\ w : R^n \to R^n$ *of solenoidal vector fields from the class* $C_0^2(R^n)$, *the identity holds:*

$$\int (w_{i,\,j} + w_{j,\,i}) c_{ki}(v) c_{kj}(u) dx = -\int w_i (c_{ki}(u) \triangle v_k + c_{ki}(v) \triangle u_k) dx.$$

Hence, it follows (one should take $u = v = w$):

$$\int u_{i,\,j} c_{ki}(u) c_{kj}(u) dx = -\int u_i c_{ki}(u) \triangle u_k dx.$$

Corollary 1. *(see [33].) If dimension* $n = 2$, *then every solenoidal vector* $u \in C_0^2(R^2)$ *satisfies the integral identity:*

$$\int u_i u_{k,\,i} \triangle u_k dx = 0. \tag{91}$$

Obviously, it implies some interesting applications to the 2D Navier–Stokes and Euler equations (see [32]).

(1) We deduce a priori estimate for a solution u, which is not independent of a viscosity:

$$\|\nabla u\|_2 \leq \|\nabla \varphi\|_2 + \int_0^t \|\nabla f\|_2 dt, \tag{92}$$

where f is an outer force. This improves essentially Ladyzhenskaya's estimate (see [34]).
(2) In the case $f = 0$, we have formula (83) and, therefore, the norm $\|\nabla u\|_2$ is a decreasing function.
(3) We give the new proof of the existence of a global weak solution for the Euler equations in plane in the case when an outer force $f = 0$. In addition, the estimate $\|\nabla u\|_2 \leq \|\nabla \varphi\|_2$ is exact and it does not follows from Judovich's results [35]. This explains "the simplicity" of a motion of an ideal fluid on plane.

Remark 6. *Let $n = 2$, $f = 0$. Then, the product $\|\nabla u\|_2 \|u\|_2$ is a decreasing function in any case.*

Remark 7. *If dimension $n = 3$, then integral from (91) may be not equal to null.*

For a simple example, there is the vector field with the following coordinates:

$$u_i(x) = \lambda_i^2 (l_i, x) e^{-\frac{1}{2}\left(\frac{x_1^2}{\lambda_1^2} + \frac{x_2^2}{\lambda_2^2} + \frac{x_3^2}{\lambda_3^2}\right)}, \ i = 1, 2, 3,$$

where l_i is the i –th vector row of the skew-symmetric matrix. Since

$$\int_{R^n} \sum_{i,\,k=1}^{n} u_i u_{k,\,i} \triangle u_k dx = -\int_{R^n} \sum_{i,\,k,\,j=1}^{n} u_{i,\,j} u_{k,\,i} u_{k,\,j} dx,$$

then simple calculations show

$$\int_{R^3} \sum_{i,\,j=1}^{3} u_i u_{j,\,i} \triangle u_j dx = c \sum_{i \neq k} \lambda_i^2 \lambda_k^4 \sum_{j}^{3} l_{ki} l_{ij} l_{kj}$$

with a constant $c \neq 0$. A coefficient $\sum_j l_{ki} l_{ij} l_{kj}$ may be not equal to zero for fixed different means k and i because there is the linear independence of polynomials $\lambda_i^2 \lambda_k^4 - \lambda_k^2 \lambda_i^4$, $i < k$, $i, k = 1, 2, 3$. It gives a distinct from zero of the integral when we choose a suitable skew-symmetric matrix. Respectively, the right side (see (83)) for dimensions $n \geq 3$ can be taken with a large value implying a positive mean of the difference

$$\int_{R^n} \sum_{i,\,j=1}^{n} u_i u_{j,\,i} \triangle u_j dx - \nu \|\triangle u\|_2^2$$

for $t \simeq 0$. It is possible because we can take a factor for initial data $\alpha\varphi$ or diminish viscosity coefficient ν. This implies a growth of the norm $\|\nabla u\|_2$ for space. Obviously, on the plane, this phenomena does not appear.

8. Conclusions

Briefly, the main achievements (see Theorems 7–10) have an obvious physical interpretation and, therefore, it may be interesting for applications. Nevertheless, they are connected with monitoring of blow up.

First of all, no phenomena blow up if parameter $\lambda \geq 1$ or kinetic energy satisfies inequality:

$$\lim_{t \uparrow T_0} \|u(t, \cdot)\|_2^2 \geq \|\varphi\|_2^2 \sqrt{1 - \lambda^4}$$

for $\lambda < 1$.

No phenomena blow up on the time interval $[0, T)$ if kinetic energy satisfies inequality:

$$\|u(t, \cdot)\|_2^2 \geq \|\varphi\|_2^2 \left(1 - \lambda^2 \sqrt{\frac{t}{T_0}}\right)$$

with condition $\lambda < 1$.

Finally, we have the importance of the exact lower estimates for kinetic energy of a fluid flow. It is possible that this is one of the new ways where the interesting problem will be studied.

Funding: This research received no external funding.

Acknowledgments: I would like to say many thanks to all my reviewers for their remarks and advices.

Conflicts of Interest: The author declares no conflict of interest.

Appendix A

Appendix A.1. About the Riesz Potentials and Integral Representations

Some technical results are given.

Lemma A1. *(Hardy–Littlewood–Sobolev's inequality ([24], p. 141). Let $I_\alpha(f)$ be Riesz's potential defined by (4). Set $0 < \alpha < n$. Then, there exists a constant $A = A(p,q)$ where $\frac{1}{q} = \frac{1}{p} - \frac{\alpha}{n}$, $1 < p < q$, such that the following inequality holds:*

$$\|I_\alpha(f)\|_q \leq A\|f\|_p.$$

In a special case, we give an estimate for operator norm.

Corollary A1. *The inequality $A(6,2) \leq \sqrt[3]{\frac{4}{\pi}}$ is true, i.e., $\|u\|_6 \leq \sqrt[3]{\frac{4}{\pi}}\|\nabla u\|_2$.*

Proof. It is sufficient to verify this inequality for smooth and finite mappings. From Riesz's formula, we have:

$$|u(x)|^4 = \frac{1}{\pi}\int_{R^3} \frac{|u(y)|^2 u_i(y) u_{i,j}(y)(x_j - y_j)dy}{|x-y|^3}.$$

Multiply it by $|u(x)|^2$. Then, we make a simple estimate and integrate over space. Hence,

$$\|u\|_6^6 \leq \frac{1}{\pi}\int |u(y)|^3 |\nabla u(y)| \int \frac{|u(x)|^2}{|x-y|^2} dxdy.$$

The interior integral we estimate applying Leray's inequality

$$\int \frac{|u(x)|^2}{|x-y|^2} dx \leq 4\|\nabla u\|_2^2$$

(see [4], also [7], p. 24), thereupon we use Hölder's inequality. Then,

$$\|u\|_6^6 \leq \frac{4}{\pi}\|u\|_6^3 \|\nabla u\|_2^3.$$

It gives the required estimate. □

Let us make more precise well-known integral representations as Poisson's formula and Riesz's formula for smooth functions with compact support.

Lemma A2. *Let $w \in C^2(R^3) \bigcap L_p(R^3)$, $p \geq 1$, be a mapping and its Laplacian $\triangle w$ has a compact support. Then, the equalities hold:*

$$w(x) = -\frac{1}{4\pi}\int_{R^3} \frac{\triangle w(y)dy}{|x-y|}, \; w_{,j}(x) = \frac{1}{4\pi}\int_{R^3} \frac{\triangle w(y)(x_j - y_j)dy}{|x-y|^3}, \tag{A1}$$

$$w(x) = \frac{1}{4\pi}\int_{R^3} \frac{w_{,j}(y)(x_j - y_j)dy}{|x-y|^3}, \tag{A2}$$

(In (A2), repeated indices give summation.)

Proof. To integral

$$\int_{\varepsilon \leq |x-y| \leq r} \frac{\triangle w(y)dy}{|x-y|},$$

we apply twice the Stokes formula removing integrals over spherical layer and derivatives of the mapping w. As the result, we have two integrals over sphere $\mid x - y \mid = \varepsilon$ and two integrals over sphere $\mid x - y \mid = r$. They are:

$$\int_{|x-y|=\varepsilon} \frac{w(y)dS}{\varepsilon^2}, \int_{|x-y|=r} \frac{w(y)dS}{r^2}, \tag{A3}$$
$$\int_{|x-y|=\varepsilon} \frac{w,_j(y)(x_j - y_j)dS}{\varepsilon^2}, \int_{|x-y|=r} \frac{w,_j(y)(x_j - y_j)dS}{r^2}.$$

The third and the fourth integrals we transform again applying the Stokes formula and getting integrals over balls $\mid x - y \mid \leq \varepsilon$, and $\mid x - y \mid \leq r$, respectively. Every integral must contain Laplacian. Since support of $\triangle w$ is a compact set, then these integrals tend to zero as $\varepsilon \to 0, r \to \infty$.

The second integral in (A3) we denote by a symbol I. Then,

$$I = r^{-3} \int_{|x-y|=r} w(y)(x_j - y_j)\frac{(x_j - y_j)}{r} dS.$$

The Stokes formula application gives the equality:

$$I = r^{-3} \int_{|x-y|\leq r} (3w(y) + w,_j(y)(x_j - y_j))dy. \tag{A4}$$

The second term in (A4) we integrate by parts. Therefore,

$$\int_{|x-y|\leq r} (w,_j(y)(x_j - y_j))dy = \frac{1}{2} \int_{|x-y|\leq r} (\triangle w(y)(|x-y|^2 - r^2)dy.$$

The integral from the first term in (A4) we estimate applying the Hölder's inequality. Then,

$$|I| \leq 3r^{-3} \|w\|_p (\sigma_3 r^3)^{1-1/p} + \frac{1}{2r} \int_{|x-y|\leq r} |\triangle w(y)| dy,$$

where σ_3—is the volume of a unit ball. From compactness of Laplacian support and lemma condition, we obtain that integral $I \to 0$ as $r \to \infty$. The first integral in (A3) tends to the mean $4\pi w(x)$ as $\varepsilon \to 0$. formula (A2) we prove by the same way. □

Corollary A2. *A mapping w from Lemma A2 satisfies inequalities: $|\nabla w(x)| \leq C_1 x^{-2}$, $|w(x)| \leq C_2 x^{-1}$ with some constants C_1 and C_2 .*

Proof. The first inequality follows from the second representation of Lemma A2 and compactness of Laplacian support. The second estimate follows from the third representation of Lemma A2 because the first estimate from the corollary gives:

$$|w(x)| \leq \frac{C_1}{4\pi} \int_{R^3} \frac{dy}{|y|^2 |x-y|^2}.$$

A change of variables $y = |x|z$ proves the second estimate. □

Corollary A3. *Let $v,w : R^3 \to R^3$ be mappings which satisfy conditions from Lemma A2. Then,*

$$\int_{R^3} v_k \triangle w_k dy = -\int v_{k,\,j} w_{k,\,j} dy.$$

Proof. We apply the Stokes formula to the integral from the left side of this equality. From Corollary A2 on a sphere $|y| = r$, we get the following formula: $v_k w_{k,\,j} = O(r^{-3})$. A passage to the limit as $r \to \infty$ gives the required equality. □

Lemma A3. *Let $P : R^3 \to R$ be a function and $P \in L_r(R^3)$ with some exponent $r > 1$ and distributions $P_{,ij} \in L_1((R^3) \bigcap L_s(R^3)$ for some exponent $s \in (1, 3/2)$. Then, for function P, Poisson's formula (the first equality from (A2)) is true.*

Proof. For any smooth function P, we verify the integral identity the same way as in Lemma A2 with application of Lemma A4 (see below). A density of smooth functions and Lemma A1 prove the statement in a general case because there is continuity of the Riesz potentials in spaces L_q. □

Lemma A4. *Suppose that a continuous mapping $w : R^n \to R^n$ belongs to the class $W^1_p(R^n)$, $p > 1$. Then, for any point x, an exponent α, where $\alpha > (n-1)(1 - 1/p)$,*

$$r^{-\alpha} \int_{|x-y|=r} w(y) dS \to 0$$

as $r \to \infty$.

Proof. Hölder's inequality implies an estimate:

$$|\int_{|x-y|=r} w(y) dS| \le (\omega_{n-1} r^{n-1})^{1/q} \Big(\int_{|x-y|=r} |w(y)|^p dS \Big)^{1/p}. \tag{A5}$$

Here, ω_{n-1} – is the surface measure of an unit sphere, $q = \frac{p}{p-1}$. Let $J = \int_{|x-y|=r} |w(y)|^p dS$. Then,

$$J = \int_{|x-y|=r} |w(y)|^p \frac{y_j - x_j}{r} \frac{y_j - x_j}{r} dS =$$

$$= \frac{n}{r} \int_{|x-y| \le r} |w(y)|^p dy + p \int_{|x-y| \le r} |w(y)|^{p-2} w_k(y) w_{k,\,j}(y) \frac{y_j - x_j}{r} dy.$$

The second integral on the right-hand side is estimated by application of Hölder's inequality. Furthermore, we replace the integration over a ball by the integration over the whole space. Hence,

$$J \le \frac{n}{r} \|w(y)\|_p^p + p \|w(y)\|_p^{p-1} \|\nabla w\|_p$$

and $J = O(1)$ as $r \to \infty$. Therefore, from (A5), we have the statement. □

Appendix A.2. Logarithmic Convexity Inequalities and Its Corollaries

Lemma A5. *([36], p. 21). A function $\beta(p) = \|w\|_p$ is a logarithmic convex function. That is, for exponents $r \ge 1$, $s \ge 1$ with condition $\frac{1}{p} = \frac{1-t}{r} + \frac{t}{s}$, where $t \in [0,1]$, the inequality $\|w\|_p \le \|w\|_r^{1-t} \|w\|_s^t$ is fulfilled.*

Corollary A4. *Let $u, v, w : R^3 \to R^3$ be a triple of mappings satisfying conditions of Lemma A2. Then, the inequality holds:*

$$|\int_{R^3} u_i v_{k,\,i} \triangle w_k dy| \le a \|\nabla u\|_2 \|\nabla v\|_2^{1/2} \|\triangle v\|_2^{1/2} \|\triangle w\|_2$$

with a constant $a = \sqrt{\frac{4}{\pi}}$. In addition, for a solenoidal vector field u, there is a more exact estimate:

$$|\int_{R^3} u_i u_{k,\,i} \triangle u_k dy| \le a_1 \|\nabla u\|_2^{3/2} \|\triangle u\|_2^{3/2}, \; k, i = 1,\, 2,\, 3,$$

where $a_1 = \frac{8\sqrt[4]{12}}{27}$.

Proof. We have estimates:

$$|u_i v_{k,\,i} \triangle w_k| \le |u| \cdot |\nabla v_k| \cdot |\triangle w_k| \le |u| \cdot |\nabla v| \cdot |\triangle w|,$$

which follow from the Cauchy–Bunyakovskii's inequality. Apply Hölder's inequality for three factors. Then,

$$|\int_{R^3} u_i v_{k,\,i} \triangle w_k dy| \le \|\nabla u\|_6 \|\nabla v\|_3 \|\triangle w\|_2. \tag{A6}$$

For each coordinate $u_i \in C_0^\infty$, we have $\|u_i\|_6 \le A\|\nabla u_i\|_2$ where $A = \sqrt[3]{\frac{4}{\pi}}$ (see Corollary A1). A density of smooth functions and Lemma A1 give the required estimate in a general case.

Since $\|u\|_6^2 \le \sum_i \|u_i\|_6^2$ (we apply the Minkovskii's inequality with exponent 3), then $\|u\|_6 \le A\|\nabla u\|_2$. Respectively, we have $\|\nabla v\|_6^2 \le \sum_i \|v_{,\,i}\|_6^2$ and $\|v_{,\,i}\|_6^2 \le \|\nabla v_{,\,i}\|_2^2$, $\sum_i \|\nabla v_{,\,i}\|_2^2 = \sum_i \|\triangle v\|_2^2$.

From Lemma A5 with exponents $p = 3,\ r = 2,\ s = 6$ and number $t = 0,5$, we obtain: $\|\nabla v\|_3 \le \|\nabla v\|_2^{1/2} |\nabla v\|_6^{1/2}$. Then, from inequalities above and formula (A6), we prove the estimate with a constant $a = A^{3/2}$.

Now, we verify the other inequality. For solenoidal vector fields, we get (see Corollary A2):

$$\int_{R^3} u_i u_{k,\,i} \triangle u_k dy = -\int_{R^3} u_{i,\,j} u_{k,\,i} u_{k,\,j} dy = -\int_{R^3} u_{i,\,j}\Big(u_{k,\,i}u_{k,\,j} - \frac{1}{3}\delta_{ij}|\nabla u|_2^2\Big)dy,$$

where δ_{ij} is Kronecker's delta. Applying Hölder's inequality to a pair

$$u_{i,\,j},\ u_{k,\,i}u_{k,\,j} - \frac{1}{3}\delta_{ij}|\nabla u|_2^2,$$

we obtain:

$$|\int_{R^3} u_i u_{k,\,i} \triangle u_k dy| \le \sqrt{\frac{2}{3}}\|\nabla u\|_2 \|\nabla u\|_4^2.$$

Since

$$\Big(\int_{R^3}\Big(\sum_i |u_{,\,i}|^2\Big)^2 dy\Big)^{1/2} \le \sum_i \Big(\int_{R^3} |u_{,\,i}|^4 dy\Big)^{1/2},$$

then, from the inequality

$$\|f\|_4^2 \le \Big(\frac{4}{3\sqrt{3}}\Big)^{3/2} \|f\|_2^{1/2} \|\nabla f\|_2^{3/2}$$

for vector fields (see [23], Chapter 2 and [27]), we get the second part of the lemma comparing all estimates from above. □

Appendix A.3. Vanishing of Harmonic and Biharmonic Functions

Lemma A6. *If a harmonic function* $h : R^3 \to R$ *is represented by sum* $h = h_s + h_3 + h_6$ *where functions* $h_p \in L_p(R^3)$, $p = s,\ 3,\ 6,\ 1 < s \le 2$, *then* $h \equiv 0$.

Proof. Without loss of generality, we assume that functions h_p are smooth. Otherwise, we take its average defined by a formula

$$h^\tau(x) = \int h(x + \tau y)\omega(y)dy$$

with a kernel $\omega \in C_0^\infty(R^3)$. In the equality,

$$0 = \int_{|y-x| \le r} \triangle h(y)|x - y|^\beta dy = I,$$

$-1 < \beta < -0,5$, we transform the integral applying the Stokes theorem. Let $x = 0$. Then,

$$I = -\beta \int_{|y|\le r} h_{,j}(y) y_j |y|^{\beta-2} dy + \int_{|y|=r} h_{,j}(y) |y|^{\beta} \frac{y_j}{r} dS = \beta J_1 + J_2.$$

The integral over surface $J_2 = 0$ since $J_2 = r^{\beta} \int_{|y|\le r} \triangle h(y) dy = 0$. Hence, $J_1 = 0$. This integral is transformed in the same way as the integral I. From the equality,

$$J_1 = \int_{|y|\le r} h(y) \frac{\partial}{\partial y_j} \left(y_j |y|^{\beta-2} \right) dy - r^{\beta-1} \int_{|y|=r} h(y) dS,$$

by application of the theorem on the mean value of a harmonic function, we conclude the formula:

$$h(0) = \frac{\beta+1}{4\pi r^{\beta+1}} \left(\int_{|y|\le r} \frac{h_s(y)}{|y|^{2-\beta}} dy + \int_{|y|\le r} \frac{h_3(y)}{|y|^{2-\beta}} dy + \int_{|y|\le r} \frac{h_6(y)}{|y|^{2-\beta}} dy \right). \tag{A7}$$

For chosen means β, each potential $I^{\beta+1}(h_q)(0)$, $q = s, 3, 6$ is finite (see Lemma A1). The passage to the limit in (A7) as $r \to \infty$ yields the equality: $h(0) = 0$. □

Lemma A7. *If a biharmonic function $h : R^3 \to R$ has a decomposition $h = h_s + h_3 + h_6$ where functions $h_p \in L_p(R^3)$, $p = s,\ 3,\ 6,\ 1 < s \le 2$, then $h \equiv 0$.*

Proof. Without loss of the generality, we can replace functions h_p by its averages (see above). Then, every average $h_p^{\tau} \in W_p^1(R^3)$. Now, we fix the averaging parameter τ. Let $x = 0$ and $1 < \beta < 1,5$. Let us show that function h is a harmonic function. It is sufficient to apply the theorem about the mean value of a harmonic function to $\triangle h$ and use the spherical coordinates. Then, for the average, we have:

$$\triangle h^{\tau}(0) = \frac{\beta+1}{4\pi r^{\beta+1}} \int_{|y|\le r} \triangle h^{\tau}(y) |y|^{\beta-2} dy = \frac{\beta+1}{4\pi r^{\beta+1}} J_3. \tag{A8}$$

The integral J_3 is transformed by applying three times of the Stokes theorem: twice to the integrals over volume and once to the integral over surface. As a result, we obtain:

$$J_3 = (\beta^2 - 3\beta + 2) \int_{|y|\le r} h^{\tau}(y) |y|^{\beta-4} dy - (\beta-2) r^{\beta-3} \int_{|y|=r} h^{\tau}(y) dS + r^{\beta-2} \int_{|y|\le r} \triangle h^{\tau}(y) dy.$$

Furthermore, we apply again the theorem about a mean value for a harmonic function to the third integral. After that, we input the mean of integral J_3 in (A8). Then, we conclude:

$$\frac{\triangle h^{\tau}(0)}{3} = \frac{1-\beta^2}{4\pi r^{\beta+1}} \int_{|y|\le r} h^{\tau}(y) |y|^{\beta-4} dy + \frac{\beta+1}{4\pi r^4} \int_{|y|=r} h^{\tau}(y) dS.$$

Here, the integral over the volume set tends to a finite mean as $r \to \infty$. The finiteness of this mean is proved in the same way as in Lemma A6. This implies

$$\frac{\triangle h(0)}{3} = \lim_{r\to\infty} \frac{\beta+1}{4\pi r^{\beta+1}} \int_{|y|=r} h(y) dS.$$

An exponent mean β belongs to the interval $(1, 3/2)$. Hence, and from Lemma A4, we obtain $\triangle h^{\tau}(0) = 0$. Taking assumption about a ball center, we obtain that the function h^{τ} is a harmonic function. Then, from Lemma A6, $h^{\tau} \equiv 0$. Let $\tau \to 0$. Then, $h \equiv 0$. □

References

1. Oseen, C. *Neuere Methoden und Ergebnisse der Hydrodynamik*; Akademische Verlagsgesellschaft: Leipzig, Germany, 1927.
2. Oldquist, F.K.G. Über die Randwertaufgaben der Hydrodynamik zäher Flüssigkeiten. *Math. Z.* **1930**, *32*, 329–375.
3. Leray, J. Sur le mouvement d'un liquide visqueux emplissant l'espace. *Acta Math.* **1934**, *63*, 193–248. [CrossRef]
4. Leray, J. Etude de diverses equations integrales non lineaires et de quelques problemes que pose l'hydrodynamique. *J. Math. Pure Appl.* **1933**, *12*, 1–82.
5. Leray, J. Essai sur les mouvements plans d'un liquide visqueux que limitent des parois. *J. Math. Pures Appl.* **1933**, *9*, 331–418.
6. Hopf, E. Über die Anfangswertaufgabe für die Hydrodynamischen Grundgleichungen. *Math. Nachr.* **1951**, *4*, 213–231. [CrossRef]
7. Ladyzhenskaya, O.A. *Mathematical Questions Of Dynamics of Viscous Incompressible Fluid*, 2nd ed.; Nauka: Moscow, Russia, 1970.
8. Prodi, G. Un teorema di unicita per le equazioni di Navier–Stokes. *Ann. Mat. Pura Appl.* **1959**, *48*, 173–182. [CrossRef]
9. Ladyzhenskaya, O.A. The sixth problem of millenium: Navier–Stokes equations, existence and regularity. *Uspekhi Matematicheskih Nauk* **2003**, *58*, 45–78.
10. Caffarelli, L.; Kohn, R.; Nirenberg, L. Partial regularity of suitable weak solutions of the Navier–Stokes equations. *Commun. Pure Appl. Math.* **1982**, *35*, 771–831. [CrossRef]
11. Serrin, J. On the interior regularity of weak solutions of the Navier–Stokes equations. *Arch. Ration. Mech. Anal.* **1962**, *9*, 187–195. [CrossRef]
12. Escauriaza, L.; Seregin, G.A.; Sverak, V. $L_{3,\infty}$ solutions to the Navier–Stokes equations and backward uniqieness. *Uspekhi Matematicheskih Nauk* **2003**, *58*, 3–44.
13. Fabes, E.; Jones, B.; Riviere, N. The initial value problem for the Navier–Stokes equations with data in R^n. *Arch. Ration. Mech. Anal.* **1972**, *45*, 222–240. [CrossRef]
14. Solonnikov, V.A. On differential properties of solution in the first boundary value problem for nonstationary equations system. *Trudy MIAN* **1964**, *73*, 221–291.
15. Temam, R. *Navier–Stokes Equations. Theory and Numerical Analysis*; AMS Chelsea Publishing: Amsterdam, The Netherlands, 1977.
16. Kato, T. Strong L^p –solutions of the Navier–Stokes equations in R^m with applications to weak solutions. *Math. Z.* **1984**, *187*, 471–480. [CrossRef]
17. Galdi, G.P. *An Introduction to the Mathematical Theory of the Navier–Stokes Equations. V. I, II*; Springer: New York, NY, USA, 1994.
18. Fefferman, C. *Existence and Smoothness of the Navier–Stokes Equation*; Clay Mathematics Institute: Cambridge, MA, USA, 2000; pp. 1–5.
19. Kiselev, A.A.; Ladyzhenskaya, O.A. On existence and uniqueness of solution in nonstationary problem for viscous incompressible fluid. *Izv. Akad. Nauk SSSR Ser. Mat.* **1957**, *21*, 655–680.
20. Semenov, V.I. *On Determinism of Fluid Dynamics and Navier–Stokes Equations, Differential Equations. Functional Spaces. Approximation Theory*; Dedicated to S.L. Sobolev; Institut Matematiki: Novosibirsk, Russia, 2008; p. 204.
21. Semenov, V.I. *Necessary and Sufficient Conditions Of Existence of Global Solutions of Navier–Stokes Equations in Space, Contemporary Problems of Approximative Mathematics and Mathematical Physics*; Dedicated to A. A. Samarskii; Moscow State University: Moscow, Russia, 2009; pp. 257–258.
22. Semenov, V.I. *On Smoothness Property of Solutions Of Navier–Stokes Equations in Nonlinear Nonstationary the Cauchy Problem in Space*; Kuzbassvuzizdat: Kemerovo, Russia, 2007; pp. 1–40.
23. Sobolev, S.L. *Some Applications of Functional Analysis in Mathematical Physics*; SO AN: Novosibirsk, Russia, 1962.
24. Stein, E. *Singular Integrals and Differential Functions Properties*; Prinston Univ. Press: Prinston, NJ, USA, 1970.
25. Gol'dstein, V.M.; Reshetnyak, J.G. *Introduction in the Function Theory With Generalized Derivatives and Quasiconformal Mappings*; Nauka: Moscow, Russia, 1983.

26. Ladyzhenskaya, O.A. On uniqueness and smoothness of weak solutions of the Navier–Stokes equations. *J. Math. Sci.* **2005**, *130*, 4884–4892.
27. Serrin, J. *The Initial Value Problem for the Navier–Stokes Equations, Nonlinear Problems*; Langer, R., Ed.; Univ. of Wisconsin Press: Madison, WI, USA, 1963; pp. 69–98.
28. Michlin, S.G. *Many-Dimensional Singular Integrals and Integral Equations*; Fizmatgiz: Moscow, Russia, 1962. Hille, E.; Fillips, R. *Functional Analysis and Semi–Groups*; American Mathematical Society: Providence, RI, USA, 1957.
29. Sobolev, S.L. On some new problem of mathematical physics. *Izv. Akad. Nauk SSSR Ser. Mat.* **1954**, *18*, 3–50.
30. Lewis, J.E. Mixed estimates for singular integrals and an application to initial value problems in parabolic differential equations. *Proc. Symp. Pure Math.* **1963**, *10*, 218–231.
31. Maslennikova, V.N.; Bogovskii, M.E. Approximation of potential and solenoidal vector fields. *Sib. Math. J.* **1983**, *24*, 149–171. [CrossRef]
32. Semenov, V.I. Certain general properties of solenoidal vector fields and its applications to the 2d Navier–Stokes and Euler equations. *Nauchnye Vedomosti Belgorodskogo gos. univ., ser. Matem.* **2008**, *15*, 109–129.
33. Semenov, V.I. Some New integral identities for solenoidal fields and applications. *Mathematics* 2014, *2*, 29–36. [CrossRef]
34. Ladyzhenskaya, O.A. Solution in global of boundary value problem for the Navier–Stokes equations in case of two spatial variables. *Doklady Akad. Nauk SSSR* **1958**, *123*, 427–429.
35. Judovich, V.I. Periodic solutions of viscous incompressible fluids. *Doklady Akad. Nauk SSSR* **1960**, *130*, 1214–1217.
36. Bourbaki, N. *Integration, Measure, Measure Integration*; Nauka: Moscow, Russia, 1967.

axioms

MDPI

Article

Sumudu Decomposition Method for Solving Fuzzy Integro-Differential Equations

Shin Min Kang [1,2], **Zain Iqbal** [3], **Mustafa Habib** [3] **and Waqas Nazeer** [4,*]

1 Department of Mathematics and Research Institute of Natural Science, Gyeongsang National University, Jinju 52828, Korea; smkang@gnu.ac.kr
2 Center for General Education, China Medical University, Taichung 40402, Taiwan
3 Department of Mathematics, University of Engineering and Technology, Lahore 54000, Pakistan; zainiqbal3778@gmail.com (Z.I.); mustafa@uet.edu.pk (M.H.)
4 Division of Science and Technology, University of Education, Lahore 54000, Pakistan
* Correspondence: nazeer.waqas@ue.edu.pk; Tel.: +92-321-470-7379

Received: 24 March 2019; Accepted: 12 June 2019; Published: 20 June 2019

Abstract: Different results regarding different integro-differentials are usually not properly generalized, as they often do not satisfy some of the constraints. The field of fuzzy integro-differentials is very rich these days because of their different applications and functions in different physical phenomena. Solutions of linear fuzzy Volterra integro-differential equations (FVIDEs) are more generalized and have better applications. In this report, the Sumudu decomposition method (SDM) was used to find the solution to some linear and nonlinear fuzzy integro-differential equations (FIDEs). Some examples are given to show the validity of the presented method.

Keywords: integro-differentials; Sumudu decomposition method; dynamical system

1. Introduction

Linear and nonlinear phenomena are a fundamental part of science and construction. Nonlinear equations are seen in an alternative way when dealing with physical problems such as liquid elements, plasma material science, strong mechanics, quantum field hypotheses, the proliferation of shallow water waves, and numerous other models, all of which are to found within the field of incomplete differential equations. The wide use of these equations is the key to why they have drawn the attention of mathematicians. Regardless of this, they are difficult to solve, either numerically or theoretically. Previously, dynamic examinations were much examined for the potential of finding exact or approximate solutions to these sorts of equations [1,2].

In the recent years, the area of FIDEs has developed a lot and plays a key role in the field of engineering. The elementary impression and arithmetic of fuzzy sets were first introduced by Zadeh. Later, the area of fuzzy derivative and fuzzy integration was studied, and some general results were developed. Fuzzy differential equations (FDEs) and FIDEs are very important in the study of fuzzy theory and have many beneficial consequences related to different problems. Modeling of different physical systems in the differential way gives us different FIDEs [2,3]. Furthermore, FIDEs in a fuzzy setting are a natural way to model the ambiguity of dynamic systems. Consequently, different scientific fields, such as physics, geography, medicine, and biology, pay much importance to the solution of different FIDEs. Solutions to these equations can be utilized in different engineering problems. Seikkala first defined fuzzy derivatives, while the concept of integration of fuzzy functions was first introduced by Dubois and Prade. However, analytic solutions to nonlinear FIDE types are often difficult to find. Therefore, most of the time, an approximate solution is required. There are also useful numerical schemes that can produce a numerical approximation to solutions for some problems [4,5].

The literature on numerical solutions of integro-differential equations (IDEs) is vast. We used the Sumudu decomposition method [6–8] to solve linear and nonlinear fuzzy integral equations (FIEs). The method gives more realistic series solutions that converge very rapidly in physical problems. Sumudu transforms are also used for solving IDEs, which can be seen in [4,5]. IDEs are transforms to FIDEs that are more general and give better results. After applying a Sumudu transform, a decomposition method is used for the approximate solution [8,9].

2. Preliminaries

Integral Equation (IE)

The obscure function $Y(\xi)$ that shows up under an integral symbol is known as an integral equation. Usually, we write an integral equation as follow:

$$Y(\xi) = f(\xi) + \int_{g(\xi)}^{h(\xi)} K(\xi, t)Y(t)dt, \tag{1}$$

where $k(\xi, t)$ and λ are the kernel and constant parameter, respectively. The kernel is identified as the function of dual variables ξ and t, whereas $g(\xi)$ and $h(\xi)$ are recognized as the limitations for integration. The function $Y(\xi)$ to be resolved shows up under the integral symbol; it has the property of appearing in both the outside as well as inside of the integral symbol. The functions that will be specified in progressive are $f(\xi)$ and $k(\xi, t)$. Limitations of integration can adopt both forms, either as the variable, constant, or blended [10].

Types of Integral Equation

IEs show up in numerous forms. Different sorts are generally contingent on the limitations of antiderivatives as well as the kernel of equality. In this content, we focus on the following sorts of IE [11]:

i. Fredholm IE;
ii. Volterra IE;
iii. Volterra-Fredholm IE;
iv. Singular.

Volterra Integral Equations (VIEs)

There is a restriction for the VIE, which is that at least one limit should be a variable. Likewise, in FIEs, there are two varieties of VIEs, which are more easily described through the following:

$$f(\xi) = \int_{0}^{\xi} K(\xi, t)Y(t)dt. \tag{2}$$

Equation (2) is a VIE of the first kind.
That is:

$$\xi e^{-\xi} = \int_{0}^{\xi} e^{t-\xi}Y(t)dt$$

$$Y(\xi) = f(\xi) + \lambda \int_{0}^{\xi} K(\xi, t)Y(t)dt \tag{3}$$

Equality (3) is a VIE of 2nd type.

For illustration,

$$Y(\xi) = 1 - \int_0^{\xi} Y(t)dt$$

Classification of Integro-Differential Equations

Different types of dynamical physical problems possess integro-differential equations, specifically during the conversion of initial value problems (IVPs) and boundary value problems (BVPs). Differential operators as well as integral operators are involved in an integro-differential equation. There could be any order for the presence of derivatives of the unknown function. In characterizing integro-differential equations, we pursued a similar class as used previously. The following are well-known types of integro-differential equations:

i. Fredholm integro-differential equations;
ii. Volterra integro-differential equations;
iii. Volterra–Fredholm integro-differential equations.

Volterra Integro-Differential Equation

The Volterra integro-differential equation appears during the conversion of IVPs into the integral equation. In the Volterra integro-differential equation, the unidentified function and its derivatives appear inside as well as outside of the integral operator. For VIE, at least one limit of integration is variable. In order to obtain the exact solution, we need initial conditions in the Volterra integro-differential equation (VIDE). Consider the following VIDE:

$$Y^n(\xi) = f(\xi) + \lambda \int_0^{\xi} K(\xi, t)Y(t)dt \tag{4}$$

where Y^n denotes the derivative of order n of $Y(\xi)$. The VIDE given in Equation (4) can be written as:

$$Y'(\xi) = 3 + \frac{1}{4}\xi^2 - \xi e^{\xi} - \int_0^{\xi} tY(t)dt, \quad Y(0) = 0, \tag{5}$$

3. Theorems and Definitions Interrelated to Fuzzy Perceptions

Fuzzy Number

A fuzzy number is a generalization of a regular, real number in the sense that it does not refer to one single value but rather to a connected set of possible values, where each possible value has its own weight between 0 and 1. This weight is called the membership function.

Let E be the set of all fuzzy numbers which upper semicontinuous and compact. The α level set $[Y]_\rho$ where Y is the collection of fuzzy numbers, $0 < \rho \leq 1$, is defined as:

$$[Y]_\rho = \{t \in R, Y(t) \geq \rho\}$$

The set E is convex if $Y(t) \geq Y(s) \wedge Y(r) = \min(Y(s), Y(r))$, where $s < t < r$.

If $\exists\, t_o \in R$ such that $Y(t_o) = 1$, then E becomes normal. E is said to be upper semicontinuous if for every $\varepsilon > 0$, *such that* $Y^{-1}([0, a + \varepsilon))$, $\forall\, a \in [0, 1]$ is open in the typical topology of R [12,13].

Absolute value $|Y|$ of $Y \in E$ is defined as:

$$\begin{aligned} |Y|(t) &= max\{Y(t), Y(-t)\},\ if\ t \geq 0 \\ &= 0,\ if\ t < 0 \end{aligned}$$

Consider the mapping $\overline{d} : L(R) \times L(R) \to R$ defined as:

$$d(Y, V) = \sup_{0 \leq \rho \leq 1} max\{|\underline{Y}(\rho) - \underline{V}(\rho)|, |\overline{Y}(\rho) - \overline{V}(\rho)|\}$$

where:

$$\overline{Y} = [\underline{Y}(\rho), \overline{Y}(\rho)] \text{ and } \overline{V} = [\underline{V}(\rho), \overline{V}(\rho)].$$

Then,d is a metric on $L(R)$ satisfying the properties:

1. $d(Y + w, V + w) = d(Y, V)$ *for all* $Y, V, w \in L(R)$;
2. $d(kY, kV) = |k|d(Y, V)$ for all $Y, V, \in L(R)$;
3. $d(Y + w, w + e) \leq d(Y, w) + d(V, e)$ *for all* $Y, V, w, e \in L(R)$;
4. $(d, L(R))$ is a complete metric space.

Definition 1. *Let $f : R \to L(R)$ be a fuzzy valued function, then f is continuous if for $t_o \in R$ and for each $\varepsilon > 0$, there exists $\delta > 0$ such that:*

$$d((f(t), f(t_0))) < \varepsilon \text{ whenever } |t - t_o| < \delta$$

Definition 2. *Let $f : R \to L(R)$ be a fuzzy valued function and $\xi_o \in R$, then f is differentiable at ξ_o. If $\exists$ $f'(\xi_o) \in L(R)$ such that:*

(a) $\lim_{h \to 0+} \frac{f(\xi_o+h)-f(\xi_o)}{h} = \lim_{h \to 0+} \frac{f(\xi_o)-f(\xi_o-h)}{h} = f^{(1)}(\xi_o)$

(b) $\lim_{h \to 0-} \frac{f(\xi_o+h)-f(\xi_o)}{h} = \lim_{h \to 0-} \frac{f(\xi_o)-f(\xi_o-h)}{h} = f^{(1)}(\xi_o)$

Theorem 1. *Consider $f : R \to L(R)$ as a fuzzy valued function defined as $f(t) = [\underline{f}(t, \rho), \overline{f}(t, \rho)]$ for each $0 \leq \alpha \leq 1$, and - f is differentiable, then $\underline{f}(t, \rho)$ and $\overline{f}(t, \rho)$ are differentiable and $f^{(1)}(t) = \left[\underline{f}^{(1)}(t, \rho), \overline{f}^{(1)}(t, \rho)\right]$.*

Theorem 2. *Let $f : R \to L(R)$ be the fuzzy valued function defined as $f(t) = [\underline{f}(t, \rho), \overline{f}(t, \rho)]$ for each $0 \leq \rho \leq 1$. If f and $f^{(1)}$ have the property of differentiability, then $\overline{f}^{(1)}(t, \rho)$ and $\underline{f^{(1)}}(t, \rho)$ are differentiable and:*

$$f^{(2)}(t) = \left[\underline{f^{(2)}}(t, \rho), \overline{f}^{(2)}(t, \rho)\right]$$

Theorem 3. *Consider a real valued function $f(\xi)$ defined on $[0, \infty]$ such that $\underline{f}(\xi, \rho), \overline{f}(\xi, \rho)$ are Riemann-integrable on $[a, b]$, for each $b \geq a$ and there exist positive constants $\underline{M}(\rho), \overline{M}(\rho)$ such that:*

$$\int_a^b |\underline{f}(\xi, \rho)| d\xi \leq \underline{M}(\rho) \text{ and } \int_a^b |\overline{f}(\xi, \rho)| d\xi \leq \overline{M}(\rho)$$

for every $b \geq a$. Then, $f(\xi)$ is an improper fuzzy Riemann integrable on $[0, \infty]$, and $f(\xi)$ is a fuzzy number. Additionally, we we have:

$$\int_a^\infty f(\xi) d\xi = \int_a^b \underline{f}(\xi, \rho) d\xi, \int_a^b \overline{f}(\xi, \rho) d\xi$$

Theorem 4. *The sum of two fuzzy Riemann integrable functions is a Riemann integrable.*

Definition 3. *The fuzzy Laplace transform (FLT) of a fuzzy function f is defined as:*

$$f(s) = L\{f(t)\} = \int_0^\infty e^{-st} f(t)dt = \lim_{T\to\infty} \int_0^T e^{-st} f(t)dt,$$

where L denotes FLT. In addition, the fuzzy Laplace transform for $f(t)$ can be as follows:

$$f(s,\rho) = L\{f(t,\rho)\} = [l\{\underline{f}(t,\rho\}, l\{\overline{f}\{t,\rho\}]$$
$$l\{\underline{f}(t,\rho\} = \int_0^\infty e^{-st}\underline{f}(t)dt = \lim_{T\to\infty} \int_0^T e^{-st}\underline{f}(t)dt$$
$$0 \le \rho \le 1$$
$$l\{\overline{f}\{t,\rho\} = \int_0^\infty e^{-st}\overline{f}(t)dt = \lim_{T\to\infty} \int_0^T e^{-st}\overline{f}(t)dt$$
$$0 \le \rho \le 1$$

Theorem 5 (Fuzzy Convolution Theorem).

Let f and g be two fuzzy real valued functions. Then, the convolution of f and g is defined as:

$$(f * g)(t) = \int_0^T f(T)g(t-T)dT.$$

Theorem 6. *Consider f and g defined on R are two continuous (piecewise) functions defined on $[0,\infty]$ having exponential order p, then:*

$$L\{(f * g)(t)\} = L\{f(t)\}L\{g(t)\} = F(s).G(s)$$

Definition 4. (Sumudu transform) [14–16]

The Sumudu transform of the function $f(t)$ is defined as:

$$F(u) = \mathcal{S}[f(t)] = \int_0^\infty \frac{1}{u} e^{(-\frac{t}{u})} f(t)dt,$$
$$F(u) = \mathcal{S}[f(t)] = \int_0^\infty f(ut)e^{-t}dt,$$

for any function $f(t)$ and $-\tau_1 < u < \tau_2$.

Theorem 7. *If $c_1 \ge 0$, $c_2 \ge 0$ and $c \ge 0$ are any constant and $f_1(t)$, $f_2(t)$, and $f(t)$ any functions having the Sumudu transform $G_1(u)$, $G_2(u)$, and $G(u)$, respectively, then:*

i. $\mathcal{S}[c_1 f_1(t) + c_2 f_2(t)] = c_1\mathcal{S}[f_1(t)] + c_2\mathcal{S}[f_2(t)] = c_1G_1(u) + c_2G_2(u)$;
ii. $\mathcal{S}[f(ct)] = G(cu)$;
iii. $\lim_{t\to\infty} f(t) = f(0) = \lim_{u\to 0} G(u)$.

For more details, we refer the readers to [17,18].

Fuzzy Sumudu Transfom

Let $f : R \to f(R)$ be a continuous fuzzy function, then the fuzzy Sumudu transform (FST) can be defined as:

$$\begin{aligned} F(u) &= \mathcal{S}[f(\xi)] = \int_0^\infty f(u\xi) \odot e^{-\xi} d\xi,\ u \in [\tau_1, \tau_2] \\ &= \mathcal{S}[f(\xi)] = \left[\mathcal{S}\left[\underline{f}_\alpha(\xi)\right], \mathcal{S}\left[\overline{f}_\alpha(\xi)\right]\right] \end{aligned}$$

Theorem 7. *Let $f : R \to f(R)$ be a continuous fuzzy valued function. If $F(u) = \mathcal{S}[f(\xi)]$, then:*

$$\mathcal{S}\left[f^{(1)}(\xi)\right] = \begin{cases} \frac{F(u)}{u} - \frac{f(0)}{u} & if\ f\ is\ (i) differentiable\ and\ u > 0 \\ -\frac{f(0)}{u} - \frac{(-F(u))}{u} & if\ f\ is\ (ii) differentiable\ and\ u > 0 \end{cases}$$

Proof. Case (i) Let f be differentiable, then:

$$\begin{aligned} \frac{F(u)}{u} - \frac{f(0)}{u} &= \left[\frac{\mathcal{S}\left[\underline{f}_\rho(\xi)\right] - \underline{f}_\rho(0)}{u}, \frac{\mathcal{S}\left[\overline{f}_\rho(x)\right] - \overline{f}_\rho(0)}{u}\right] \\ &= \mathcal{S}\left[\left[\underline{f}_\rho(\xi)\right]\mathcal{S}\left[\overline{f}_\rho(\xi)\right]\right] = \mathcal{S}\left[f^{(1)}(\xi)\right] \end{aligned}$$

Case (ii) Let f be differentiable, then:

$$\begin{aligned} -\frac{f(0)}{u} - \frac{(-F(u))}{u} &= \left[-\frac{\underline{f}_\rho(0) + \mathcal{S}\left[\underline{f}_\rho(\xi)\right]}{u}, -\frac{\overline{f}_\rho(0) + \mathcal{S}\left[\overline{f}_\rho(\xi)\right]}{u}\right] \\ &= \mathcal{S}\left[\left[\underline{f}_\rho(\xi)\right]\mathcal{S}\left[\overline{f}_\rho(\xi)\right]\right] \\ &= \mathcal{S}\left[f^{(1)}(\xi)\right] \end{aligned}$$

□

Theorem 8. *Let $f : R \to f(R)$ be a continuous fuzzy valued function, and if $F(u) = \mathcal{S}[f(x)]$, then:*

$$\mathcal{S}\left(e^{-a\xi} \odot f(t)\right) = \frac{1}{1+au} F\left(\frac{u}{1+au}\right), \quad au \neq -1\ and \frac{1}{1+au} > 0$$

Proof.

$$\begin{aligned} \mathcal{S}\left(e^{-a\xi} \odot f(t)\right) &= \left[\int_0^\infty \underline{f}_\rho(u\xi) e^{-a(\xi u)} e^{-\xi} d\xi, \int_0^\infty \overline{f_\rho}(u\xi) e^{-a(\xi u)} e^{-\xi} d\xi\right] \\ &= \left[\int_0^\infty \underline{f}_\rho(u\xi) e^{-a(1+u)\xi} d\xi, \int_0^\infty \overline{f_\rho}(\xi) e^{-a(1+u)\xi} d\xi\right] \end{aligned}$$

Now, let $v = (1+au)\xi$ *and* $d\xi = \frac{v}{1+au}$.
Thus, we have:

$$\begin{aligned} \mathcal{S}\left(e^{-a\xi} \odot f(t)\right) &= \left[\frac{1}{1+au} \int_0^\infty \underline{f}_\rho\left(\frac{uv}{1+au}\right) e^{-v} dv, \frac{1}{1+au} \int_0^\infty \overline{f_\rho}\left(\frac{uv}{1+au}\right) e^{-v} dv,\right] \\ &= \frac{1}{1+au} \int_0^\infty f_\alpha\left(\frac{uv}{1+au}\right) e^{-v} dv. \end{aligned}$$

Hence:

$$\mathcal{S}\left(e^{-a\xi} \odot f(t)\right) = \frac{1}{1+au} F\left(\frac{u}{1-au}\right).$$

In the same way, we can prove that:

$$\left(e^{a\xi}\odot f(t)\right)=\frac{1}{1-au}F\left(\frac{u}{1-au}\right).$$

□

Theorem 9. *Let $f:R\to f(R)$ be a continuous fuzzy valued function, and if $F(u)=\mathcal{S}[f(\xi)]$, then:*

$$\mathcal{S}\left[\int_0^{\xi} f(\xi)d\xi\right]=uF(u)$$

Proof. Assume function h is differentiable, and:

$$\underline{h}_{\rho}(\xi)=\int_0^{\xi}\underline{f}_{\rho}(\xi)d\xi,\ \overline{h}_{\rho}(\xi)=\int_0^{\xi}\overline{f}_{\rho}(\xi)d\xi \quad \underline{h}_{\rho}(0)=0=\overline{h}_{\rho}(0), \qquad h^{(1)}(\xi)=f(\xi).$$

Then:

$$\begin{aligned}\mathcal{S}\left(h^{(1)}(\xi)\right)=\frac{H(u)}{u}-\frac{h(o)}{u} &=\left[\frac{\mathcal{S}\left[\underline{h}_{\rho}(\xi)\right]}{u}-\frac{\underline{h}_{\rho}(0)}{u},\ \frac{\mathcal{S}\left[\overline{h}_{\rho}(\xi)\right]}{u}-\frac{\overline{h}_{\rho}(0)}{u}\right]\\ &=\left[\frac{\mathcal{S}\left[\underline{h}_{\rho}(\xi)\right]}{u},\ \frac{\mathcal{S}\left[\overline{h}_{\rho}(\xi)\right]}{u}\right]\\ &=\left[\frac{1}{u}\,\mathcal{S}\int_0^{\xi}\underline{f}_{\rho}(\xi)d\xi,\ \frac{1}{u}\mathcal{S}\int_0^{\xi}\overline{f}_{\rho}(\xi)d\xi\right]\end{aligned}$$

Thus, we have:

$$\mathcal{S}\left[\int_0^{\xi} f(\xi)d\xi\right]=uF(u)$$

□

4. Sumudu Decomposition Method for Fuzzy Integro-Differential Equation (Analysis of Method)

Consider a Volterra integro-differential equation:

$$\Upsilon^{n}(\xi,\rho)=f(\xi,\rho)+\int_0^{\xi}k(\xi-t)\Upsilon(t,\rho)dt \tag{6}$$

$$\Upsilon^{k}(0)=\overline{p}=(\underline{p}_{k},\overline{p}_{k});\ _{0\le k\le n-1}$$

By taking sumudu transform on Equation (6), we have:

$$\mathcal{S}[\Upsilon^{n}(\xi,\rho)]=\mathcal{S}[f(\xi,\rho)]+\mathcal{S}\left[\int_0^{\xi}k(\xi-t)\Upsilon(t,\rho)dt\right]. \tag{7}$$

This will give us:

$$\begin{aligned}&\tfrac{1}{u^{n}}\mathcal{S}[\Upsilon(\xi,\rho)]-\tfrac{1}{u^{n}}\Upsilon(0,\rho)-\tfrac{1}{u^{n-1}}\Upsilon^{(1)}(0,\rho)-\cdots-\tfrac{\Upsilon^{(n-1)}(0,\rho)}{u}\\ &=\mathcal{S}[f(\xi,\rho)]+\mathcal{S}\left[\int_0^{\xi}k(\xi-t)\Upsilon(t,\rho)dt\right]\end{aligned} \tag{8}$$

$$\frac{1}{u^n}\mathcal{S}[\underline{Y}(\xi,\rho)] - \frac{1}{u^n}\underline{Y}(0,\rho) - \frac{1}{u^{n-1}}\underline{Y}'(0,\rho) - \cdots - \frac{\underline{Y}^{n-1}(0,\rho)}{u} = \mathcal{S}\left[\underline{f}(\xi,\rho)\right] + u\underline{\mathcal{S}[k(\xi-t)]\mathcal{S}[Y(\xi,t)]} \quad (9)$$

$$\frac{1}{u^n}\mathcal{S}\left[\overline{Y}(\xi,\rho)\right] - \frac{1}{u^n}\overline{Y}(0,\rho) - \frac{1}{u^{n-1}}\overline{Y}^{(1)}(0,\rho) - \cdots - \frac{\overline{Y}^{(n-1)}(0,\rho)}{u} = \mathcal{S}\left[\overline{f}(\xi,\rho)\right] + u\overline{\mathcal{S}[k(\xi-t)]\mathcal{S}[Y(\xi,t)]} \quad (10)$$

Note that:

$$\underline{Y}(0,\rho) = \underline{p}_0,\ \underline{Y}^{(1)}(0,\rho) = \underline{p}_1 \cdots\cdots\cdots \underline{Y}^{n-1}(0,\rho) = \underline{p}_{n-1}$$
$$\overline{Y}(0,\rho) = \overline{p}_0,\ Y^{(1)}(0,\rho) = \overline{p}_1 \cdots\cdots\cdots \overline{Y}^{n-1}(0,\rho) = \overline{p}_{n-1}$$

Thus, we have:

$$\frac{1}{u^n}\mathcal{S}[\underline{Y}(\xi,\rho)] - \frac{1}{u^n}\underline{p}_0 - \frac{1}{u^{n-1}}\underline{p}_1 - \cdots - \frac{\underline{p}_{n-1}}{u} = \mathcal{S}\left[\underline{f}(\xi,\rho)\right] + u\mathcal{S}[\underline{k}(\xi-t)]\mathcal{S}[\underline{Y}(\xi,\rho)] \quad (11)$$

$$\frac{1}{u^n}\mathcal{S}\left[\overline{Y}(\xi,\rho)\right] - \frac{1}{u^n}\overline{p}_0 - \frac{1}{u^{n-1}}\overline{p}_1 - \cdots - \frac{\overline{p}_{n-1}}{u} = \mathcal{S}\left[\overline{f}(\xi,\rho)\right] + u\mathcal{S}\left[\overline{k}(\xi-t)\right]\mathcal{S}\left[\overline{Y}(\xi,\rho)\right] \quad (12)$$

The following cases can be discussed:

(i) if $Y(\xi;\rho)$ and $k(\xi;\rho)$ both are positive:

$$\mathcal{S}[\underline{k}(\xi,\rho)]\mathcal{S}[\underline{Y}(\xi,\rho)] = \mathcal{S}[\underline{k}(\xi,\rho)]\mathcal{S}[\underline{Y}(\xi,\rho)]$$
$$\mathcal{S}\left[\overline{k}(\xi,\rho)\right]\mathcal{S}\left[\overline{Y}(\xi,\rho)\right] = \mathcal{S}\left[\overline{k}(\xi,\rho)\right]\mathcal{S}\left[\overline{Y}(\xi,\rho)\right]$$

(ii) if $Y(\xi;\rho)$ is negative and $k(\xi;\rho)$ is positive:

$$\mathcal{S}[\underline{k}(\xi,\rho)]\mathcal{S}[\underline{Y}(\xi,\rho)] = \mathcal{S}\left[\overline{k}(\xi,\rho)\right]\mathcal{S}[\underline{Y}(\xi,\rho)]$$
$$\mathcal{S}\left[\overline{k}(\xi,\rho)\right]\mathcal{S}\left[\overline{Y}(\xi,\rho)\right] = \mathcal{S}[\underline{k}(\xi,\rho)]\mathcal{S}\left[\overline{Y}(\xi,\rho)\right]$$

(iii) if $Y(\xi;\rho)$ is positive and $k(\xi;\rho)$ is negative:

$$\mathcal{S}[\underline{k}(\xi,\rho)]\mathcal{S}[\underline{Y}(\xi,\rho)] = \mathcal{S}[\underline{k}(\xi,\rho)]\mathcal{S}\left[\overline{Y}(\xi,\rho)\right]$$
$$\mathcal{S}\left[\overline{k}(\xi,\rho)\right]\mathcal{S}\left[\overline{Y}(\xi,\rho)\right] = \mathcal{S}\left[\overline{k}(\xi,\rho)\right]\mathcal{S}[\underline{Y}(\xi,\rho)]$$

(iv) if $Y(\xi;\rho)$ and $k(\xi;\rho)$ both are negative:

$$\mathcal{S}[\underline{k}(\xi,\rho)]\mathcal{S}[\underline{Y}(\xi,\rho)] = \mathcal{S}\left[\overline{k}(\xi,\rho)\right]\mathcal{S}\left[\overline{Y}(\xi,\rho)\right]$$
$$\mathcal{S}\left[\overline{k}(\xi,\rho)\right]\mathcal{S}\left[\overline{Y}(\xi,\rho)\right] = \mathcal{S}[\underline{k}(\xi,\rho)]\mathcal{S}[\underline{Y}(\xi,\rho)]$$

Exploring Case (i), we can see that is remains are same.
After simplification, (11) and (12) become:

$$\mathcal{S}[\underline{Y}(\xi,\rho)] - \underline{p}_0 - u\underline{p}_1 - \cdots - u^{n-1}\underline{p}_{n-1} = u^n\mathcal{S}\left[\underline{f}(\xi,\rho)\right] + u^{n+1}\,\mathcal{S}[\underline{k}(\xi,\rho)]\mathcal{S}[\underline{Y}(\xi,\rho)] \quad (13)$$

$$\mathcal{S}\left[\overline{Y}(\xi,\rho)\right] - \overline{p}_0 - u\overline{p}_1 - \cdots - u^{n-1}\overline{p}_{n-1} = u^n\mathcal{S}\left[\overline{f}(\xi,\rho)\right] + u^{n+1}\mathcal{S}\left[\overline{k}(\xi,\rho)\right]\mathcal{S}\left[\overline{Y}(\xi,\rho)\right] \quad (14)$$

After simplification:

$$\mathcal{S}[\underline{Y}(\xi,\rho)] - u^{n+1}\,\mathcal{S}[\underline{k}(\xi,\rho)]\mathcal{S}[\underline{Y}(\xi,\rho)] = u^n\mathcal{S}\left[\underline{f}(\xi,\rho)\right] + \underline{p}_0 + u\underline{p}_1 + \cdots + u^{n-1}\underline{p}_{n-1} \quad (15)$$

$$\mathcal{S}[Y(\xi,\rho)] - u^{n+1}\mathcal{S}\left[\overline{k}(\xi,\rho)\right]\mathcal{S}\left[\overline{Y}(\xi,\rho)\right] = u^n\mathcal{S}\left[\overline{f}(\xi,\rho)\right] + u\overline{p}_1 + \cdots + u^{n-1}\overline{p}_{n-1} \quad (16)$$

Equations (15) and (16) give us:

$$\mathcal{S}[\underline{Y}(\xi,\rho)] = \frac{u^n\mathcal{S}\left[\underline{f}(\xi,\rho)\right] + \underline{p}_0 + u\underline{p}_1 + \cdots + u^{n-1}\underline{p}_{n-1}}{(1-u^{n+1}\mathcal{S}[\underline{k}(\xi-t)])} \tag{17}$$

$$\mathcal{S}\left[\overline{Y}(\xi,\rho)\right] = \frac{u^n\mathcal{S}\left[\overline{f}(\xi,\rho)\right] + u\overline{p}_1 + \cdots + u^{n-1}\overline{p}_{n-1}}{(1-u^{n+1}\mathcal{S}\left[\left(\overline{k}(\xi-t)\right]\right)} \tag{18}$$

By taking the inverse Sumudu transforms, we can get the value of $\overline{Y}(\xi,\rho)$ and $Y(\xi,\rho)$. Now, using the decomposition method:

$$\sum_{i=0}^{\infty}\underline{Y}_i(\xi,\rho) = \underline{Y}_0(\xi,\rho) + \underline{Y}_1(\xi,\rho) + \underline{Y}_2(\xi,\rho) + \cdots \underline{Y}_n(\xi,\rho)$$

and:

$$\sum_{i=0}^{\infty}\overline{Y}_i(\xi,\rho) = \overline{Y}_0(\xi,\rho) + \overline{Y}_1(\xi,\rho) + \overline{Y}_2(\xi,\rho) + \cdots \overline{Y}_n(\xi,\rho),$$

we can write as:

$$\begin{aligned}
\mathcal{S}\left[\underline{Y}_0(\xi,\rho)\right] &= u^n\mathcal{S}\left[\underline{f}(\xi,\rho)\right] + \underline{p}_0 + u\underline{p}_1 + \cdots + u^{n-1}\underline{p}_{n-1}\\
\mathcal{S}\left[\underline{Y}_1(\xi,\rho)\right] &= u^{n+1}\underline{\mathcal{S}[k(\xi-t)]\mathcal{S}[Y_0(\xi,\rho)]}\\
\mathcal{S}\left[\underline{Y}_1(\xi,\rho)\right] &= u^{n+1}\underline{\mathcal{S}[k(\xi-t)]\mathcal{S}[Y_1(\xi,\rho)]}\\
&\vdots\\
\mathcal{S}\left[\underline{Y}_n(\xi,\rho)\right] &= u^{n+1}\underline{\mathcal{S}[k(\xi-t)]\mathcal{S}[Y_{n-1}(\xi,\rho)]}
\end{aligned} \tag{A}$$

Similarly:

$$\begin{aligned}
\mathcal{S}\left[\overline{Y}_0(\xi,\rho)\right] &= u^n\mathcal{S}\left[\underline{f}(\xi,\rho)\right] + \underline{p}_0 + u\underline{p}_1 + \cdots + u^{n-1}\underline{p}_{n-1}\\
\mathcal{S}\left[\overline{Y}_1(\xi,\rho)\right] &= u^{n+1}\overline{\mathcal{S}[k(\xi-t)]\mathcal{S}[Y_0(\xi,\rho)]}\\
\mathcal{S}\left[\overline{Y}_2(\xi,\rho)\right] &= u^{n+1}\overline{\mathcal{S}[k(\xi-t)]\mathcal{S}[Y_1(\xi,\rho)]}\\
&\vdots\\
\mathcal{S}\left[\overline{Y}_n(\xi,\rho)\right] &= u^{n+1}\overline{\mathcal{S}[k(\xi-t)]\mathcal{S}[Y_{n-1}(\xi,\rho)]}
\end{aligned} \tag{B}$$

For nonlinear equations, we use the adomian polynomials:

$$\begin{aligned}
A_0 = {Y_O}^2,\quad & A_1 = 2Y_O\,Y_1\\
A_2 = 2Y_O\,Y_2 + {Y_1}^2,\quad & A_3 = 2Y_O\,Y_1 + 2Y_1Y_2
\end{aligned}$$

Then, Equation (B) becomes:

$$\mathcal{S}\left[\sum_{i=0}^{\infty}\underline{Y}_i(\xi,\rho)\right] - \underline{p}_0 - u\underline{p}_1 - \cdots - u^{n-1}\underline{p}_{n-1} = u^n\mathcal{S}\left[\underline{f}(\xi,\rho)\right] + u^{n+1}\underline{\mathcal{S}\left[k(x-t)]\mathcal{S}[\sum_{j=1}^{\infty}A_j\right]} \tag{19}$$

$$\mathcal{S}\left[\sum_{i=0}^{\infty}\overline{Y}_i(\xi,\rho)\right] - \overline{p}_0 - u\overline{p}_1 - \cdots - u^{n-1}\overline{p}_{n-1} = u^n\mathcal{S}\left[\overline{f}(\xi,\rho)\right] + u^{n+1}\overline{\mathcal{S}\left[k(\xi-t)]\mathcal{S}[\sum_{i=j}^{\infty}A_j\right]}$$

5. Numerical Examples

Example 1. *A linear fuzzy integro-differential equation is:*

$$\mathrm{Y}^{(1)}(\xi,\rho) = f(\xi,\rho) - \int_0^{\xi} \mathrm{Y}(t,\rho)dt,$$

with conditions, $\mathrm{Y}(0,\rho) = (0,0)$, *where:*

$$\lambda = 1,\ 0 \le t \le \xi,\ 0 \le \rho \le 1,\ K(\xi,t) = 1,$$

i.e.,

$$f(\xi,\rho) = \left(\left(\rho^2+\rho\right),(5-\rho)\right)$$

To solve this fuzzy integro-differential, we proceed as follows:

$$\begin{cases} \underline{\mathrm{Y}^{(1)}}(\xi,\rho) = f(\xi,\rho) - \int_0^{\xi} \underline{\mathrm{Y}(t,\rho)dt} \\ \overline{\mathrm{Y}^{(1)}}(\xi,\rho) = f(\xi,\rho) - \int_0^{\xi} \overline{\mathrm{Y}(t,\rho)dt} \end{cases} \tag{20}$$

$$\begin{cases} \underline{\mathrm{Y}^{(1)}}(\xi,\rho) = \left(\rho^2+\rho\right) - \int_0^{\xi} \underline{\mathrm{Y}(t,\rho)dt} \\ \overline{\mathrm{Y}^{(1)}}(\xi,\rho) = (5-\rho) - \int_0^{\xi} \overline{\mathrm{Y}(t,\rho)dt} \end{cases} \tag{21}$$

Applying Sumudu transform on (21) and using Equations (A) and (B), we have:

$$\begin{cases} \mathcal{S}\left[\underline{\mathrm{Y}^{(1)}}(\xi,\rho)\right] = \mathcal{S}(\left(\rho^2+\rho\right) - \mathcal{S}\int_0^{\xi} \underline{\mathrm{Y}(t,\rho)dt} \\ \mathcal{S}[\overline{\mathrm{Y}^{(1)}}(\xi,\rho)] = \mathcal{S}((5-\rho) - \mathcal{S}\int_0^{\xi} \overline{\mathrm{Y}(t,\rho)dt} \end{cases} \tag{22}$$

$$\begin{cases} \underline{\mathrm{Y}}(\xi,\rho) = \mathcal{S}^{-1}\left(u\left(\rho^2+\rho\right)\right) - \mathcal{S}^{-1}(u^2\mathcal{S}(\underline{\mathrm{Y}}(\xi,\rho)) \\ \overline{\mathrm{Y}}(\xi,\rho) = \mathcal{S}^{-1}\ (u(5-\rho)) - \mathcal{S}^{-1}(u^2\mathcal{S}\left(\overline{\mathrm{Y}}(\xi,\rho)\right) \end{cases} \tag{23}$$

For $\underline{\mathrm{Y}}(\xi,\rho)$*:*

$$\begin{aligned} &\mathrm{Y}_0 = \xi\left(\alpha^2+\alpha\right) \\ &\mathrm{Y}_1 = \mathcal{S}^{-1}(u^2\mathcal{S}\left(\xi\left(\rho^2+\rho\right)\right), && \mathrm{Y}_1 = -\tfrac{\xi^3}{3!}\left(\rho^2+\rho\right) \\ &\mathrm{Y}_2 = \mathcal{S}^{-1}(u^2\mathcal{S}\left(\tfrac{\xi^3}{3!}\left(\rho^2+\rho\right)\right), && \mathrm{Y}_2 = \tfrac{\xi^5}{5!}\left(\rho^2+\rho\right) \\ &\mathrm{Y}_3 = \mathcal{S}^{-1}(u^2\mathcal{S}\left(\tfrac{\xi^5}{5!}\left(\rho^2+\rho\right)\right), && \mathrm{Y}_3 = -\tfrac{\xi^7}{7!}\left(\rho^2+\rho\right) \end{aligned}$$

Similarly, for $\overline{\mathrm{Y}(\xi,\rho)}$*:*

$$\begin{aligned} &\mathrm{Y}_0 = \xi(5-\rho) \\ &\mathrm{Y}_1 = \mathcal{S}^{-1}(u^2\mathcal{S}(\xi(5-\rho)), && \mathrm{Y}_1 = -\tfrac{\xi^3}{3!}(5-\rho), \\ &\mathrm{Y}_2 = \mathcal{S}^{-1}(u^2\mathcal{S}\left(-\tfrac{\xi^3}{3!}(5-\rho)\right), && \mathrm{Y}_2 = \tfrac{\xi^5}{5!}(5-\rho), \\ &\mathrm{Y}_3 = \mathcal{S}^{-1}(u^2\mathcal{S}\left(\tfrac{\xi^5}{5!}(5-\rho)\right), && \mathrm{Y}_3 = -\tfrac{\xi^7}{7!}(5-\rho) \end{aligned}$$

Thus, using above iterative results, the series form solution is given as:

$$\begin{cases} \underline{Y(\xi,\rho)} = \xi(5-\rho) - \frac{\xi^3}{3!}(5-\rho) + \frac{\xi^5}{5!}(5-\rho) - \frac{\xi^7}{7!}(5-\rho) + \ldots. \\ \overline{Y(\xi,\rho)} = \xi(5-\rho) - \frac{\xi^3}{3!}(5-\rho) + \frac{\xi^5}{5!}(5-\rho) - \frac{\xi^7}{7!}(5-\rho) + \ldots.. \end{cases} \tag{24}$$

Using (17), we get the exact solution:

$$\begin{cases} \underline{Y(\xi,\rho)} = sin\xi\left(\rho^2+\rho\right) \\ \overline{Y(\xi,\rho)} = sin\xi(5-\rho) \end{cases} \tag{25}$$

A graphical representation of the solution is given in Figure 1.

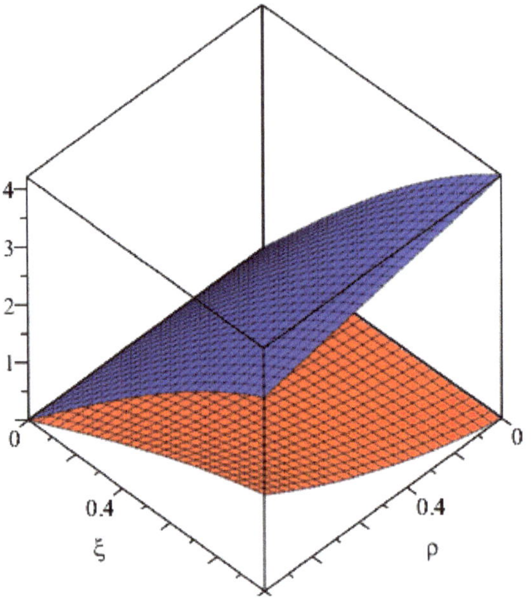

Figure 1. Graphical interpolation of Example 1.

Example 2. *Consider the following fuzzy Volterra integro-differential equation:*

$$\begin{cases} \underline{Y}^{(1)}(\xi,\rho) = (\rho-1) + \int_0^{\xi} \underline{Y}(t,\rho)dt \\ \overline{Y}^{(1)}(\xi,\rho) = (1-\rho) + \int_0^{\xi} \overline{Y}(t,\rho)dt \end{cases} \tag{26}$$

$$\underline{Y}^{(1)}(0) = 0 = \overline{Y}^{(1)}(0); 0 \le \rho \le 1,\ 0 \le t \le \xi, \quad \xi \in [0,1]$$

Using (A) and (B) on both sides and taking the inverse:

$$\begin{cases} \underline{Y}(\xi,\rho)) = \mathcal{S}^{-1}(u(\rho-1)) + \mathcal{S}^{-1}\left(u^2\mathcal{S}[\underline{Y}(\xi,\rho)]\right) \\ \overline{Y}(\xi,\rho)) = \mathcal{S}^{-1}(u(1-\rho)) + \mathcal{S}^{-1}\left(u^2\mathcal{S}\left[\overline{Y}(\xi,\rho)\right]\right) \end{cases} \tag{27}$$

$$\begin{cases} \underline{Y}(\xi,\rho)) = \xi(\rho-1)) + \mathcal{S}^{-1}\left(u^2\mathcal{S}[\underline{Y}(\xi,\rho)]\right) \\ \overline{Y}(\xi,\rho)) = \xi(1-\rho)) + \mathcal{S}^{-1}\left(u^2\mathcal{S}\left[\overline{Y}(\xi,\rho)\right]\right) \end{cases}$$

Then, the solution in the series form will be:

$$\sum_{i=0}^{\infty}\underline{Y}_i(\xi,\rho) = \underline{Y}_0(\xi,\rho) + \underline{Y}_1(\xi,\rho) + \underline{Y}_2(\xi,\rho) + \underline{Y}_3(\xi,\rho) + \cdots$$

$$= \xi(\rho-1) + \frac{\xi^3}{3!}(\rho-1) + \frac{\xi^5}{3!}(\rho-1) + \frac{\xi^7}{3!}(\rho-1) + \cdots$$

Similarly, for $\overline{Y}(\xi,\rho)$*:*

$$= \xi(1-\rho) + \frac{\xi^3}{3!}(1-\rho) + \frac{\xi^5}{5!}(1-\rho) + \frac{\xi^7}{5!}(1-\rho) + \cdots$$

and the exact solution is given using (17).

$$\overline{Y}(\xi,\rho) = (\rho-1)sinht \text{ and } \overline{Y}(\xi,\rho) = (1-\rho)sinht$$

The graphical representation of the solution is given in Figure 2.

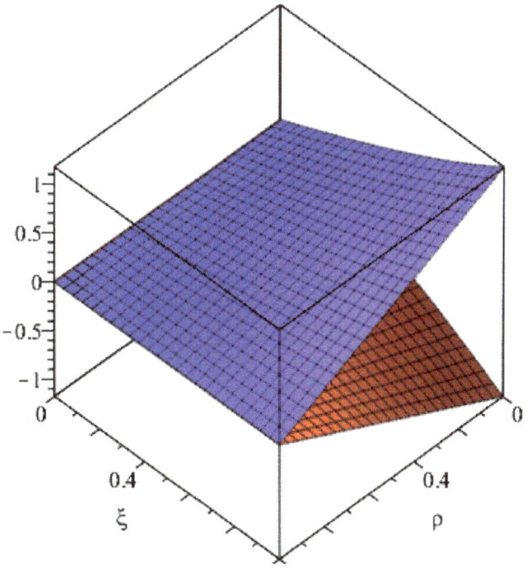

Figure 2. Graphical interpolation of Example 2.

Example 3. *Consider the following fuzzy Volterra integro-differential equation:*

$$\begin{cases} \underline{Y}^{(1)}(\xi,\rho) = (\rho+1)(1+\xi) + \int_0^{\xi}\underline{Y}(t,\rho)dt \\ \overline{Y}^{(1)}(\xi,\rho) = (\rho-2)(1+\xi) + \int_0^{\xi}\overline{Y}(t,\rho)dt \end{cases} \tag{28}$$

$$\underline{Y}^{(1)}(0) = 0 = \overline{Y}^{(1)}(0); 0 \le \rho \le 1,\ 0 \le t \le \xi, \quad \xi \in [0,1]$$

Using (A) and (B) on (28), we get:

$$\begin{cases} \mathcal{S}(\underline{Y}^{(1)}(\xi,\rho)) = \mathcal{S}((\rho+1)(1+\xi)) + \mathcal{S}[\int_0^{\xi} \underline{Y}(t,\rho)]dt \\ \mathcal{S}(\overline{Y}^{(1)}(\xi,\rho)) = \mathcal{S}((\rho-2)(1+\xi)) + \mathcal{S}[\int_0^{\xi} \overline{Y}(t,\rho)]dt \end{cases}$$
$$\begin{cases} \underline{Y}(\xi,\rho)) = \mathcal{S}^{-1}(u(\rho+1)) + \mathcal{S}^{-1}(u^2(\rho+1) + \mathcal{S}^{-1}\big(u^2\mathcal{S}[\underline{Y}(\xi,\rho)]\big) \\ \overline{Y}(\xi,\rho)) = \mathcal{S}^{-1}(u(\rho-2)) + \mathcal{S}^{-1}(u^2(\rho-2) + \mathcal{S}^{-1}\big(u^2\mathcal{S}\big[\overline{Y}(\xi,\rho)\big]\big) \end{cases} \quad (29)$$

Then, the solution in the series form will be:

$$\sum_{i=0}^{\infty} \underline{Y}_i(\xi,\rho) = \underline{Y}_0(\xi,\rho) + \underline{Y}_1(\xi,\rho) + \underline{Y}_2(\xi,\rho) + \underline{Y}_3(\xi,\rho) + \cdots$$

$$\begin{aligned} &= \xi(\rho+1) + \tfrac{\xi^2}{2!}(\rho+1) + \tfrac{\xi^3}{3!}\,(\rho+1) + \tfrac{\xi^4}{4!}(\rho+1) + \tfrac{\xi^5}{5!}\,(\rho+1) \\ &\quad + \tfrac{\xi^6}{6!}(\rho+1)\tfrac{\xi^7}{7!}\,(\rho+1) + \tfrac{\xi^8}{8!}(\rho+1) + \cdots \end{aligned} \quad (30)$$

Similarly, for $\overline{Y}(\xi,\rho)$*:*

$$\begin{aligned} &= \xi(\rho-2) + \tfrac{\xi^2}{2!}(\rho-2) + \tfrac{\xi^3}{3!}\,(\rho-2) + \tfrac{\xi^4}{4!}(\rho-2) + \tfrac{\xi^5}{5!}\,(\rho-2) \\ &\quad + \tfrac{\xi^6}{6!}(\rho-2)\tfrac{\xi^7}{7!}\,(\rho-2) + \tfrac{\xi^8}{8!}(\rho-2) + \cdots \end{aligned} \quad (31)$$

Using (17), we get the exact solution.

$$\underline{Y}(\xi,\rho) = (\rho+1)(e^x-1) \text{ and } \overline{Y}(\xi,\rho) = (\rho-2)(e^x-1)$$

The graphical representation of the solution is given in Figure 3.

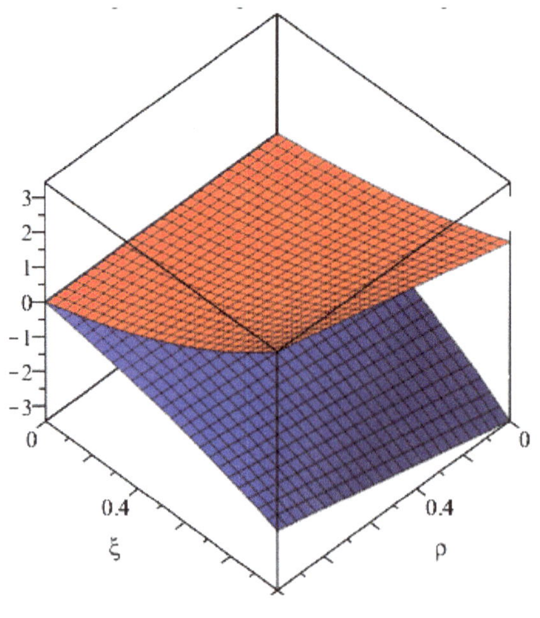

Figure 3. Graphical interpretation of Example 3.

Example 4. *Consider a Volterra integro-differential equation:*

$$Y^{(2)}(\xi,\rho) = f(\xi,\rho) - \int_0^{\xi} (\xi - t)Y(t,\rho)dt \tag{32}$$

with conditions:

$$u(0,\rho) = (\rho+1, 3-\rho); \quad u^{(1)}(0,\rho) = (\rho, 2-\rho)$$
$$\lambda = 1,\ 0 \le t \le \xi,\ 0 \le \rho \le 1,\ K(\xi,t) = (\xi - t),$$

To solve Equation (32), we proceed as follows:

$$\begin{cases} \underline{u^{(2)}(\xi,\rho)} = f(\xi,\rho) - \int_0^{\xi}(\xi-t)\underline{Y(t,\rho)}dt \\ \overline{u^{(2)}(\xi,\rho)} = f(\xi,\rho) - \int_0^{\xi}(\xi-t)\overline{Y(t,\rho)}dt \end{cases} \tag{33}$$

$$\begin{cases} \underline{u^{(2)}(\xi,\rho)} = (\rho+2)\xi - \int_0^{\xi}(\xi-t)\underline{Y(t,\rho)}dt \\ \overline{u^{(2)}(\xi,\alpha)} = (4-\rho)\xi - \int_0^{\xi}(\xi-t)\overline{Y(t,\rho)}dt \end{cases} \tag{34}$$

Using (A) and (B), we get:

$$\begin{cases} \underline{Y(\xi,\rho)} = \mathcal{S}^{-1}(u^3(\rho+2) + \mathcal{S}^{-1}(u\rho) + \mathcal{S}^{-1}(\rho+1) - \mathcal{S}^{-1}(u^4\mathcal{S}\big(\underline{Y(\xi,\rho)}\big) \\ \overline{Y(\xi,\rho)} = \mathcal{S}^{-1}(u^3(4-\rho) + \mathcal{S}^{-1}(2-\rho) + \mathcal{S}^{-1}(3-\rho) - \mathcal{S}^{-1}(u^4\mathcal{S}\big(\overline{Y(\xi,\rho)}\big) \end{cases} \tag{35}$$

Now, applying the decomposition method for $\underline{Y(\xi,\rho)}$*:*

$$\begin{aligned} &Y_0 = \tfrac{\xi^3}{3!}(\rho+2) + \xi\rho + (\rho+1) \\ &Y_1 = \mathcal{S}^{-1}\Big(u^4\mathcal{S}\Big(\tfrac{\xi^3}{3!}(\rho+2) + \xi\rho + (\rho+1)\Big)\Big), && Y_1 = \tfrac{\xi^7}{7!}(\rho+2) + \tfrac{\xi^5}{5!}\rho + \tfrac{\xi^4}{4!}(\rho+1) \\ &Y_2 = \mathcal{S}^{-1}\Big(u^4\mathcal{S}(\tfrac{\xi^7}{7!}(\rho+2) + \tfrac{\xi^5}{5!}\rho + \tfrac{\xi^4}{4!}(\rho+1)\Big) && Y_2 = \tfrac{\xi^{11}}{11!}(\rho+2) + \tfrac{\xi^9}{9!}\rho + \tfrac{\xi^8}{8!}(\rho+1) \end{aligned}$$

Similarly, we can find $\underline{Y_3}(\xi,\rho), \underline{Y_4}(\xi,\rho), \ldots.$*:*

$$\begin{aligned} \sum_{i=0}^{\infty} \underline{Y_i}(\xi,\rho) &= \underline{Y_0}(\xi,\rho) + \underline{Y_1}(\xi,\rho) + \underline{Y_2}(\xi,\rho) + \ldots. \\ &= \Big(\xi + \tfrac{\xi^5}{5!} + \tfrac{\xi^9}{9!} + \ldots..\Big)\rho + \Big(1 + \tfrac{\xi^4}{4!} + \tfrac{\xi^8}{8!} + \ldots..\Big)(\rho+1) \\ &+\Big(\tfrac{\xi^3}{3!} + \tfrac{\xi^7}{7!} + \tfrac{\xi^{11}}{11!}\ldots..\Big)(\rho+2) + \cdots \end{aligned} \tag{36}$$

Now, for $\overline{\mathrm{Y}(\xi,\rho)}$:

$$\mathrm{Y}_0 = \frac{\xi^3}{3!}(4-\rho) + \xi(2-\rho) + (3-\rho)$$
$$\mathrm{Y}_1 = \mathcal{S}^{-1}\left(u^4\mathcal{S}\left(\frac{\xi}{3!}(4-\rho) + \xi(2-\rho) + (3-\rho)\right)\right),$$

$$\mathrm{Y}_1 = \frac{\xi^7}{7!}(4-\rho) + \frac{\xi^5}{5!}(2-\rho) + \frac{\xi^4}{4!}(3-\rho)$$
$$\mathrm{Y}_2 = \mathcal{S}^{-1}\left(u^4\mathcal{S}\left(\frac{\xi^7}{7!}(4-\rho) + \frac{\xi^5}{5!}(2-\rho) + \frac{\xi^4}{4!}(3-\rho)\right)\right),$$
$$\mathrm{Y}_2 = \frac{\xi^{11}}{11!}(4-\rho) + \frac{\xi^9}{9!}(2-\rho) + \frac{\xi^8}{8!}(3-\rho)$$

Similarly, for $\overline{\mathrm{Y}_i}(\xi,\rho)$:

$$= \left(\xi + \frac{\xi^5}{5!} + \frac{\xi^9}{9!} + \ldots..\right)(2-\rho) + \left(1 + \frac{\xi^4}{4!} + \frac{\xi^8}{8!} + \ldots..\right)(3-\rho)$$
$$+\left(\frac{\xi^3}{3!} + \frac{\xi^7}{7!} + \frac{\xi^{11}}{11!}\ldots..\right)(4-\rho) + \cdots$$

and the exact solution is given as:

$$\underline{\mathrm{Y}}(\xi;\rho) = (\rho+2).\frac{1}{2}(sinh\xi - sin\xi) + (\rho+1).\frac{1}{2}(cos\xi + cosh\xi) + (\rho)(sin\xi + sinh\xi)$$

$$\overline{\mathrm{Y}}(t;\rho) = (4-\rho).\frac{1}{2}(sinh\xi - sin\xi) + (3-\rho).\frac{1}{2}(cos\xi + cosh\xi)$$
$$+ (2-\rho)(sin\xi + sinh\xi) \qquad 0 \le \rho \le 1$$

The graphical representation of the solution is given in Figure 4.

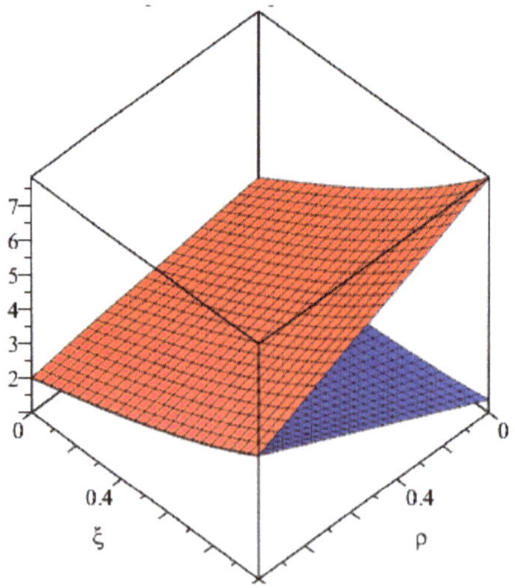

Figure 4. Graphical interpretation of Example 4.

Example 5. *Consider a nonlinear fuzzy Volterra integro-differential equation:*

$$\mathrm{Y}^{(1)}(\xi,\rho) = f(\xi,\rho) - \int_0^{\xi} \mathrm{Y}^2(t,\rho)dt \tag{37}$$

with conditions $\mathrm{Y}(0,\rho) = (0,0)$, *where:*

$$\lambda = 1,\ 0 \le t \le \xi,\ 0 \le \rho \le 1,\ K(\xi,t) = 1, \text{ i.e.,}$$
$$f(\xi,\rho) = (\rho, 7-\rho)$$

To solve Equation (37), we proceed as follows:

$$\begin{cases} \underline{\mathrm{Y}^{(1)}(\xi,\rho)} = f(\xi,\rho) - \int_0^{\xi} \underline{\mathrm{Y}^2(t,\rho)dt} \\ \overline{\mathrm{Y}^{(1)}(\xi,\rho)} = f(\xi,\rho) - \int_0^{\xi} \overline{\mathrm{Y}^2(t,\rho)dt} \end{cases} \tag{38}$$

$$\begin{cases} \underline{\mathrm{Y}^{(1)}(\xi,\rho)} = \rho - \int_0^{\xi} \underline{\mathrm{Y}^2(t,\rho)dt} \\ \overline{\mathrm{Y}^{(1)}(\xi,\rho)} = 7 - \rho - \int_0^{\xi} \overline{\mathrm{Y}^2(t,\rho)dt} \end{cases} \tag{39}$$

Applying the Sumudu transform on both sides of the equation, we get:

$$\begin{cases} \mathcal{S}(\underline{\mathrm{Y}^{(1)}(\xi,\rho)} = \mathcal{S}(\rho) - \mathcal{S}\int_0^{\xi} \underline{\mathrm{Y}^2(t,\rho)dt} \\ \mathcal{S}(\overline{\mathrm{Y}^{(1)}(\xi,\rho)} = \mathcal{S}(7-\rho) - \mathcal{S}\int_0^{\xi} \overline{\mathrm{Y}^2(t,\rho)dt} \end{cases} \tag{40}$$

Applying the inverse Sumudu transform and using (19), we get:

$$\begin{cases} \underline{\mathrm{Y}(\xi,\rho)} = \mathcal{S}^{-1}(u\rho) - \mathcal{S}^{-1}(u^2\mathcal{S}\big(\mathrm{Y}^2\underline{(\xi,\rho)}\big) \\ \overline{\mathrm{Y}(\xi,\rho)} = \mathcal{S}^{-1}\ (u(7-\rho)) - \mathcal{S}^{-1}(u^2\mathcal{S}\big(\overline{\mathrm{Y}^2(\xi,\rho)}\big) \end{cases}$$

$$\begin{aligned} \sum_{i=0}^{\infty} \overline{\mathrm{Y}_i}(\xi,\rho) &= \overline{\mathrm{Y}_1}(\xi,\rho) + \overline{\mathrm{Y}_2}(\xi,\rho) + \overline{\mathrm{Y}_3}(\xi,\rho) + \ldots. \\ &= \xi(7-\rho) + \tfrac{\xi^4}{12}(7-\rho)^2 + \tfrac{\xi^7}{252}(7-\rho)^3 + \ldots \end{aligned}$$

The graphical representation of the solution is given in Figure 5.

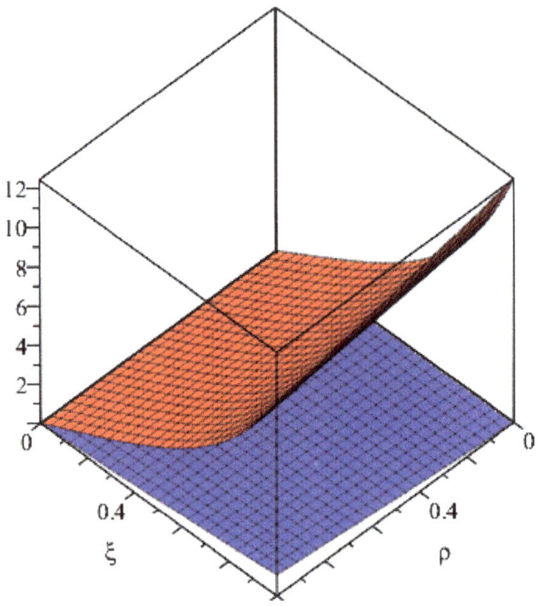

Figure 5. Graphical interpretation of Example 5.

6. Conclusions

Usually, it is difficult to solve fuzzy integro-differential equations analytically. Most probably, it is required to obtain the approximate solutions. In this paper, we developed a numerical technique (Sumudu decomposition method) to find the solution to linear and nonlinear fuzzy Volterra integro-differential equations. A general method for solving VIDE was developed. This technique proved reliable and effective based on the achieved results. It gives fast convergence because by utilizing a lower number of iterations, we get approximate as well as exact solutions.

Author Contributions: The problem was proposed by S.M.K. The results were proven and the paper was drafted by Z.I. M.H. supervised this work, and W.N. verified and analyzed the results.

Funding: Funding is not available for this research.

Conflicts of Interest: The authors declare no conflict of interest.

References

1. Ullah, S.; Farooq, M.; Ahmad, L.; Abdullah, S. Application of fuzzy Laplace transforms for solving fuzzy partial Volterra integro-differential equations. *arXiv* **2014**, arXiv:1405.1895.
2. Mikaeilv, N.; Khakrangin, S.; Allahviranloo, T. Solving fuzzy Volterra integro-differential equation by fuzzy differential transform method. In Proceedings of the 7th Conference of the European Society for Fuzzy Logic and Technology, Aix-Les-Bains, France, 18–22 July 2011.
3. Ahmad, J.; Nosher, H. Solution of Different Types of Fuzzy Integro-Differential Equations Via Laplace Homotopy Perturbation Method. *J. Sci. Arts* **2017**, *17*, 5.
4. Mohmmed, S.E.A.A. Solution of Linear and Nonlinear Partial Differential Equations by Mixing Adomian Decomposition Method and Sumudu Transform. Ph.D. Thesis, Sudan University of Science and Technology, Kashmu, Sudan, 2016.
5. Eltayeb, H.; Kılıçman, A. Application of Sumudu decomposition method to solve nonlinear system of partial differential equations. In *Abstract and Applied Analysis*; Eltayeb, H., Kılıçman, A., Eds.; Hindawi: London, UK, 2012.

6. Gomes, L.T.; de Barros, L.C.; Bede, B. *Fuzzy Differential Equations in Various Approaches*; Springer: Berlin, Germany, 2015.
7. Das, M.; Talukdar, D. Method for solving fuzzy integro-differential equations by using fuzzy Laplace transformation. *Int. J. Sci. Tech.* **2014**, *3*, 291–295.
8. Kumar, D.; Singh, J.; Rathore, S. Sumudu decomposition method for nonlinear equations. *Int. Math. Forum* **2012**, *7*, 515–521.
9. Bildik, N.; Deniz, S. The use of Sumudu decomposition method for solving predator-prey systems. *Math. Sci. Lett.* **2016**, *5*, 285–289. [CrossRef]
10. Burton, A.J.; Miller, G.F. The application of integral equation methods to the numerical solution of some exterior boundary-value problems. *Proc. R. Soc. Lond. A Math. Phys. Sci.* **1971**, *323*, 201–210. [CrossRef]
11. Wazwaz, A.M. *A First Course in Integral Equations*; World Scientific Publishing Company: Singapore, 2015.
12. Zeinali, M.; Shahmorad, S.; Mirnia, K. Fuzzy integro-differential equations: Discrete solution and error estimation. *Iran. J. Fuzzy Syst.* **2013**, *10*, 107–122.
13. Rajkumar, A.; Mohammed Shapique, A.; Jesuraj, C. Solving Fuzzy Linear Volterra Intergro-Differential Equation Using Fuzzy Sumudu Transform. *Int. J. Pure Appl. Math.* **2018**, *119*, 3173–3184.
14. Prakash, A.; Kumar, M.; Baleanu, D. A new iterative technique for a fractional model of nonlinear Zakharov–Kuznetsov equations via Sumudu transform. *Appl. Math. Comput.* **2018**, *334*, 30–40. [CrossRef]
15. Singh, J.; Kumar, D.; Baleanu, D.; Rathore, S. An efficient numerical algorithm for the fractional Drinfeld–Sokolov–Wilson equation. *Appl. Math. Comput.* **2018**, *335*, 12–24. [CrossRef]
16. Ziane, D.; Baleanu, D.; Belghaba, K.; Cherif, M.H. Local fractional Sumudu decomposition method for linear partial differential equations with local fractional derivative. *J. King Saud Univ. Sci.* **2017**, *31*, 83–88. [CrossRef]
17. Rathore, S.; Kumar, D.; Singh, J.; Gupta, S. Homotopy analysis Sumudu transform method for nonlinear equations. *Int. J. Ind. Math.* **2012**, *4*, 301–314.
18. Li, K.; Xie, Y. A brief introduction of Sumudu transform and comparison with other integral transforms. In Proceedings of the 6th Asia-Pacific Conference on Environmental Electromagnetics (CEEM), Shanghai, China, 6–9 November 2012; pp. 285–287.

axioms

MDPI

Article

On the Polynomial Solution of Divided-Difference Equations of the Hypergeometric Type on Nonuniform Lattices

Mama Foupouagnigni [1,2,*,†] **and Salifou Mboutngam** [3,†]

1 Department of Mathematics, Higher Teachers' Training College, University of Yaounde 1, Yaounde, Cameroon
2 The African Institute for Mathematical Sciences, Limbe, Cameroon
3 Department of Mathematics, Higher Teachers' Training College, University of Maroua, Maroua, Cameroon; mbsalif@gmail.com
* Correspondence: foupouagnigni@gmail.com
† These authors contributed equally to this work.

Received: 31 January 2019; Accepted: 10 April 2019; Published: 21 April 2019

Abstract: In this paper, we provide a formal proof of the existence of a polynomial solution of fixed degree for a second-order divided-difference equation of the hypergeometric type on non-uniform lattices, generalizing therefore previous work proving existence of the polynomial solution for second-order differential, difference or q-difference equation of hypergeometric type. This is achieved by studying the properties of the mean operator and the divided-difference operator as well as by defining explicitly, the right and the "left" inverse for the second operator. The method constructed to provide this formal proof is likely to play an important role in the characterization of orthogonal polynomials on non-uniform lattices and might also be used to provide hypergeometric representation (when it does exist) of the second solution—non polynomial solution—of a second-order divided-difference equation of hypergeometric type.

Keywords: second-order differential/difference/q-difference equation of hypergeometric type; non-uniform lattices; divided-difference equations; polynomial solution

1. Introduction

Classical orthogonal polynomials of a continuous variable (P_n) are known to satisfy a second-order differential equation of hypergeometric type

$$\sigma(x)\,y''(x) + \tau(x)\,y'(x) + \lambda\,y(x) = 0, \tag{1}$$

where σ is a polynomial of degree at most 2, τ is a first degree polynomial and λ is a constant with respect to x.

In [1,2], it is shown that Equation (1) has a polynomial solution of exactly n degree for a specific given constant $\lambda = \lambda_n$. This is achieved mainly by showing that :

- the nth derivative $y^{(n)}$ of any solution y of (1) satisfies an equation of the same type (hypergeometric aspect), that is, an equation of the form

$$\sigma(x)\,y''(x) + \tau_n(x)\,y'(x) + \lambda_n\,y(x) = 0, \tag{2}$$

where τ_n is a first degree polynomial and λ_n is a constant given by

$$\tau_n(x) = \tau + n\,\sigma'(x),\ \lambda_n = \lambda + n\,\tau' + \frac{n(n-1)}{2}\,\sigma''. \tag{3}$$

- Any solution of (2) can be written as the nth derivative of a solution of (1), provided that $\lambda_j \neq 0,\ j = 1...n-1$.

The fact that the constant solution of (2) when $\lambda_n = 0$ is the nth derivative of a solution of (1) leads to the existence of a polynomial solution of (1), of exactly n degree, when

$$\lambda = \lambda_n = -n\,\tau' - \frac{n(n-1)}{2}\,\sigma''.$$

This result proves not only the existence of a polynomial solution for Equation (1) but also allows for establishing the Rodrigues formula expressing the polynomial solution in the term of the nth derivative:

$$P_n(x) = \frac{B_n}{\rho(x)}\,[\sigma^n(x)\,\rho(x)]^{(n)},$$

where B_n is a constant and ρ is the weight function satisfying the Pearson equation

$$(\sigma(x)\,\rho(x))' = \tau(x)\,\rho(x).$$

It is worth mentioning that Hermite, Laguerre, Jacobi and Bessel polynomials are the polynomial eigenfunctions of the second-order linear differential operation given in (1).

Using the same approach, similar results have been established in [3] (See also [4]) for the classical orthogonal polynomials of a discrete variable satisfying instead a second-order difference equation of hypergeometric type

$$\sigma(x)\,\Delta\nabla y(x) + \tau(x)\,\Delta y(x) + \lambda_n\,y(x) = 0, \tag{4}$$

where Δ and ∇ are the forward and the backward operators defined by

$$\Delta f(s) = f(s+1) - f(s),\ \nabla f(s) = f(s) - f(s-1).$$

Furthermore, it should be noticed that Charlier, Krawtchuk, Meixner and Hahn polynomials are the polynomial eigenfunctions of the second-order linear difference operation given in (4).

The same result can be established in the same way to for the classical orthogonal polynomials of a q-discrete variable satisfying a second-order q-difference equation of hypergeometric type [5] (See also [6,7])

$$\sigma(x)\,D_q^2 y(x) + \tau(x)\,D_q y(x) + \lambda_n\,y(x) = 0, \tag{5}$$

where D_q is the Hahn operator [8] defined by

$$D_q(f(x)) = \frac{f(qx) - f(x)}{(q-1)\,x},\ x \neq 0,\ D_q f(0) := f'(0),$$

provided that $f'(0)$ exists. Orthogonal polynomials which are eigenfunctions of the second-order q-difference operator given defined by (5) are [5]: Big q-Jacobi, Big q-Laguerre, Little q-Jacobi, Little q-Laguerre (Wall), q-Laguerre, Alternative q-Charlier, Al-Salam–Carlitz I, Al-Salam–Carlitz II, Stieltjes–Wigert, Discrete q–Hermite I, Discrete q–Hermite II, q-Hahn, q-Meixner, Quantum q-Krawtchouk, q-Krawtchouk, Affine q-Krawtchouk, the q-Charlier and the q-Charlier II polynomials.

Classical orthogonal polynomials on non-uniform lattices (including but not limited to Askey–Wilson polynomials, Racah and q-Racah polynomials), are known to satisfy a second-order divided-difference equation of the form [9,10] (see also [11])

$$\phi(x(s))\,\frac{\Delta}{\Delta x(s-\frac{1}{2})}\left[\frac{\nabla y(x(s))}{\nabla x(s)}\right] + \frac{\psi(x(s))}{2}\left[\frac{\Delta y(x(s))}{\Delta x(s)} + \frac{\nabla y(x(s))}{\nabla x(s)}\right] + \lambda_n\,y(x(s)) = 0, \tag{6}$$

where ψ and ϕ are polynomials of degree 1 and at most 2, respectively; λ_n is a constant depending on n and on the leading coefficients of ϕ and ψ. The lattice $x(s)$ is defined by [9,10]

$$x(s) = \begin{cases} c_1 q^{-s} + c_2 q^s + c_3 & \text{if} \quad q \neq 1, \\ c_4 s^2 + c_5 s + c_6 & \text{if} \quad q = 1, \end{cases} \tag{7}$$

is known as non-uniform lattice and fulfills various important properties.

Equation (6) can be transformed into equation [12]

$$\phi(x(s))\mathbb{D}_x^2 y(x(s)) + \psi(x(s))\mathbb{S}_x\mathbb{D}_x y(x(s)) + \lambda_n y(x(s)) = 0, \tag{8}$$

called divided-difference equation of the hypergeometric type by means of the two companion operators $\mathbb{D}_x$ (called divided-difference operator) and $\mathbb{S}_x$ (called mean operator) defined as [9,10,12,13]

$$\mathbb{D}_x f(x(s)) = \frac{f(x(s+\frac{1}{2})) - f(x(s-\frac{1}{2}))}{x(s+\frac{1}{2}) - x(s-\frac{1}{2})}, \quad \mathbb{S}_x f(x(s)) = \frac{f(x(s+\frac{1}{2})) + f(x(s-\frac{1}{2}))}{2}. \tag{9}$$

Using appropriate bases, computer algebra software has been used to solve divided-difference Equation (8) for specific families of classical orthogonal polynomials on a non-uniform lattice. For some special values of the parameter for the specific case of Askey–Wilson polynomials, non-polynomial solution has been recovered together with the polynomial one [14] (see page 15, Equations (62) and (63)). In addition, the operators $\mathbb{D}_x$ and $\mathbb{S}_x$ have played a decisive role not only for establishing the functional approach of the characterization theorem of classical orthogonal polynomials on non-uniform lattices, but also for providing algorithmic solution to linear homogeneous divided-difference equations with polynomial coefficients, allowing to solve explicitly [13] the first-order divided-difference equations satisfied by the basic exponential function

$$\mathbb{D}_x y(x(s)) = \frac{2wq^{\frac{1}{4}}}{1-q} y(x(s)),$$

and the second-order divided-difference equation satisfied by the basic trigonometric functions

$$\mathbb{D}_x^2 y(x(s)) = -\left(\frac{2wq^{\frac{1}{4}}}{1-q}\right)^2 y(x(s)),$$

where w is a given constant.

The aim of this work is:

1. redto define the right and the "left" inverses of the operator $\mathbb{D}_x$;
2. to provide a formal proof of the existence of a polynomial solution of a preassigned degree of the divided-difference equation of hypergeometric type (8), extending and generalising therefore—by means of specialisation and limiting situations on the lattice $x(s)$—similar results obtained for second-order differential, difference or q-difference equation of hypergeometric type.

2. Preliminary Results: Known and New Properties

Since the main result of this paper is based on the operators $\mathbb{D}_x$ and $\mathbb{S}_x$ which are defined by using the lattice $x(s)$, we will provide in this section some known and basic properties of $x(s)$, $\mathbb{D}_x$ and $\mathbb{S}_x$. We will also derive new properties such as the right and the "left" inverses of the operator $\mathbb{D}_x$, required in the next section.

2.1. Known Properties of the Lattice $x(s)$

Taking into account the notation

$$x_\mu(s) = x(s + \frac{\mu}{2}),$$

the non-uniform lattice $x(s)$ defined by Equation (7) satisfies

$$\begin{aligned} x(s+k) - x(s) &= \gamma_k \nabla x_{k+1}(s), && (10) \\ \frac{x(s+k)+x(s)}{2} &= \alpha_k x_k(s) + \beta_k, && (11) \end{aligned}$$

for $k = 0, 1, \dots,$ with

$$\alpha_0 = 1,\ \alpha_1 = \alpha,\ \beta_0 = 0,\ \beta_1 = \beta, \gamma_0 = 0,\ \gamma_1 = 1, \tag{12}$$

where the sequences (α_k), (β_k), (γ_k) satisfy the following relations:

$$\begin{aligned} \alpha_{k+1} - 2\,\alpha\,\alpha_k + \alpha_{k-1} &= 0, \\ \beta_{k+1} - 2\,\beta_k + \beta_{k-1} &= 2\,\beta\,\alpha_k, \\ \gamma_{k+1} - \gamma_{k-1} &= 2\,\alpha_k, \end{aligned} \tag{13}$$

and are given explicitly by [9,10]

$$\alpha_n = 1,\ \beta_n = \beta\, n^2,\ \gamma_n = n,\ \text{for } \alpha = 1, \tag{14}$$

and

$$\alpha_n = \frac{q^{\frac{n}{2}} + q^{-\frac{n}{2}}}{2},\ \beta_n = \frac{\beta(1-\alpha_n)}{1-\alpha},\ \gamma_n = \frac{q^{\frac{n}{2}} - q^{-\frac{n}{2}}}{q^{\frac{1}{2}} - q^{-\frac{1}{2}}},\ \text{for } \alpha = \frac{q^{\frac{1}{2}} + q^{-\frac{1}{2}}}{2}. \tag{15}$$

2.2. Known Properties of the Operators $\mathbb{D}_x$ and $\mathbb{S}_x$

The operators $\mathbb{S}_x$ and $\mathbb{D}_x$ fulfil the so-called Product rules I [13,14]:

$$\begin{aligned} \mathbb{D}_x\left(f(x(s))g(x(s))\right) &= \mathbb{S}_x f(x(s))\,\mathbb{D}_x g(x(s)) + \mathbb{D}_x f(x(s))\,\mathbb{S}_x g(x(s)), && (16) \\ \mathbb{S}_x\left(f(x(s))g(x(s))\right) &= U_2(x(s))\,\mathbb{D}_x f(x(s))\,\mathbb{D}_x g(x(s)) + \mathbb{S}_x f(x(s))\,\mathbb{S}_x g(x(s)), && (17) \end{aligned}$$

where U_2 is a polynomial of degree 2

$$U_2(x(s)) = (\alpha^2 - 1)\,x^2(s) + 2\,\beta\,(\alpha+1)\,x(s) + \eta_x, \tag{18}$$

and η_x is a constant given by [14]

$$\eta_x = \frac{x^2(0)+x^2(1)}{4\alpha^2} - \frac{(2\alpha^2-1)}{2\alpha^2}x(0)\,x(1) - \frac{\beta\,(\alpha+1)}{\alpha^2}(x(0)+x(1)) + \frac{\beta^2\,(\alpha+1)^2}{\alpha^2}. \tag{19}$$

The operators $\mathbb{D}_x$ and $\mathbb{S}_x$ also satisfy the so-called Product Rules II [13,14]:

$$\mathbb{D}_x\,\mathbb{S}_x = \alpha\,\mathbb{S}_x\,\mathbb{D}_x + U_1(x(s))\,\mathbb{D}_x^2;\ \ \mathbb{S}_x^2 = U_1(x(s))\,\mathbb{S}_x\,\mathbb{D}_x + \alpha\,U_2(x(s))\,\mathbb{D}_x^2 + \mathbb{I}, \tag{20}$$

where $\mathbb{I}$ is the identity operator $\mathbb{I}f(x) = f(x)$, and

$$U_1(x(s)) = (\alpha^2 - 1)\,x(s) + \beta\,(\alpha+1).$$

2.3. An Appropriate Basis for the Operators $\mathbb{D}_x$ and $\mathbb{S}_x$

Searching for a polynomial basis on which the action of the companion operators will give a linear combination of at most two elements of the basis, Foupouagnigni et al. proved in [14] that the polynomial F_n defined by

$$F_n(x(s)) = F_n(x(s), x(\varepsilon)), \text{ with } F_n(x(s), x(\varepsilon)) = \prod_{j=1}^{n} [x(s) - x_j(\varepsilon)], \tag{21}$$

where ε is the unique solution (provided that the lattice $x(s)$ is quadratic or q-quadratic: i.e., the constants c_j in (7) satisfy $c_1 c_2 \neq 0$ or $c_4 \neq 0$) in the variable t of the equation $x_1(t) = x(t)$, is the right basis for the operators $\mathbb{S}_x$ and $\mathbb{D}_x$ because it satisfies the following properties

$$\mathbb{D}_x F_n(x(s)) = \gamma_n F_{n-1}(x(s)), \tag{22}$$

$$\mathbb{S}_x F_n(x(s)) = \alpha_n F_n(x(s)) + \frac{\gamma_n}{2} \nabla x_{n+1}(z_x) F_{n-1}(x(s)), \tag{23}$$

where the constants α_n and γ_n are given in (14) and (15).

After reviewing some properties of the operators $\mathbb{D}_x$ and $\mathbb{S}_x$, we now state and prove the following proposition providing the left and right inverse of the operator $\mathbb{D}_x$, to be used in the next section to complete the proof of the main theorem of this paper.

Proposition 1.
Let $\mathbb{F}_x$ be a linear operator defined on the basis $(F_n)_n$ by

$$\mathbb{F}_x F_n = \frac{F_{n+1}}{\gamma_{n+1}}, \; n \geq 0, \; \mathbb{F}_x 0 := 0. \tag{24}$$

Then, $\mathbb{F}_x$ satisfies the following relations:

$$\mathbb{D}_x \mathbb{F}_x = \mathbb{I}, \; \mathbb{F}_x \mathbb{D}_x = \mathbb{I} - \delta_{x(\varepsilon)}, \tag{25}$$

where $\mathbb{I}$ is the identity operator and $\delta_{x(\varepsilon)}$ is the Dirac delta distribution defined by

$$\langle \delta_{x(\varepsilon)}, P \rangle = P(x(\varepsilon)), \; \forall P,$$

with ε is defined in (21).

Proof. For all positive integer n, F_n defined by (21) is a polynomial of degree exactly n. (F_n) is therefore a basis of $\mathbb{C}[x]$. Letting $f \in \mathbb{C}_n[x]$, there exist $f_0, \ldots, f_n \in \mathbb{C}[x]$ such that

$$f(x(s)) = \sum_{j=0}^{n} f_j F_j(x(s)).$$

We have

$$\langle \delta_{x(\varepsilon)}, f \rangle = f(x(\varepsilon)) = \sum_{j=1}^{n} f_j F_j(x(\varepsilon)) + f_0 = f_0. \tag{26}$$

$$\begin{aligned}
\mathbb{D}_x\mathbb{F}_x f(x(s)) &= \sum_{j=0}^{n} f_j \mathbb{D}_x\mathbb{F}_x F_j(x(s)) \\
&= \sum_{j=0}^{n} \frac{f_j \mathbb{D}_x F_{j+1}(x(s))}{\gamma_{j+1}} \\
&= \sum_{j=0}^{n} f_j F_j(x(s)) = f(x(s)).
\end{aligned}$$

Hence, the first part of Relation (24) holds. Using (26), we have

$$\begin{aligned}
\mathbb{F}_x\mathbb{D}_x f(x(s)) + \langle \delta_\varepsilon, f\rangle &= \mathbb{F}_x\mathbb{D}_x f(x(s)) + f_0 \\
&= \sum_{j=0}^{n} f_j \mathbb{F}_x\mathbb{D}_x F_j(x(s)) + f_0 \\
&= \sum_{j=1}^{n} f_j \gamma_j \mathbb{F}_x F_{j-1}(x(s)) + f_0 \\
&= \sum_{j=1}^{n} f_j F_j(x(s)) + f_0 \\
&= f(x(s)).
\end{aligned}$$

The second part of (25) is therefore satisfied. □

3. Existence of the Polynomial Solution of the Divided-Difference Equation of the Hypergeometric Type

Having stated and proved required properties of the operators $\mathbb{D}_x$ and $\mathbb{S}_x$, we will now state and prove the main theorem of this paper.

Theorem 1. *Let n be a nonnegative integer, ψ and ϕ be two polynomials of degree 1 and at most 2, respectively, such that*

$$\forall k \in \mathbb{N},\ \eta_k := \phi_2\,\gamma_k + \psi_1\,\alpha_k \neq 0. \tag{27}$$

Then, the divided-difference equation

$$\phi(x(s))\,\mathbb{D}_x^2 y(x(s)) + \psi(x(s))\,\mathbb{S}_x\mathbb{D}_x y(x(s)) + \lambda_{n,0}\,y(x(s)) = 0, \tag{28}$$

with

$$\lambda_{n,0} = -\gamma_n\,(\phi_2\,\gamma_{n-1} + \psi_1\,\alpha_{n-1}) = -\gamma_n\eta_{n-1}, \tag{29}$$

where ψ_1 and ϕ_2 are leading coefficients of polynomials ψ and ϕ respectively, has a polynomial solution of exactly n degree.

The proof of Theorem 1 will be organized as follows: we split the proof in five lemmas which we first state, prove, and then put these lemmas together in combination with Proposition 1 to deduce the proof of this theorem.

Lemma 1.
If the function y_0 is a solution of (28), then the function $y_1 = \mathbb{D}_x\,y_0$ satisfies

$$\phi^{[1]}(x(s))\,\mathbb{D}_x^2 y(x(s)) + \psi^{[1]}(x(s))\,\mathbb{S}_x\mathbb{D}_x y(x(s)) + \lambda_{n,1}\,y(x(s)) = 0, \tag{30}$$

where

$$\begin{cases} \phi^{[1]}(x(s)) = \mathbb{S}_x\phi(x(s)) + \alpha\, U_2(x(s))\, \mathbb{D}_x\psi + U_1(x(s))\, \mathbb{S}_x\psi(x(s)), \\ \psi^{[1]}(x(s)) = \mathbb{D}_x\phi(x(s)) + U_1(x(s))\, \mathbb{D}_x\psi + \alpha\, \mathbb{S}_x\psi(x(s)), \\ \lambda_{n,1} = \lambda_{n,0} + \mathbb{D}_x\psi. \end{cases} \tag{31}$$

Proof. Assume that y_0 satisfies (28). Applying the operator $\mathbb{D}_x$ to (28) in which y is replaced by y_0 and using the product rule I in (16) and (17), we obtain

$$\mathbb{D}_x\phi(x(s))\, \mathbb{S}_x\mathbb{D}_x^2 y_0(x(s)) + \mathbb{S}_x\phi(x(s))\, \mathbb{D}_x^3 y_0(x(s)) + \mathbb{D}_x\psi(x(s))\, \mathbb{S}_x^2\mathbb{D}_x y_0(x(s))$$
$$+\mathbb{S}_x\psi(x(s))\, \mathbb{D}_x\mathbb{S}_x\mathbb{D}_x y_0(x(s)) + \lambda_{n,0}\, \mathbb{D}_x y_0(x(s)) = 0.$$

Using the product rules II in (20) to replace $\mathbb{S}_x^2$ and $\mathbb{D}_x\mathbb{S}_x$ in the previous equation, we have

$$\phi^{[1]}(x(s))\, \mathbb{D}_x^3 y_0(x(s)) + \psi^{[1]}(x(s))\, \mathbb{S}_x\mathbb{D}_x^2 y_0(x(s)) + \lambda_{n,1}\, \mathbb{D}_x y_0(x(s)) = 0,$$

where $\phi^{[1]}(x(s))$, $\psi^{[1]}(x(s))$ and $\lambda_{n,1}$ are defined by (31). Therefore, $y_1 = \mathbb{D}_x\, y_0$ is a solution of Equation (30). □

Lemma 2.
If the function y_0 is a solution of (28), then the function $y_k = \mathbb{D}_x^k\, y_0$ is a solution of the equation

$$\phi^{[k]}(x(s))\, \mathbb{D}_x^2 y(x(s)) + \psi^{[k]}(x(s))\, \mathbb{S}_x\mathbb{D}_x y(x(s)) + \lambda_{n,k}\, y(x(s)) = 0, \tag{32}$$

where the polynomials $\phi^{[k]}$, $\psi^{[k]}$ and the constant $\lambda_{n,k}$ satisfy

$$\begin{cases} \phi^{[k+1]}(x(s)) = \mathbb{S}_x\phi^{[k]}(x(s)) + \alpha\, U_2(x(s))\, \mathbb{D}_x\psi^{[k]} + U_1(x(s))\, \mathbb{S}_x\psi^{[k]}(x(s)), \\ \psi^{[k+1]}(x(s)) = \mathbb{D}_x\phi^{[k]}(x(s)) + U_1(x(s))\, \mathbb{D}_x\psi^{[k]} + \alpha\, \mathbb{S}_x\psi^{[k]}(x(s)), \\ \lambda_{n,k+1} = \lambda_{n,k} + \mathbb{D}_x\psi^{[k]}, \end{cases} \tag{33}$$

with the following initial values: $\phi^{[0]} := \phi$, $\psi^{[0]} := \psi$.

Proof. Lemma 1 assures the validity of the result for $k = 1$.

Let k be a positive integer. Assume that y_k is solution of Equation (32). Applying the operator $\mathbb{D}_x$ to (32) in which y is replaced by y_k and using the Product Rules I, we obtain

$$\mathbb{D}_x\phi^{[k]}(x(s))\, \mathbb{S}_x\mathbb{D}_x^2 y_k(x(s)) + \mathbb{S}_x\phi^{[k]}(x(s))\, \mathbb{D}_x^3 y_k(x(s)) + \mathbb{D}_x\psi^{[k]}(x(s))\, \mathbb{S}_x^2\mathbb{D}_x y_k(x(s))$$
$$+\mathbb{S}_x\psi^{[k]}(x(s))\, \mathbb{D}_x\mathbb{S}_x\mathbb{D}_x y_k(x(s)) + \lambda_{n,k}\, \mathbb{D}_x y_k(x(s)) = 0.$$

Using the products rule II to replace $\mathbb{S}_x^2$ and $\mathbb{D}_x\mathbb{S}_x$ in the previous equation, we have

$$\phi^{[k+1]}(x(s))\, \mathbb{D}_x^3 y_k(x(s)) + \psi^{[k+1]}(x(s))\, \mathbb{S}_x\mathbb{D}_x^2 y_k(x(s)) + \lambda_{n,k+1}\, \mathbb{D}_x y_k(x(s)) = 0,$$

where $\phi^{[k+1]}(x(s))$, $\psi^{[k+1]}(x(s))$ and $\lambda_{n,k+1}$ are defined by (33). Thus, $y_{k+1} = \mathbb{D}_x\, y_k$ satisfies

$$\phi^{[k+1]}(x(s))\, \mathbb{D}_x^2 y_{k+1}(x(s)) + \psi^{[k+1]}(x(s))\, \mathbb{S}_x\mathbb{D}_x y_{k+1}(x(s)) + \lambda_{n,k+1}\, y_{k+1}(x(s)) = 0.$$

□

Lemma 3.
If a given function y_1 satisfies (30) with $\lambda_{n,0} \neq 0$, then there exists a function y_0 satisfying (28) such that

$$y_1 = \mathbb{D}_x y_0. \tag{34}$$

Proof. Let y_1 be a solution of (30) with $\lambda_{n,0} \neq 0$. If there would exist a solution v_0 of (28) such that $y_1 = \mathbb{D}_x v_0$, then from (28) we can express v_0 as:

$$v_0(x(s)) = -\frac{1}{\lambda_{n,0}} \left[\phi(x(s))\, \mathbb{D}_x y_1(x(s)) + \psi(x(s))\, \mathbb{S}_x y_1(x(s))\right]. \tag{35}$$

Now, it remains to verify that the function v_0 defined in terms of y_1 by (35) satisfies Equation (28) with

$$\mathbb{D}_x v_0 = y_1. \tag{36}$$

By applying $\mathbb{D}_x$ to (35) and using product rules I, II and the fact that y_1 is solution of (30), we get

$$\begin{aligned} -\lambda_{n,0}\, \mathbb{D}_x v_0(x(s)) &= \mathbb{D}_x \left[\phi(x(s))\, \mathbb{D}_x y_1(x(s)) + \psi(x(s))\, \mathbb{S}_x y_1(x(s))\right] \\ &= \phi^{[1]}(x(s))\, \mathbb{D}_x^2 y_1(x(s)) + \psi^{[1]}(x(s))\, \mathbb{S}_x \mathbb{D}_x y_1(x(s)) + (\lambda_{n,1} - \lambda_{n,0})\, y_1(x(s)) \\ &= -\lambda_{n,0}\, y_1(x(s)). \end{aligned}$$

Therefore, $\mathbb{D}_x v_0 = y_1$ since $\lambda_{n,0} \neq 0$.

We prove that v_0 is solution of (28) by replacing y_1 in the Equation (35) by $\mathbb{D}_x v_0$. □

Lemma 4.
For any positive integer n, the coefficients $\lambda_{n,k}$ defined by relation (33) satisfy

$$\begin{aligned} \lambda_{n,k} &= \lambda_{n,0} - \lambda_{k,0},\ 0 \leq k \leq n, && (37) \\ \lambda_{n,k} &\neq 0, \qquad \text{for } 0 \leq k \leq n-1, \text{ and } (n,k) \neq (0,0), && (38) \end{aligned}$$

where

$$\lambda_{n,0} = -\gamma_n(\phi_2 \gamma_{n-1} + \psi_1 \alpha_{n-1}).$$

Proof. If we denote by $\phi^{[k]}(x(s)) = \phi_2^{[k]}\, F_2(x(s)) + \phi_1^{[k]}\, F_1(x(s)) + \phi_0^{[k]}$ and $\psi^{[k]}(x(s)) = \psi_1^{[k]}\, F_1(x(s)) + \psi_0^{[k]}$, then from (33), we have the following system of recurrence equation

$$\left\{ \begin{aligned} \phi_2^{[k+1]} &= \alpha_2\, \phi_2^{[k]} + \alpha\, \gamma_1(\alpha^2 - 1)\, \psi_1^{[k]} + \alpha_1\, (\alpha^2 - 1)\, \psi_1^{[k]}, \\ \psi_1^{[k+1]} &= \gamma_2\, \phi_2^{[k]} + (\alpha^2 - 1)\, \gamma_1\, \psi_1^{[k]} + \alpha\, \alpha_1\, \psi_1^{[k]}, \\ \lambda_{n,k+1} &= \lambda_{n,k} + \psi_1^{[k]}. \end{aligned} \right.$$

Using relations

$$\alpha_2 = 2\,\alpha^2 - 1,\ \gamma_2 = 2\,\alpha,$$

derived from Equations (12) and (13), the previous system of equations becomes

$$\left\{ \begin{aligned} \phi_2^{[k+1]} &= (2\,\alpha^2 - 1)\, \phi_2^{[k]} + 2\,\alpha\, (\alpha^2 - 1)\, \psi_1^{[k]}, \\ \psi_1^{[k+1]} &= 2\,\alpha\, \phi_2^{[k]} + (2\,\alpha^2 - 1)\, \psi_1^{[k]}, \\ \lambda_{n,k+1} &= \lambda_{n,k} + \psi_1^{[k]}. \end{aligned} \right.$$

Solving this system of recurrence equations with the initial values $\phi_2^{[0]} = \phi_2$, $\psi_1^{[0]} = \psi_1$, we obtain for the q-quadratic lattice

$$\lambda_{n,k} = \frac{(q^k - q^n)}{(q-1)^2 q^k q^n} \left[\sqrt{q}(q^k q^n - q)\phi_2 + \frac{1}{2}(q-1)(q + q^k q^n)\psi_1\right]. \tag{39}$$

Using the definition of $\lambda_{n,0}$ which of course coincides with the one of $\lambda_{n,k}$ for $k = 0$, we derive (37) from (39).

Solving the following the equation

$$\lambda_{n,k} = 0,$$

in terms of the unknown k keeping in mind (27), gives a unique solution $k = n$. Thus, relation (38) is satisfied. It can easily be proved in the same way that relation (38) is satisfied for the quadratic lattice. □

Lemma 5.
Let n be a fixed positive integer and let k be an integer such that $0 \leq k \leq n$. Then, if y_k is a solution of Equation (32), then there exists y_0 solution of Equation (28) such that

$$y_k = \mathbb{D}_x^k\, y_0.$$

Proof. Let k be a nonnegative integer with $k \leq n$. Assume that y_k satisfies (32). Then, we obtain that there exists a function y_{k-1} solution of the equation obtained by replacing k in (32) by $k-1$, namely,

$$\phi^{[k-1]}(x(s))\, \mathbb{D}_x^2 y(x(s)) + \psi^{[k-1]}(x(s))\, \mathbb{S}_x \mathbb{D}_x y(x(s)) + \lambda_{n,k-1}\, y(x(s)) = 0, \tag{40}$$

such that

$$y_k = \mathbb{D}_x\, y_{k-1}.$$

This is achieved using the fact that $\lambda_{n,k-1} \neq 0$ thanks to Lemma 4, and also using Lemma 3 but with the functions y_0 and y_1 replaced, respectively, by the functions y_{k-1} and y_k while Equation (28) is replaced by Equation (40). In addition, Equation (30) is replaced by the Equation (32). □

The proof is completed by repeating the same process for $y_{k-1}, y_{k-2}, \ldots, y_1$ and using Lemmas 3 and 4.

Proof of Theorem 1. Since, for $k = n$, $\lambda_{n,k} = \lambda_{n,n} = 0$ thanks to (37), Equation (32) admits a constant solution, namely $F_0(x(s)) = 1$. We therefore deduce from Lemma 5 that there exists a function v_0 solution of (28) such that

$$F_0(x(s)) = \mathbb{D}_x^n\, v_0(x(s)). \tag{41}$$

Next, we apply the operator $\mathbb{F}_x$ on both members of the previous equation and deduce by applying the second relation of Equation (25) of Proposition 1 that

$$\mathbb{F}_x\, F_0(x(s)) = \mathbb{F}_x \mathbb{D}_x\, \mathbb{D}_x^{n-1} v_0(x(s)) = \mathbb{D}_x^{n-1} v_0(x(s)) - \mathbb{D}_x^{n-1} v_0(x(s))\,|_{s=\varepsilon}\,.$$

Hence,

$$\mathbb{D}_x^{n-1} v_0(x(s)) = \mathbb{F}_x\, F_0(x(s)) + C_{n-1}\, F_0(x(s)),$$

where $C_{n-1} = \mathbb{D}_x^{n-1} v_0(x(s))\,|_{s=\varepsilon}$.

By applying again the operator $\mathbb{F}_x$ on both members of the previous equation and using the second relation of Equation (25), we get

$$\mathbb{D}_x^{n-2} v_0(x(s)) = \mathbb{F}_x^2\, F_0(x(s)) + C_{n-1}\, \mathbb{F}_x\, F_0(x(s)) + C_{n-2}\, F_0(x(s)),$$

where $C_{n-2} = \mathbb{D}_x^{n-2} v_0(x(s))\,|_{s=\varepsilon}$. Repeating the same process, we express v_0 as

$$v_0(x(s)) = \mathbb{F}_x^n\, F_0(x(s)) + \sum_{j=0}^{n-1} C_j\, \mathbb{F}_x^j\, F_0(x(s)) = \frac{F_n(x(s))}{\prod_{l=1}^{n} \gamma_l} + \sum_{j=0}^{n-1} C_j\, \mathbb{F}_x^j\, F_0(x(s)),$$

where $C_j = \mathbb{D}_x^j v_0(x(s))\,|_{s=\varepsilon}$. Therefore, $v_0(x(s))$ is a polynomial of degree exactly n in $x(s)$. □

4. Conclusions and Perspectives

In this work, we have derived the right and the "left" inverse of the operator $\mathbb{D}_x$ and used the properties of the inverse operators, as well as those of the operators $\mathbb{D}_x$ and $\mathbb{S}_x$, to provide a formal proof that the divided-difference equation of hypergeometric type (28) has a polynomial solution of degree exactly n.

The novelty of our work is the formal proof of the existence of this polynomial solution, confirming therefore the fact that, in [14], by solving divided-difference (8) on a case by case basis and using most appropriate polynomial basis for each case, we have obtained for each family of classical orthogonal polynomials on non-uniform lattice, a hypergeometric or q-hypergeometric solution which happens to be a polynomial because of the form of one of the upper parameters obtained in the hypergeometric (or q-hypergeometric) representation of the obtained solution.

Finding hypergeometric representation of the non polynomial solution of (8) is not obvious and this was obtained unexpectedly for the Askey–Wilson polynomials when the parameters fulfill $b = a\,q^{\frac{1}{2}}$, $d = a\,q^{\frac{1}{2}}$ [14] (see page 15, Equations (62) and (63)). The method developed here might help to understand when and why such a hypergeometric representation exists for non-polynomial solutions.

As an additional potential application of our paper, the right and the "left" inverse of the operator $\mathbb{D}_x$ are likely to play important role in the study of the properties of orthogonal polynomials on the non-uniform latices, and on the search of the solutions of divided-difference equations on non-uniform lattices, as well as on the hypergeometric representation (when they exist) of the second-solution—non polynomial solution—of Equation (28).

Author Contributions: These authors contributed equally to this work, in terms of investigation, conceptualization and methodology.

Funding: The Research of the first author was partially supported by the AIMS-Cameroon Research Allowance 2018–2019.

Acknowledgments: We would like to thank Gabriel Nguetseng from the Mathematics Department of the University of Yaounde I, for asking the right question whose answer led to this result. In addition, we would like to thank the anonymous reviewers whose comments and suggestions helped to improve the quality of our work.

Conflicts of Interest: The authors declare no conflict of interest.

References

1. Nikiforov, A.F.; Uvarov, V.B. *Special Functions of Mathematical Physics*; Birkhäuser: Basel, Switzerland; Boston, MA, USA, 1988.
2. Bochner, S. Über Sturm-Liouvillesche Polynomsysteme. *Math. Z.* **1929**, *29*, 730–736. [CrossRef]
3. Nikiforov, A.F.; Suslov, S.K.; Uvarov, V.B. *Classical Orthogonal Polynomials of a Discrete Variable*; Springer Series in Computational Physics; Springer: Berlin, Germany, 1991.
4. Lancaster, O.E. Orthogonal polynomials defined by difference equations. *Am. J. Math.* **1941**, *63*, 185–207. [CrossRef]
5. Koekoek, R.; Lesky, P.A.; Swarttouw, R.F. *Hypergeometric Orthogonal Polynomials and Their q-Analogues*; Springer Monographs in Mathematics; Springe: Berlin, Germany, 2010.
6. Lesky, P. Über Polynomsystem, die Sturm-Liouvilleschen Diefferenzengleichungen gegen. *Math. Z.* **1982**, *78*, 656–663.
7. Lesky, P. Zweigliedrige Rekursionen für die Koeffizienten von polynomlösungen Sturm-Liouvillescher q-Differezengleichungen. *Z. Angew. Math. Mech.* **1994**, *74*, 497–500. [CrossRef]
8. Hahn, W. Über Orthogonalpolynome, die q-Differenzengleichungen genugen. *Math. Nachr.* **1949**, *2*, 4–34. [CrossRef]
9. Suslov, S.K. The theory of difference analogues of special functions of hypergeometric type. *Russ. Math. Surv.* **1989**, *44*, 227–278. [CrossRef]
10. Atakishiyev, N.M.; Rahman, M.; Suslov, S.K. On Classical Orthogonal Polynomials. *Constr. Approx.* **1995**, *11*, 181–226. [CrossRef]

11. Costas-Santos, R.S.; Marcellán, F. q-Classical Orthogonal Polynomials: A General Difference Calculus Approach. *Acta Appl. Math.* **2010**, *111*, 107–128. [CrossRef]
12. Foupouagnigni, M. On difference equations for orthogonal polynomials on non-uniform lattices. *J. Differ. Equ. Appl.* **2008**, *14*, 127–174. [CrossRef]
13. Foupouagnigni, M.; Kenfack-Nangho, M.; Mboutngam, S. Characterization theorem of classical orthogonal polynomials on non-uniform lattices: The functional approach. *Integral Transforms Spec. Funct.* **2011**, *22*, 739–758. [CrossRef]
14. Foupouagnigni, M.; Koepf, W.; Kenfack-Nangho, M.; Mboutngam, S. On Solutions of Holonomic Divided-Difference Equations on non-uniform Lattices. *Axioms* **2013**, *3*, 404–434. [CrossRef]

axioms

MDPI

Article

Solution Estimates for the Discrete Lyapunov Equation in a Hilbert Space and Applications to Difference Equations

Michael Gil'

Department of Mathematics, Ben Gurion University of the Negev, P.0. Box 653, Beer-Sheva 84105, Israel; gilmi@bezeqint.net

Received: 15 January 2019; Accepted: 2 February 2019; Published: 6 February 2019

Abstract: The paper is devoted to the discrete Lyapunov equation $X - A^*XA = C$, where A and C are given operators in a Hilbert space $\mathcal{H}$ and X should be found. We derive norm estimates for solutions of that equation in the case of unstable operator A, as well as refine the previously-published estimates for the equation with a stable operator. By the point estimates, we establish explicit conditions, under which a linear nonautonomous difference equation in $\mathcal{H}$ is dichotomic. In addition, we suggest a stability test for a class of nonlinear nonautonomous difference equations in $\mathcal{H}$. Our results are based on the norm estimates for powers and resolvents of non-self-adjoint operators.

Keywords: discrete Lyapunov equation; difference equations; Hilbert space; dichotomy; exponential stability

1. Introduction and Notations

Let $\mathcal{H}$ be a complex separable Hilbert space with a scalar product $(.,.)$, the norm $\|.\| = \sqrt{(.,.)}$, and unit operator $I = I_{\mathcal{H}}$. By $\mathcal{B}(\mathcal{H})$, we denote the set of all bounded linear operators in $\mathcal{H}$. In addition, Ω denotes the unit circle: $\Omega = \{z \in \mathbb{C} : |z| = 1\}$. An operator A is said to be Schur–Kohn stable, or simply stable, if its spectrum $\sigma(A)$ lies inside Ω. Otherwise, A will be called an unstable operator.

Consider the discrete Lyapunov equation:

$$X - A^*XA = C, \tag{1}$$

where $A, C \in \mathcal{B}(\mathcal{H})$ are given operators and X should be found. That equation arises in various applications, cf. [1]. Sharp norm estimates for solutions of (1) with Schur–Kohn stable finite dimensional and some classes of infinite dimensional operators have been derived in [2,3]. At the same time, to the best of our knowledge, norm estimates for solutions of (1) with unstable A have not been obtained in the available literature.

Our aim in the present paper is to establish sharp norm estimates for solutions of Equation (1) with an unstable operator A. In addition, we refine and complement estimates for (1) with stable operator coefficients from [2,3].

The point estimates enable us to suggest new dichotomy conditions for nonautonomous linear difference equations and explicit stability conditions for the nonautonomous nonlinear difference equations in a Hilbert space.

The dichotomy of various abstract difference equations has been investigated by many mathematicians, cf. [4] and [5–11] and the references therein. In particular, the main result of the paper [8] gives a decomposition of the dichotomy spectrum considering the upper dichotomy spectrum, lower dichotomy spectrum, and essential dichotomy spectrum. In addition, in [8], it is proven that the dichotomy spectrum is a disjoint union of closed intervals. In [9,11], an approach concerning the characterization of the exponential dichotomy of difference equations by means of an admissible pair of sequence Banach spaces has been developed. The paper [12] considers two general concepts of dichotomy for noninvertible and nonautonomous linear discrete-time systems in Banach spaces. These concepts use two types of dichotomy projection sequences and generalize some well-known dichotomy concepts.

Certainly, we could not survey here all the papers in which in the general situation the dichotomy conditions are formulated in terms of the original norm. We formulate the dichotomy conditions in terms of solutions of Lyapunov's equation. In appropriate situations, that fact enables us to derive upper and lower solution estimates. In addition, traditionally, the existence of dichotomy projections is assumed. We obtain the existence of these projections via perturbations of operators.

The stability theory for abstract nonautonomous difference equations has a long history, but mainly linear equations have been investigated, cf. [13–15] and the references therein. Regarding the stability of nonlinear autonomous difference equations in a Banach space, see [16]. The stability theory for nonlinear nonautonomous difference equations in a Banach space is developed considerably less than the one for linear and autonomous nonlinear equations. Here, we should point out the paper [17], in which the author studied the local exponential stability of difference equations in a Banach space with slowly-varying coefficients and nonlinear perturbations. Besides, he established the robustness of the exponential stability. Regarding other results of the stability of nonlinear nonautonomous difference equations in an infinite dimensional space, see for instance [2], Chapter 12.

In this paper, we investigate semilinear nonautonomous difference equations in a Hilbert space and do not require that the coefficients are slowly varying.

Introduce the notations. For an $A \in \mathcal{B}(\mathcal{H})$, $\sigma(A)$ is the spectrum; $r_s(A)$ is the (upper) spectral radius; $r_l(A) = \inf \{|s| : s \in \sigma(A)\}$ is the lower spectral radius; A^* is adjoint to A; $R_\lambda(A) = (A - \lambda I)^{-1}$ $(\lambda \notin \sigma(A))$ is the resolvent; $\|A\|_{\mathcal{B}(\mathcal{H})} = \|A\| := \sup_{h \in \mathcal{H}} \|Ah\|/\|h\|$; $A_I = \Im A = (A - A^*)/2i$;

$$1.7em(A, \lambda) := \inf_{s \in \sigma(A)} |\lambda - s| \quad (\lambda \in \mathbb{C}).$$

The Schatten–von Neumann ideal of compact operators A in $\mathcal{H}$ with the finite Schatten–von Neumann norm $N_p(A) := (\text{trace}\,(A^*A)^{p/2})^{1/p}$ $(1 \le p < \infty)$ is denoted by SN_p. In particular, SN_2 is the Hilbert–Schmidt ideal and $N_2(.)$ is the Hilbert–Schmidt norm.

2. Auxiliary Results

In the present section, we have collected norm estimates for powers and resolvents of some classes of operators and estimates for the powers of their inverses. They give us bounds for the solution of Equation (1).

2.1. Operators in Finite Dimensional Spaces

Let $\mathcal{H} = \mathbb{C}^n$ $(n < \infty)$ be the complex n-dimensional Euclidean space and $\mathbb{C}^{n \times n}$ be the set of complex $n \times n$ matrices. In this subsection, $A \in \mathbb{C}^{n \times n}$; $\lambda_k(A), k = 1, ..., n$, are the eigenvalues of A, counted with their multiplicities. Introduce the quantity (the departure from normality of A):

$$g(A) = [N_2^2(A) - \sum_{k=1}^{n} |\lambda_k(A)|^2\,]^{1/2}.$$

The following relations are checked in [3], Section 3.1:

$$g^2(A) \le N_2^2(A)(A) - |\text{trace } A^2| \text{ and } g^2(A) \le \frac{N_2(A - A^*)}{2} = 2N_2^2(A_I).$$

If A is a normal matrix: $AA^* = A^*A$, then $g(A) = 0$.

Due to Example 3.3 from [3]:

$$\|A^m\| \le \sum_{k=0}^{n-1} \frac{m! r_s^{m-k}(A) g^k(A)}{(m-k)!(k!)^{3/2}} \quad (m = 1, 2, ...). \tag{2}$$

Recall that $\frac{1}{(m-k)!} = 0$ if $k > m$. Inequality (2) is sharp. It is attained for a normal operator A, since $g(A) = 0$, $0^0 = 1$, and $\|A^m\| = r_s^m(A)$ in this case.

By Theorem 3.2 from [3]:

$$\|(A - \lambda I)^{-1}\| \le \sum_{k=0}^{n-1} \frac{g^k(A)}{(1.7em(A,\lambda))^{k+1}\sqrt{k!}} \quad (\lambda \notin \sigma(A)). \tag{3}$$

This inequality is also attained for a normal operator.

Now, let $r_l > 0$. Then, by Corollary 3.6 from [3],

$$\|A^{-m}\| \le \sum_{k=0}^{n-1} \frac{g^k(A^m)}{r_l^{mk}(A)(k!)^{1/2}} \quad (A \in \mathbb{C}^{n\times n};\ m = 1, 2, ...). \tag{4}$$

Inequality (4) is equality if A is a normal operator. In addition, by Theorem 3.3 of [3] for any invertible $A \in \mathbb{C}^{n\times n}$ and $1 \le p < \infty$, one has:

$$\|A^{-1} \det A\| \le \frac{N_p^{n-1}(A)}{(n-1)^{(n-1)/p}}$$

and:

$$\|A^{-1} \det A\| \le \|A\|^{n-1}.$$

Hence,

$$\|A^{-m}\| \le \frac{N_p^{n-1}(A^m)}{(n-1)^{(n-1)/p} |\det A|^m} \tag{5}$$

and:

$$\|A^{-m}\| \le \frac{\|A^m\|^{n-1}}{|\det A|^m}.$$

Now, (2) and (5) imply:

$$\|A^{-m}\| \le \frac{1}{|\det A|^m} \left(\sum_{k=0}^{n-1} \frac{m! r_s^{m-k}(A) g^k(A)}{(m-k)!(k!)^{3/2}} \right)^{n-1} \quad (m = 1, 2, ...). \tag{6}$$

2.2. Hilbert–Schmidt Operators

In the sequel, $\mathcal{H}$ is infinite dimensional. In this subsection, A is in SN_2 and:

$$g(A) = [N_2^2(A) - \sum_{k=1}^{\infty} |\lambda_k(A)|^2\,]^{1/2},$$

where $\lambda_k(A)$ $(k = 1, 2, ...)$ are the eigenvalues of $A \in \mathcal{B}(\mathcal{H})$, counted with their multiplicities and enumerated in the nonincreasing order of their absolute values.

Since:

$$\sum_{k=1}^{\infty} |\lambda_k(A)|^2 \ge |\sum_{k=1}^{\infty} \lambda_k^2(A)| = |\text{trace } A^2|,$$

one can write:

$$g^2(A) \le N_2^2(A) - |\text{trace } A^2|.$$

If A is a normal Hilbert–Schmidt operator, then $g(A) = 0$, since:

$$N_2^2(A) = \sum_{k=1}^{\infty} |\lambda_k(A)|^2$$

in this case. Moreover,

$$g^2(A) \leq \frac{N_2^2(A - A^*)}{2} = 2N_2^2(A_I), \tag{7}$$

cf. [3], Section 7.1. Due to Corollary 7.4 from [3], for any $A \in SN_2$, we have:

$$\|A^m\| \leq \sum_{k=0}^{m} \frac{m! r_s^{m-k}(A) g^k(A)}{(m-k)!(k!)^{3/2}} \quad (m = 1, 2, ...). \tag{8}$$

This inequality and Inequality (9) below are attained for a normal operator.

Furthermore, by Theorem 7.1 from [3], for any $A \in SN_2$, we have:

$$\|R_\lambda(A)\| \leq \sum_{k=0}^{\infty} \frac{g^k(A)}{(1.7em(A,\lambda))^{k+1}\sqrt{k!}} \quad (\lambda \notin \sigma(A)). \tag{9}$$

By the Schwarz inequality:

$$\left(\sum_{j=0}^{\infty} \frac{(cg(A))^j}{c^j \sqrt{j!} x^j} \right)^2 \leq \sum_{k=0}^{\infty} c^{2k} \sum_{j=0}^{\infty} \frac{g^{2j}(A)}{c^{2j} j! x^{2j}}$$

$$= \frac{1}{1-c^2} \exp\left[\frac{g^2(A)}{c^2 x^2}\right] \quad (x > 0, c \in (0,1)).$$

Taking $c^2 = 1/2$, from (9), we arrive at the inequality:

$$\|R_\lambda(A)\| \leq \frac{\sqrt{2}}{1.7em(A,\lambda)} \exp\left[\frac{g^2(A)}{(1.7em(A,\lambda))^2}\right] \quad (\lambda \notin \sigma(A)). \tag{10}$$

2.3. *Schatten–von Neumann Operators*

In this subsection, $A \in SN_{2p}$ for an integer $p \geq 1$. Making use of Theorems 7.2 and 7.3 from [3], we have:

$$\|R_\lambda(A)\| \leq \sum_{m=0}^{p-1} \sum_{k=0}^{\infty} \frac{(2N_{2p}(A))^{pk+m}}{(1.7em(A,\lambda))^{pk+m+1}\sqrt{k!}} \quad (\lambda \notin \sigma(A)) \tag{11}$$

and:

$$\|R_\lambda(A)\| \leq \sqrt{e} \sum_{m=0}^{p-1} \frac{(2N_{2p}(A))^m}{(1.7em(A,\lambda))^{m+1}} \, exp \left[\frac{(2N_{2p}(A))^{2p}}{2(1.7em(A,\lambda))^{2p}}\right] \quad (\lambda \notin \sigma(A)). \tag{12}$$

Since, the condition $A \in SN_{2p}$ implies $A - A^* \in SN_{2p}$, and one can use estimates for the resolvent presented in the next two subsections.

Furthermore, if $A \in SN_{2p}$, then $A^p \in SN_2$. For any $m = p\nu + i$ $(i = 1, ..., p-1; \nu = 1, 2, ...)$, we have:

$$\|A^m\| \leq \|A^i\| \|(A^p)^\nu\|.$$

Now, (8) implies:

$$\|A^{p\nu+i}\| \leq \|A^i\| \sum_{k=0}^{\nu} \frac{\nu! r_s^{p(\nu-k)}(A) g^k(A^p)}{(\nu-k)!(k!)^{3/2}} \quad (\nu = 1, 2, ...; \ i = 1, ..., p-1). \tag{13}$$

2.4. Noncompact Operators with Hilbert–Schmidt Hermitian Components

In this subsection, we suppose that:

$$A_I = (A - A^*)/(2i) \in SN_2. \tag{14}$$

To this end, introduce the quantity:

$$g_I(A) := \sqrt{2}\left[N_2^2(A_I) - \sum_{k=1}^{\infty}(\Im\,\lambda_k(A))^2\right]^{1/2}.$$

Obviously, $g_I(A) \le \sqrt{2}N_2(A_I)$. If A is normal, then $g_I(A) = 0$ by Lemma 9.3 of [3]. Due to Example 10.2 [3],

$$\|A^m\| \le \sum_{k=0}^{m} \frac{m! r_s^{m-k}(A) g_I^k(A)}{(m-k)!(k!)^{3/2}} \quad (m = 1, 2, \ldots). \tag{15}$$

Furthermore, by Theorem 9.1 from [3], under Condition (14), we have,

$$\|R_\lambda(A)\| \le \sum_{k=0}^{\infty} \frac{g_I^k(A)}{(1.7em(A,\lambda))^{k+1}\sqrt{k!}} \tag{16}$$

and:

$$\|R_\lambda(A)\| \le \frac{\sqrt{e}}{1.7em(A,\lambda)} \exp\left[\frac{g_I^2(A)}{2(1.7em(A,\lambda))^2}\right] \quad (\lambda \notin \sigma(A)). \tag{17}$$

Now, let $r_l > 0$. Then, by (16):

$$\|A^{-1}\| \le \sum_{k=0}^{\infty} \frac{g_I^k(A)}{r_l^{k+1}(A)(k!)^{1/2}}. \tag{18}$$

Similarly, by (17):

$$\|A^{-1}\| \le \frac{\sqrt{e}}{r_l(A)} \exp\left[\frac{g_I^2(A)}{2r_l^2(A)}\right]. \tag{19}$$

Let us point out an additional estimate for $\|A^{-m}\|$.

Lemma 1. *Let Condition (14) hold and A be invertible. Then:*

$$\|A^{-m}\| \le \sum_{k=0}^{m} \frac{m!(\|A^{-1}\|^2 N_2(A - A^*))^k}{2^{k/2} r_l^{m-k}(A)(m-k)!(k!)^{3/2}} \quad (m = 1, 2, \ldots). \tag{20}$$

Proof. Put $B = A^{-1}$. By (15):

$$\|B^m\| \le \sum_{k=0}^{m} \frac{m! r_s^{m-k}(B) g_I^k(B)}{(m-k)!(k!)^{3/2}} \quad (m = 1, 2, \ldots).$$

However,

$$N_2(B - B^*) = N_2(A^{-1} - (A^{-1})^*) = N_2(A^{-1}(A - (A)^*)(A^{-1})^*) \le \|A^{-1}\|^2 N_2(A - (A^{-1})^*).$$

Thus,

$$g_I(A^{-1}) \le \frac{1}{\sqrt{2}} N_2(A^{-1} - (A^{-1})^*) \le \frac{1}{\sqrt{2}}\|A^{-1}\|^2 N_2(A - (A^{-1})^*).$$

This proves the lemma. □

Note that $\|A^{-1}\|$ can be estimated by (18) and (19).

2.5. Noncompact Operators with Schatten–von Neumann Hermitian Components

In this subsection, it is assumed that:

$$A_I = (A - A^*)/2i \in SN_{2p} \text{ for an integer } p \geq 2. \tag{21}$$

By Theorem 9.5 of [3], for any quasinilpotent operator $V \in SN_p$, there is a constant b_p dependent on p only, such that $N_p(V + V^*) \leq b_p N_p(V - V^*)$. According to Lemma 9.5 from [3], $b_p \leq \frac{p}{2} e^{1/3}$. Put:

$$\tau_p(A) = (1 + b_{2p})(N_{2p}(A_I) + N_{2p}(D_I)).$$

Therefore,

$$\tau_p(A) \leq (1 + pe^{1/3})(N_{2p}(A_I) + N_{2p}(D_I)) \leq (1 + 2p)(N_{2p}(A_I) + N_{2p}(D_I)).$$

From the Weyl inequalities ([3], Lemma 8.7), we have $N_{2p}(D_I) \leq N_{2p}(A_I)$. Thus:

$$\tau_p(A) \leq 2(1 + 2p) N_{2p}(A_I). \tag{22}$$

If A has a real spectrum, then:

$$\tau_p(A) \leq (1 + 2p) N_{2p}(A_I). \tag{23}$$

We need the following result ([3], Theorem 9.5).

Theorem 1. *Let Condition (21) hold. Then:*

$$\|R_\lambda(A)\| \leq \sum_{m=0}^{p-1} \sum_{k=0}^{\infty} \frac{\tau_p^{pk+m}(A)}{(1.7em(A,\lambda))^{pk+m+1} \sqrt{k!}} \tag{24}$$

and:

$$\|R_\lambda(A)\| \leq \sqrt{e} \sum_{m=0}^{p-1} \frac{\tau_p^m(A)}{(1.7em(A,\lambda))^{m+1}} \exp\left[\frac{\tau_p^{2p}(A)}{2(1.7em(A,\lambda))^{2p}}\right] \quad (\lambda \notin \sigma(A)). \tag{25}$$

If A is self-adjoint, then Inequality (24) takes the form $\|R_\lambda(A)\| = \frac{1}{1.7em(A,\lambda)}$.

2.6. Applications of the Integral Representation for Powers

For an arbitrary $A \in \mathcal{B}(\mathcal{H})$ and an $r_0 > r_s(A)$, we have:

$$A^m = -\frac{1}{2\pi i} \int_{|\lambda| = r_0} \lambda^m R_\lambda(A) d\lambda \quad (m = 1, 2, \ldots). \tag{26}$$

Let there be a monotonically-increasing nonnegative continuous function $F(x)$ $(x \geq 0)$, such that $F(0) = 0$, $F(\infty) = \infty$, and:

$$\|(\lambda I - A)^{-1}\| \leq F(1/1.7em(A,\lambda)) \quad (\lambda \notin \sigma(A)). \tag{27}$$

Obviously, $1.7em(A,z) \geq \epsilon = r_0 - r_s(A)$ $(|z| = r_0)$ by (26):

$$\|A^m\| \leq r_0^{m+1} F(1/\epsilon) \quad (r_0 = r_s(A) + \epsilon;\ m = 1, 2, \ldots).$$

All the above estimates for the resolvent satisfy Condition (27). For example, under Condition (14), due to (17), we have (27) with:

$$F(x) = F_2(x) := x\sqrt{e}\exp\,[\,\frac{x^2 g_I^2(A)}{2}]. \tag{28}$$

Under Condition (21), due to (25), we have (27) with:

$$F(x) = \hat{F}_p(x) := \sqrt{e}\sum_{m=0}^{p-1} x^{m+1}\tau_p^m(A)\exp\left[\frac{1}{2}x^{2p}\tau^{2p}(A)\right]. \tag{29}$$

Similarly, (24) can be taken.

Furthermore, let A be invertible. With a constant $s_l > 1/r_l(A) = r_s(A^{-1})$, we can write:

$$A^{-m} = -\frac{1}{2\pi i}\int_{|\lambda|=s_l}\lambda^m R_\lambda(A^{-1})d\lambda.$$

Hence:

$$A^{-m-1} = -\frac{1}{2\pi i}\int_{|\lambda|=s_l}\lambda^m A^{-1}R_\lambda(A^{-1})d\lambda = \frac{1}{2\pi i}\int_{|\lambda|=s_l}\lambda^m(A\lambda - I)^{-1}d\lambda.$$

Under Condition (27), we get $\|I - \lambda A\| \le F(1/1.7em(\lambda A, 1))$, and therefore,

$$\|(I - \lambda A)^{-1}\| \le F(1/1.7em(\lambda A, 1)) \quad (\frac{1}{\lambda} \notin \sigma(A)). \tag{30}$$

With $s_l = \epsilon + 1/r_l(A)$, we have $1.7em(\lambda A, 1) \ge r_l(A)\epsilon$ $(|\lambda| = s_l)$. Therefore, the inequalities:

$$\|A^{-m}\| \le s_l^{m-1}\frac{1}{2\pi}\int_{|\lambda|=s_l}\|(I - \lambda A)^{-1}\||d\lambda| \le s_l^m \sup_{|\lambda|=s_l}\|(I - \lambda A)^{-1}\|$$

hold and (30) implies:

$$\|A^{-m}\| \le (\epsilon + \frac{1}{r_l(A)})^m F(1/(r_l(A)\epsilon) \quad (\epsilon > 0;\ m = 1, 2, ...). \tag{31}$$

Note that the analogous results can be found in the book [18] (see the Exercises at the end of Chapter 1).

3. The Discrete Lyapunov Equation with a Stable Operator Coefficient

Theorem 2. *Let $A \in \mathcal{B}(\mathcal{H})$ and $r_s(A) < 1$. Then, for any $C \in \mathcal{B}(\mathcal{H})$, there exists a linear operator $X = X(A, C)$, such that:*

$$X - A^*XA = C. \tag{32}$$

Moreover,

$$X(A, C) = \sum_{k=0}^{\infty}(A^*)^k C A^k. \tag{33}$$

and:

$$X(A, C) = \frac{1}{2\pi}\int_0^{2\pi}(Ie^{-i\omega} - A^*)^{-1}C(Ie^{i\omega} - A)^{-1}d\omega. \tag{34}$$

Thus, if C is strongly positive definite, then $X(A, C)$ is strongly positive definite.

For the proof of this theorem and the next lemma, for instance see [1] ([2], Section 7.1).

Lemma 2. *If Equation (32) with $C = C^* > 0$ has a solution $X(A, C) > 0$, then the spectrum of A is located inside the unit disk.*

Due to Representations (33) and (34), we have:

$$\|X(A, C)\| \le \|C\| \sum_{k=0}^{\infty} \|A^k\|^2 \tag{35}$$

and:

$$\|X(A, C)\| \le \frac{\|C\|}{2\pi} \int_0^{2\pi} \|(e^{it}I - A)^{-1}\|^2 dt,$$

respectively. From the latter inequality, it follows

$$\|X(A, C)\| \le \|C\| \sup_{|z|=1} \|(zI - A)^{-1}\|^2 \tag{36}$$

Similar results can be found in the Exercises of Chapter 1 from [18].

Again, assume that Condition (27) holds. Then, for $|z| = 1$, $1.7em(A, z) \ge 1 - r_s(A)$; therefore, $\|(Iz - A)^{-1}\| \le F(1/(1 - r_s(A)))$. Now, (36) implies:

$$\|X(A, C)\| \le \|C\| F^2 \left(\frac{1}{1 - r_s(A)} \right). \tag{37}$$

If A is normal, then $\|A^k\| = r_s^k(A)$, and (35) yields:

$$\|X(A, C)\| \le \|C\| \frac{1}{1 - r_s^2(A)}. \tag{38}$$

Example 1. *Let $A \in \mathbb{C}^{n \times n}$. Then, (2) and (35) yield:*

$$\|X(A, C)\| \le \|C\| \sum_{m=0}^{\infty} \left(\sum_{k=0}^{n-1} \frac{m! r_s^{m-k}(A) g^k(A)}{(m-k)!(k!)^{3/2}} \right)^2.$$

Note that if A is normal, then $g(A) = 0$, and Example 3.3 gives us Inequality (38). Let us point to the more compact, but less sharper estimate for $X(A, C)$. Making use of (3) and (37), we can assert that:

$$\|X(A, C)\| \le \|C\| \left(\sum_{k=0}^{n-1} \frac{g^k(A)}{\sqrt{k!}(1 - r_s(A))^{k+1}} \right)^2 \quad (A \in \mathbb{C}^{n \times n}). \tag{39}$$

Example 2. *Let $A \in SN_2$. Then, (8) and (35) yield:*

$$\|X(A, C)\| \le \|C\| \sum_{m=0}^{\infty} \left(\sum_{k=0}^{m} \frac{m! r_s^{m-k}(A) g^k(A)}{(m-k)!(k!)^{3/2}} \right)^2.$$

If A is normal, then this example gives us Inequality (38). Furthermore, (37) and (10) imply:

$$\|X(A, C)\| \le \frac{2\|C\|}{(1 - r_s(A))^2} \exp\left[\frac{2g^2(A)}{(1 - r_s(A))^2}\right] \quad (A \in SN_2).$$

Example 3. *Assume that $A_I \in SN_2$. Then, (4) and (35) yield:*

$$\|X(A,C)\| \le \|C\| \sum_{m=0}^{\infty} \left(\sum_{k=0}^{m} \frac{m! r_s^{m-k}(A) g_I^k(A)}{(m-k)!(k!)^{3/2}} \right)^2 .$$

If A is normal, hence we get (38). Inequality (37) along with (16) and (17) give us the inequalities:

$$\|X(A,C)\| \le \|C\| \left(\sum_{j=0}^{\infty} \frac{g_I^j(A)}{\sqrt{j!}(1-r_s(A))^{j+1}} \right)^2$$

and:

$$\|X(A,C)\| \le \|C\| \frac{e}{(1-r_s(A))^2} \exp\,[\frac{g_I^2(A)}{(1-r_s(A))^2}] \ \ (A_I \in SN_2),$$

respectively. For a self-adjoint operator S, we write $S \ge 0$ ($S > 0$) if it is positive definite (strongly positive definite). The inequalities $S \le 0$ and $S < 0$ have a similar sense.

Note that (33) gives a lower bound for $X(A,C)$ with $C = C^* \ge 0$. Indeed,

$$(X(A,C)x,x) \ge \sum_{k=0}^{\infty} (CA^k x, A^k x) \ge r_l(C) \sum_{k=0}^{\infty} (A^k x, A^k x)$$

$$\ge r_l(C) \sum_{k=0}^{\infty} r_l((A^*)^k A^k)(x,x) \ \ (x \in \mathcal{H}). \tag{40}$$

If C is noninvertible, then $r_l(C) = 0$, and:

$$r_l(C) = \frac{1}{\|C^{-1}\|} \text{ and } r_l((A^k)^* A^k) = \frac{1}{\|A^{-k}\|^2},$$

if the corresponding operator is invertible. Therefore, we arrive at

Lemma 3. *Let $X(A,I) = X(A)$ be a solution of (32) with $C = I$ and $r_s(A) < 1$. Then:*

$$\|X^{-1}(A)\| \le (\sum_{k=0}^{\infty} \frac{1}{\|A^{-k}\|^2})^{-1} \textit{ if } A \textit{ is invertible} .$$

Therefore, $\|X^{-1}(A)\| \le 1$ in the general case.

4. Discrete Lyapunov's Equation with $r_l(A) > 1$

Theorem 3. *If:*

$$r_l(A) > 1, \tag{41}$$

then for any $C \in \mathcal{B}(\mathcal{H})$, there exists a linear operator $X = X(A,C)$, satisfying (32). Moreover,

$$X(A,C) = -\sum_{k=0}^{\infty} (A^*)^{-k-1} C A^{-k-1} \tag{42}$$

and:

$$X(A,C) = -\frac{1}{2\pi} \int_0^{2\pi} (Ie^{-i\omega} - A^*)^{-1} C (Ie^{i\omega} - A)^{-1} d\omega. \tag{43}$$

Proof. Rewrite (32) as the equation:

$$X - (A^{-1})^* X A^{-1} = -(A^{-1})^* C A^{-1}. \tag{44}$$

Due to (41), $r_s(A^{-1}) < 1$; from (33), we obtain (42), and from (34), it follows:

$$X(A,C) = -\frac{1}{2\pi}\int_0^{2\pi}(Ie^{-i\omega} - (A^*)^{-1})^{-1}(A^*)^{-1}CA^{-1}(Ie^{i\omega} - A^{-1})^{-1}d\omega$$
$$= -\frac{1}{2\pi}\int_0^{2\pi}(e^{-i\omega}A^* - I)^{-1}C(e^{i\omega}A - I)^{-1}d\omega$$
$$= -\frac{1}{2\pi}\int_0^{2\pi}(A^* - e^{-i\omega}I)^{-1}C(A - e^{i\omega}I)^{-1}d\omega,$$

as claimed. □

Lemma 4. *If Equation (32) with $C = C^* > 0$ has a solution $X < 0$, then the spectrum of A is located outside the unit disk.*

Proof. According to Lemma 3.2 and (43), one has $r_s(A^{-1}) < 1$, since $-X > 0$ and $(A^{-1})^*CA^{-1} > 0$. Now, the required result follows from the equality $r_l(A) = 1/r_s(A^{-1})$. □

Due Representations (41) and (42), we have:

$$\|X(A,C)\| \leq \|C\| \sum_{k=0}^{\infty} \|A^{-k-1}\|^2 \tag{45}$$

and:

$$\|X(A,C)\| \leq \frac{\|C\|}{2\pi}\int_0^{2\pi} \|(e^{it}I - A)^{-1}\|^2 dt, \tag{46}$$

respectively. From the latter inequality, it follows:

$$\|X(A,C)\| \leq \|C\| \sup_{|z|=1} \|(zI - A)^{-1}\|^2. \tag{47}$$

Let Condition (27) hold. If $|z| = 1$, then $1.7em(A,z) \geq r_l(A) - 1$, and therefore, $\|(Iz - A)^{-1}\| \leq F(1/(r_l(A) - 1))$. Hence, (43) implies:

$$\|X(A,C)\| \leq \|C\|F^2\left(\frac{1}{r_l(A) - 1}\right). \tag{48}$$

Now, we can apply estimates for resolvents from Section 2. Moreover, from (42) with positive definite C and $Y = -X(A,C)$, we get:

$$(Yx,x) \geq r_l(C)\sum_{k=0}^{\infty} r_l((A^*)^{-k-1}A^{-k-1})(x,x) \quad (x \in \mathcal{H}).$$

Hence:

$$(Yx,x) = -(X(A,C)x,x) \geq r_l(C)\sum_{k=1}^{\infty}\frac{1}{\|A^k\|^2}(x,x) \quad (x \in \mathcal{H}). \tag{49}$$

Now, we can apply estimates for powers of operators from Section 2. From (49), it follows:

Lemma 5. *Let $X(A,I) = X(A)$ be a solution of (32) with $C = I$ and $r_l(A) > 1$. Then:*

$$\|X^{-1}(A)\| \leq \left(\sum_{k=1}^{\infty}\frac{1}{\|A^k\|^2}\right)^{-1}.$$

5. Operators with Dichotomic Spectra

In this section, it is assumed that $\sigma(A)$ is dichotomic. Namely,

$$\sigma(A) = \sigma_{\text{ins}} \cup \sigma_{\text{out}}, \tag{50}$$

where σ_{ins} and σ_{out} are nonempty nonintersecting sets lying inside and outside Ω, respectively: $\sup|\sigma_{\text{ins}}| < 1$ and $\inf|\sigma_{\text{out}}| > 1$. Put:

$$P = \frac{1}{2\pi i}\int_{\Omega}(zI - A)^{-1}dz. \tag{51}$$

Therefore, P is the Riesz projection of A, such that $\sigma(AP) = \sigma_{\text{ins}}$ and $\sigma(A(I-P)) = \sigma_{\text{out}}$. We have $A = A_{\text{ins}} + A_{\text{out}}$, where $A_{\text{ins}} = AP = PA$, $A_{\text{out}} = (I-P)A = A(I-P)$.

In the sequel, $(\lambda P - A_{\text{ins}})^{-1}$ means that:

$$(\lambda P - A_{\text{ins}})(\lambda P - A_{\text{ins}})^{-1} = (\lambda P - A_{\text{ins}})^{-1}(\lambda P - A_{\text{ins}}) = P.$$

The same sense has $(\lambda(I-P) - A_{\text{out}})^{-1}$. Obviously,

$$(P - zA_{\text{ins}})(A-z)^{-1}P = (A-z)^{-1}P(P - zA_{\text{ins}}) = P \ \ (z \notin \sigma(A)).$$

Therefore,

$$(zP - A_{\text{ins}})^{-1} = P(Iz - A)^{-1}.$$

Similarly, $(z(I-P) - A_{\text{out}})^{-1} = (I-P)(Iz-A)^{-1}$ $(z \notin \sigma(A))$.

Lemma 6. *Let Conditions (50) and (27) hold. Then:*

$$\sup_{|z|=1} \|(zP - A_{\text{ins}})^{-1}\| \le F^2(1/d(A)) \tag{52}$$

and:

$$\sup_{|z|=1} \|(z(I-P) - A_{\text{out}})^{-1}\| \le (1 + F(1/d(A))F(1/d(A)), \tag{53}$$

where:

$$d(A) := \min\{1 - r_s(A_{\text{ins}}), r_l(A_{\text{out}}) - 1\}.$$

Proof. We have $1.7em(A,z) \ge d(A)$ $(|z| = 1)$. Since (27) holds,

$$\|P\| \le \sup_{|z|=1} \|(zI - A)^{-1}\| \le F(1/d(A)). \tag{54}$$

Hence, $\|I - P\| \le 1 + \|P\| \le 1 + F(1/d(A))$, and

$$\sup_{|z|=1} \|(zP - A_{\text{ins}})^{-1}\| = \sup_{|z|=1} \|(zI - A)^{-1}P\| \le F^2(1/d(A)).$$

Therefore, (52) is valid. Similarly,

$$\sup_{|z|=1} \|(z(I-P) - A_{\text{out}})^{-1}\| \le \|I - P\| \sup_{|z|=1} \|(zI - A)^{-1}\| \le (1 + F(1/d(A)))F(1/d(A)).$$

This finishes the proof. □

The analogous results can be found in ([18], Exercises of Chapter 1).

6. The Lyapunov Equation with a Dichotomic Spectrum

Assume that Condition (50) holds and P is defined by (51). Multiplying Equation (32) from the left by P^* and from the right by P, we have:

$$P^*CP = P^*XP - P^*A^*P^*XPAP = P^*XP - A_{\text{ins}}^*P^*XPA_{\text{ins}}.$$

Similarly,

$$(I-P^*)C(I-P) = (I-P^*)X(I-P) - A_{\text{out}}^*(I-P^*)X(I-P)A_{\text{out}}.$$

Therefore, with the notations $X_{\text{ins}} = P^*XP$, $X_{\text{out}} = (I-P^*)X(I-P)$, we obtain the equations:

$$X_{\text{ins}} - A_{\text{ins}}^*X_{\text{ins}}A_{\text{ins}} = P^*CP \tag{55}$$

and:

$$X_{\text{out}} - A_{\text{out}}^*X_{\text{out}}A_{\text{out}} = (I-P^*)C(I-P). \tag{56}$$

Lemma 7. *Let Conditions (50) and (27) be fulfilled. Then:*

$$\|X_{\text{ins}}\| \le \|C\|F^4(1/d(A)). \tag{57}$$

and:

$$\|X_{\text{out}}\| \le \|C\|F^2(1/d(A))(1+F(1/d(A)))^2. \tag{58}$$

Proof. According to (34) and (55):

$$X_{\text{ins}} = \frac{1}{2\pi}\int_0^{2\pi}(Pe^{-i\omega} - A_{\text{ins}}^*)^{-1}PCP(Pe^{i\omega} - A_{\text{ins}})^{-1}d\omega$$

$$= \frac{1}{2\pi}\int_0^{2\pi}(Pe^{-i\omega} - A_{\text{ins}}^*)^{-1}C(Pe^{i\omega} - A_{\text{ins}})^{-1}d\omega. \tag{59}$$

and:

$$X_{\text{out}} = \frac{1}{2\pi}\int_0^{2\pi}((I-P)e^{-i\omega} - A_{\text{out}}^*)^{-1}C((I-P)(e^{i\omega} - A_{\text{out}})^{-1}d\omega. \tag{60}$$

Now, (59) and (52) imply:

$$\|X_{\text{ins}}\| \le \|C\| \sup_{|z|=1} \|(zP - A_{\text{ins}})^{-1}\|^2 \le F^4(1/d(A)).$$

Therefore, (57) is proven. From (60) and (53), it follows:

$$\|X_{\text{out}}\| \le \|C\| \sup_{|z|=1} \|(z(I-P) - A_{\text{ins}})^{-1}\|^2 \le \|C\|F^2(1/d(A))(1+F(1/d(A)))^2.$$

Therefore, (58) is also valid. □

7. Linear Autonomous Difference Equation

In this section, we illustrate the importance of solution estimates for (32) in the simple case. To this end, consider the equation:

$$u_{k+1} = Au_k \ (k = 0, 1, 2, \ldots); \ u_0 \in \mathcal{H} \text{ is given}. \tag{61}$$

Let $X = X(A)$ be a solution of the equation:

$$X - A^*XA = I \tag{62}$$

First consider the case $r_s(A) < 1$. For any $x \in \mathcal{H}$, we have:

$$(XAx, Ax) = (Xx, x) - (x, x) \leq (Xx, x) - \frac{1}{\|X\|}(Xx, x).$$

Hence,

$$(XA^k x, A^k x) \leq (1 - \frac{1}{\|X\|})^k (Xx, x)$$

and consequently,

$$(Xu_k, u_k) \leq (1 - \frac{1}{\|X\|})^k (Xu_0, u_0) \ \ (r_s(A) < 1). \tag{63}$$

Now, let $r_l(A) > 1$ and $Y = -X$. Then, $A^*YA = Y + I$,

$$(YAx, Ax) = ((Y + I)x, x) \geq (1 + \frac{1}{\|X\|})(Yx, x).$$

Therefore,

$$(YA^k x, Ax) \geq (1 + \frac{1}{\|X\|})^k (Yx, x).$$

Consequently,

$$(Yu_k, u_k) \geq (1 + \frac{1}{\|X\|})^k (Yu_0, u_0) \ \ (Y = -X, r_l(A) > 1). \tag{64}$$

Now, assume that A has a dichotomic spectrum, i.e., (50) holds. Then, $u_k = w_k + v_k$ where w_k and v_k are solutions of the equations:

$$w_{k+1} = A_{\text{ins}} w_k \ \ (w_0 \in P\mathcal{H})$$

and:

$$v_{k+1} = A_{\text{out}} v_k \ \ (k = 0, 1, 2, ...; \ v_0 \in (I - P)\mathcal{H}).$$

Making use of (63) and (64), we have:

$$(X_{\text{ins}} w_k, w_k) \leq (1 - \frac{1}{\|X_{\text{ins}}\|})^k (X_{\text{ins}} w_0, w_0). \tag{65}$$

and:

$$(Y_{\text{out}} v_k, v_k) \geq (1 + \frac{1}{\|X_{\text{out}}\|})^k (Y_{\text{out}} v_0, v_0), \tag{66}$$

where $Y_{\text{out}} = -X_{\text{out}}$. However, as is shown in Section 6, Y_{out} and X_{ins} are upper and lower bounded. Now, (65) and (66) imply:

$$\|w_k\|^2 \leq const\ (1 - \frac{1}{\|X_{\text{ins}}\|})^k \|w_0\|^2$$

and:

$$\|v_k\|^2 \geq const\ (1 + \frac{1}{\|X_{\text{out}}\|})^k \|v_0\|^2.$$

Definition 1. *We will say the equation:*

$$u_{k+1} = A_k u_k \ \ (A_k \in \mathcal{B}(\mathcal{H}); k = 0, 1, 2, ...)$$

is dichotomic, if there exist a projection $P \neq 0$, $P \neq I$ and constants $\nu \in (0,1)$, $\mu > 1$ and $a, b > 0$ such that $\|u_k\| \leq a\nu^k \|u_0\|$ if $u_0 \in P\mathcal{H}$ and $\|u_k\| \geq m\mu^k \|u_0\|$ if $u_0 \in (I-P)\mathcal{H}$.

Therefore, Equation (61) is dichotomic, if $\sigma(A)$ is dichotomic.

8. Perturbations of Operators

To investigate nonautonomous equations, in this section, we consider some perturbations of operators.

8.1. Stable Operators

Lemma 8. *Let $A, \tilde{A} \in \mathcal{B}(\mathcal{H})$, $r_s(A) < 1$, and $X = X(A)$ be a solution of (62). If:*

$$\|X\|(2\|A-\tilde{A}\|\|A\| + \|A-\tilde{A}\|^2) < 1, \tag{67}$$

then:

$$(X\tilde{A}x, \tilde{A}x) \leq (1 - \frac{c_0}{\|X\|})(Xx, x) \quad (x \in \mathcal{H}),$$

where:

$$c_0 := 1 - \|X\|(2\|A-\tilde{A}\|\|A\| + \|A-\tilde{A}\|^2).$$

Proof. Put $Z = \tilde{A} - A$. Then:

$$X - \tilde{A}^* X \tilde{A} = X - (Z+A)^* X (Z+A) = X - A^*XA - Z^*XA - A^*XZ - Z^*XZ$$

$$= I - Z^*XA - A^*XZ - Z^*XZ.$$

By (67):

$$\|I - Z^*XA - A^*XZ - ZXZ\| \geq 1 - \|Z^*XA + A^*XZ + Z^*XZ\| \geq c_0.$$

Therefore, $X - \tilde{A}^*X\tilde{A} \geq c_0 I$ and:

$$(Xx, x) - (X\tilde{A}x, \tilde{A}x) \geq c_0(x, x) \geq c_0(\frac{X}{\|X\|}x, x) = \frac{c_0}{\|X\|}(Xx, x),$$

as claimed. □

8.2. The Case $r_l(A) > 1$

Lemma 9. *Let $A, \tilde{A} \in \mathcal{B}(\mathcal{H})$, $r_l(A) > 1$, and $X = X(A)$ be the solution of (62). If, in addition,*

$$2\|X\|\|A-\tilde{A}\|\|A\| < 1, \tag{68}$$

then with $Y = -X(A)$, one has:

$$(Y\tilde{A}x, \tilde{A}x) \geq (1 + \frac{\tilde{m}}{\|X\|})(Yx, x) \quad (x \in \mathcal{H}),$$

where $\tilde{m} = 1 - 2\|X\|\|A-\tilde{A}\|\|A\|$.

Proof. With $Z = \tilde{A} - A$, one has:

$$\tilde{A}^*Y\tilde{A} = (Z+A)^*Y(Z+A) = A^*YA + Z^*YA + A^*YZ + Z^*YZ$$

$$= Y + I + Z^*Y + A^*YZ + Z^*YZ.$$

Since Y is positive definite, hence, by (68),

$$(Y\tilde{A}x, \tilde{A}x) \geq (Yx, x) + (x, x) + (Z^*YZx, x) + (YZx, Ax)$$

$$\geq (Yx, x) + (x, x)(1 - 2\|Y\|\|Z\|) = (Yx, x) + (x, x)\tilde{m} \geq (Yx, x)(1 + \frac{\tilde{m}}{\|Y\|}),$$

as claimed. □

8.3. Perturbation of Operators with Dichotomic Spectra

Let Condition (50) hold, and:

$$\|A - \tilde{A}\| \sup_{|z|=1} \|R_z(A)\| < 1,$$

then by the Hilbert identity $R_z(\tilde{A}) - R_z(A) = R_z(\tilde{A})(A - \tilde{A})R_z(\tilde{A})$, the inequality:

$$\|R_z(\tilde{A})\| \leq \psi(A) := \sup_{|z|=1} \|R_z(A)\|(1 - \|A - \tilde{A}\|\|R_z(A)\|)^{-1} \quad (|z| = 1)$$

is fulfilled and:

$$\|R_z(\tilde{A}) - R_z(A)\| \leq q\psi(A) \sup_{|z|=1} \|R_z(A)\| \quad (|z| = 1). \tag{69}$$

Therefore, $\Omega \cap \sigma(\tilde{A}) = \emptyset$. Moreover, $\tilde{A}$ has a dichotomic spectrum:

$$\sigma(\tilde{A}) = \tilde{\sigma}_{\text{ins}} \cup \tilde{\sigma}_{\text{out}} \tag{70}$$

where $\tilde{\sigma}_{\text{ins}}$ and $\tilde{\sigma}_{\text{out}}$ are nonempty nonintersecting sets lying inside and outside Ω, respectively. Indeed, let $A_t = A + t(\tilde{A} - A)$ $(0 \leq t \leq 1)$. For each t, $\Omega \cap \sigma(A_t) = \emptyset$, since $\|A - A_t\| \sup_{|z|=1} \|R_z(A)\| < 1$. Hence, (70) follows from (50) and the semi-continuity of the spectrum. Put:

$$\tilde{P} = \frac{1}{2\pi i} \int_\Omega (zI - \tilde{A})^{-1} dz,$$

$\tilde{A}_{\text{ins}} = \tilde{P}\tilde{A}$ and $\tilde{A}_{\text{out}} = (I - \tilde{P})\tilde{A}$. With the notations of Section 5,
$A_{\text{ins}} - \tilde{A}_{\text{ins}}$

$$= \frac{1}{2\pi i} \int_\Omega z[(zI - A)^{-1} - (zI - \tilde{A})^{-1}] dz = -\frac{1}{2\pi i} \int_\Omega z[(zI - A)^{-1}(A - \tilde{A})(zI - \tilde{A})^{-1}] dz.$$

According to (69) with $q = \|A - \tilde{A}\|$, we obtain:

$$q_{\text{ins}} := \|A_{\text{ins}} - \tilde{A}_{\text{ins}}\| \leq q\psi(A) \sup_{|z|=1} \|R_z(A)\|.$$

Since $A_{\text{out}} - \tilde{A}_{\text{out}} = A - \tilde{A} - (A_{\text{ins}} - \tilde{A}_{\text{ins}})$, one has:

$$q_{\text{out}} := \|A_{\text{out}} - \tilde{A}_{\text{out}}\| \leq q + q_{\text{ins}}.$$

In this section, X_{ins} and X_{out} are solutions of the equations of (55), (56), respectively, with $C = I$; i.e.,

$$X_{\text{ins}} - A^*_{\text{ins}} X_{\text{ins}} P A_{\text{ins}} = P^*P \tag{71}$$

and:

$$X_{out} - A^*_{out} X_{out} P A_{out} = (I - P^*)(I - P).$$

Lemma 8.1 yields:

Corollary 1. *If*

$$\|X_{ins}\|(2q_{ins}\|A_{ins}\| + q_{ins}^2) < 1,$$

then:

$$(X_{ins}\tilde{A}_{ins}x, \tilde{A}_{ins}x) \le (1 - \frac{c_{ins}}{\|X_{ins}\|})(X_{ins}x, x) \ \ (x \in \mathcal{H}),$$

where:

$$c_{ins} := 1 - \|X_{ins}\|(2q_{ins}\|A_{ins}\| + q_{ins}^2).$$

Making use of Lemma 8.2, we get:

Corollary 2. *If*

$$2\|X_{out}\| q_{out}\|A_{out}\| < 1,$$

then with $Y_{out} = -X_{out}$, one has:

$$(Y_{out}\tilde{A}_{out}, \tilde{A}_{out}x, x) \ge (1 + \frac{m_{out}}{\|X_{out}\|})(Y_{out}x, x) \ \ (x \in \mathcal{H}),$$

where $m_{out} = 1 - 2\|X_{out}\| q_{out}\|A_{out}\|$.

9. Nonautonomous Linear Difference Equations

9.1. Stability

Consider the equation:

$$u_{k+1} = A_k u_k \ \ (A_k \in \mathcal{B}(\mathcal{H}); k = 0, 1, 2, ...) \tag{72}$$

with given $u_0 \in \mathcal{H}$. For some $A \in \mathcal{B}(\mathcal{H})$, define the norms:

$$\|x\|_X = \sqrt{(Xx, x)} \ \ (x \in \mathcal{H}) \text{ and } \|A\|_X = \sup_{x \in \mathcal{H}} \frac{\|Ax\|_X}{\|x\|_X}.$$

where $X = X(A)$ is the solution of (62).

Throughout this section and the next one, it is assumed that $\sup_k \|A_k\| < \infty$ and denoted $q_0 := \sup_k \|A - A_k\|$.

Theorem 4. *Let there be an $A \in \mathcal{B}(\mathcal{H})$ with $r_s(A) < 1$, such that:*

$$\sup_{k=0,1,2,...} \|X(A)\|(2q_0\|A\| + q_0^2) < 1. \tag{73}$$

Then, for any solution of u_k of (72), one has:

$$\|u_k\|_X \le (1 - \frac{a_0}{\|X\|})^{k/2}\|u_0\|_X \ \ (k = 1, 2, ...) \tag{74}$$

where $a_0 := 1 - (2q_0\|X\| + q_0^2)$.

Proof. Due to Lemma 8.1 and (73), we have:

$$\|A_k\|_X \leq \sqrt{1 - \frac{a_0}{\|X\|}} \quad (k = 0, 1, 2, ...). \tag{75}$$

Since:

$$u_{k+1} = A_k A_{k-1} \cdots A_1 A_0 u_0, \tag{76}$$

we arrive at the required result. □

Certainly, we can take $A = A_k$ for some index k.

Equation (72) is said to be exponentially stable, if there are constants $m_1 \geq 1, m_2 \in [0,1)$, such that $\|u_k\| \leq m_1 m_2^k \|u_0\|$ $(k = 1, 2, ...)$.

Note that $X = I + A^* X A \geq I$. Since $a_0 < 1$, one has $\frac{a_0}{\|X\|} < 1$. In addition, the upper and lower bounds for X presented in Section 3 show that the norms $\|\cdot\|$ and $\|\cdot\|_X$ are equivalent. Consequently, under the hypothesis of Theorem 9.1, Equation (72) is exponentially stable.

Now, we can apply the results of Section 3 to concrete operators.

9.2. Lower Bounds for Solutions

Lemma 10. *For some $A \in \mathcal{B}(\mathcal{H})$, let the condition $r_l(A) > 1$ hold and $X = X(A)$ be a solution of (62). If, in addition,*

$$2q_0 \|X\| \|A\| < 1, \tag{77}$$

then solution u_k of (72) is subject to the inequality:

$$(Yu_k, u_k) \geq (1 + \frac{m_0}{\|X\|})^k (Yu_0, u_0) \quad (k = 1, 2, ...), \tag{78}$$

where $Y = -X$ and $m_0 = 1 - 2\|X\|\|A\|q_0$.

Proof. Due to Lemma 8.2, we have:

$$(YA_k x, A_k x) \geq (1 + \frac{m_0}{\|X\|})(Yx, x).$$

Hence,

$$(Yu_{k+1}, u_{k+1}) \geq (1 + \frac{\hat{d}_0}{\|X\|})(u_k, u_k) \geq (1 + \frac{\hat{d}_0}{\|X\|})^2 (u_{k-1}, u_{k-1}). \tag{79}$$

Continuing this process, we get the required result. □

9.3. Dichotomic Equations

For an $A \in \mathcal{B}(\mathcal{H})$, let Condition (50) hold, and the inequality:

$$q_0 \sup_{|z|=1} \|R_z(A)\| < 1 \tag{80}$$

is fulfilled. Then, $\Omega \cap \sigma(A_k) = \emptyset$ for all $k \geq 0$, and by the Hilbert identity:

$$\sup_{k=0,1,...;|z|=1} \|R_z(A_k)\| \leq \psi_0 := \sup_{|z|=1} \|R_z(A)\|(1 - q_0\|R_z(A)\|)^{-1} \tag{81}$$

and:

$$\sup_{k=0,1,...;|z|=1} \|R_z(A_k) - R_z(A)\| \leq q_0 \psi_0 \sup_{|z|=1} \|R_z(A)\|. \tag{82}$$

Hence, each A_k has a dichotomic spectrum:

$$\sigma(A_k) = \sigma_{\text{ins}}(A_k) \cup \sigma_{\text{out}}(A_k),$$

where $\sigma_{\text{ins}}(A_k)$ and $\sigma_{\text{out}}(A_k)$ are nonempty nonintersecting sets lying inside and outside Ω, respectively. Put:

$$P_k = \frac{1}{2\pi i} \int_\Omega (zI - A_k)^{-1} dz,$$

$A_{k,\text{ins}} = P_k A_k$ and $A_{k,\text{out}} = (I - P_k)A_k$. With A_{ins} defined as Section 5,

$$A_{\text{ins}} - A_{k,\text{ins}} = \frac{1}{2\pi i} \int_\Omega z[(zI - A)^{-1} - (zI - A_k)^{-1}] dz.$$

According to (82):

$$q_{0,\text{ins}} := \sup_k \|A_{\text{ins}} - A_{k,\text{ins}}\| \le q_0 \psi_0 \sup_{|z|=1} \|R_z(A)\|. \tag{83}$$

Since $A_{\text{out}} - A_{k,\text{out}} = A - A_k - (A_{\text{ins}} - A_{k,\text{ins}})$, one has:

$$q_{0,\text{out}} := \sup_k \|A_{\text{out}} - A_{k,\text{out}}\| \le q_0 + q_{0,\text{ins}} \le q_0(1 + \psi_0 \sup_{|z|=1} \|R_z(A)\|). \tag{84}$$

In this section, X_{ins} and X_{out} are solutions of Equation (71) and the equation $X_{\text{out}} - A^*_{\text{out}} X_{\text{out}} P A_{\text{out}} = (I - P^*)(I - P)$, respectively. If:

$$\|X_{\text{ins}}\|(2q_{0,\text{ins}}\|A_{\text{ins}}\| + q_{0,\text{ins}}^2) < 1, \tag{85}$$

then Corollary 8.3 implies:

$$(X_{\text{ins}} A_{k,\text{ins}} x, A_{k,\text{ins}} x) \le (1 - \frac{c_{0,\text{ins}}}{\|X_{\text{ins}}\|})(X_{\text{ins}} x, x) \quad (x \in \mathcal{H}), \tag{86}$$

where:

$$c_{0,\text{ins}} := 1 - \|X_{\text{ins}}\|(2q_{0,\text{ins}}\|A_{\text{ins}}\| + q_{0,\text{ins}}^2).$$

Furthermore, if:

$$2\|X_{\text{out}}\| q_{0,\text{out}} \|A_{\text{out}}\| < 1, \tag{87}$$

then with $Y_{\text{out}} = -X_{\text{out}}$, Corollary 8.4 implies:

$$(Y_{\text{out}} A_{k,\text{out}},\ A_{k,\text{out}} x, x) \ge (1 + \frac{m_{0,\text{out}}}{\|X_{\text{out}}\|})(Y_{\text{out}} x, x) \quad (x \in \mathcal{H}), \tag{88}$$

where $m_{0,\text{out}} = 1 - 2\|X_{\text{out}}\| q_{0,\text{out}} \|A_{\text{out}}\|$.

Put $w_k = P_k A_k, w_k = (I - P_k)A_k$. Then, $u_k = w_k + v_k$, where w_k and v_k are solutions of the equations:

$$w_{k+1} = A_{k,\text{ins}} w_k \quad (w_0 \in P_0 \mathcal{H}) \tag{89}$$

and:

$$v_{k+1} = A_{k,\text{out}} v_k \quad (k = 0, 1, 2, ...;\ v_0 \in (I - P_0)\mathcal{H}). \tag{90}$$

Making use of (86), under Condition (85), we have:

$$(X_{\text{ins}} w_{k+1}, w_{k+1}) = (X_{\text{ins}} A_{k,\text{ins}} w_k, A_{k,\text{ins}} w_k)(1 - \frac{c_{0,\text{ins}}}{\|X_{\text{ins}}\|})(X_{\text{ins}} w_k, w_k) \le ...$$

$$\leq (1 - \frac{c_{0,\text{ins}}}{\|X_{\text{ins}}\|})^k (X_{\text{ins}} w_0, w_0). \tag{91}$$

Furthermore, if (87) holds, then by: (88)

$$(Y_{\text{out}} v_{k+1}, v_{k+1}) = (Y_{\text{out}} A_{k,\text{out}} v_k, A_{k,\text{out}}) \geq (1 + \frac{d_{0,\text{out}}}{\|X_{\text{out}}\|})(Y_{\text{out}} v_k, v_k) \geq$$

$$... \geq (1 + \frac{d_{0,\text{out}}}{\|X_{\text{out}}\|})^k (Y_{\text{out}} v_0, v_0). \tag{92}$$

We thus have proven:

Lemma 11. *For some $A \in \mathcal{B}(\mathcal{H})$, let Conditions (50), (85), and (87) hold. Then, (72) is a dichotomic equation. Moreover, its solution satisfies Inequalities (91) and (92).*

Let Condition (27) hold and $d(A)$ be defined as in Section 5. For brevity, put $d(A) = d$. Then, as is shown in Section 5, $\sup_{|z|=1} \|R_z(A)\| \leq F(1/d)$, $\|P\| \leq F(1/d)$, $\|I - P\| \leq 1 + F(1/d)$. By Lemma 6.1, $\|X_{\text{ins}}\| \leq F^4(1/d)$ and $\|X_{\text{out}}\| \leq F^2(1/d)(1 + F(1/d))^2$. Condition (80) takes the form:

$$q_0 F(1/d) < 1. \tag{93}$$

Therefore,

$$\psi_0 \leq \psi_1 := F(1/d)(1 - q_0 F(1/d))^{-1}$$

and $q_{0,\text{ins}} \leq q_0 \psi_1 F(1/d)$. In addition, by (84) $q_{0,\text{out}} \leq q_0(1 + \psi_1 F(1/d))$. Condition (85) is provided by:

$$F^4(1/d)(2q_0\psi_1 F(1/d))\|AP\| + q_0^2\psi_1^2 F^2(1/d)) \leq F^6(1/d) q_0 \psi_1 (2\|A\| + q_0\psi_1) < 1.$$

Condition (87) is provided by:

$$2F^2(1/d)(1 + F(1/d))^2 q_0(1 + \psi_1 F(1/d))\|A(I - P)\| \leq 2F^2(1/d)(1 + F(1/d))^3 q_0(1 + \psi_1 F(1/d))\|A\| < 1,$$

Now, Lemma 9.3 yields:

Theorem 5. *For some $A \in \mathcal{B}(\mathcal{H})$, let the Conditions (50), (27), (93), and:*

$$q_0 F^2(1/d) \max\{F^4(1/d)\psi_1(2\|A\| + q_0\psi_1), 2(1 + F(1/d))^3 \hat{(}1 + \psi_1 F(1/d))\|A\|\} < 1$$

be fulfilled. Then, (72) is a dichotomic equation. Moreover, its solution satisfies Inequalities (91) and (92).

Similar results for the periodic equations in the finite-dimensional space were established in the article [19].

10. Nonlinear Nonautonomous Equations

For a positive $\varrho \leq \infty$, put $\omega(\varrho) = \{x \in \mathcal{H} : \|x\| \leq \varrho\}$.
Let $A_k \in \mathcal{B}(\mathcal{H})$ and $G_k : \omega(\varrho) \to \mathcal{H}$. Consider the equation:

$$u_{k+1} = A_k u_k + G_k(u_k) \quad (k = 0, 1, 2, ...) \tag{94}$$

with given $u_0 \in \mathcal{H}$, assuming that:

$$\|G_k(x)\| \leq \nu_k \|x\| \quad (x \in \omega(\varrho);\ k = 0, 1, 2, ...) \tag{95}$$

with nonnegative constants ν_k.

Lemma 12. *Let Condition (95) hold with* $\varrho = \infty$. *Let there be an* $A \in \mathcal{B}(\mathcal{H})$ *with* $r_s(A) < 1$ *and:*

$$\gamma := \|X\| \sup_k [2\|A\|\|A - A_k\| + \|A - A_k\|^2 + 2\|A_k\|v_k + v_k^2] < 1, \tag{96}$$

where X *is the solution of (62). Then:*

$$(Xu_k, u_k) \le (1 - \frac{1-\gamma}{\|X\|})^k (Xu_0, u_0) \ \ (k = 1, 2, ...) \tag{97}$$

for any solution u_k *of (94).*

Proof. Multiplying (94) by X and doing the scalar product, we have.

$$(Xu_{k+1}, u_{k+1}) = (X(A_k u_k + G_k(u_k)), A_k u_k + G_k(u_k)) = (XA_k u_k, A_k u_k) + \Phi_k(u_k), \tag{98}$$

where:

$$\Phi_k(x) = (XG_k(x), A_k x) + (XA_k x, G_k(x)) + (XG_k(x), G_k(x)) \ \ (x \in \mathcal{H}).$$

However,

$$A_k^* X A_k = (A + Z_k)^* X (A + Z_k) = A^* X A + W_k = X - I + W_k,$$

where $Z_k = A_k - A$ and $W_k = Z_k^* XA + A_k^* XZ + Z_k^* XZ_k$. Thus,

$$(Xu_{k+1}, u_{k+1}) = (Xu_k, u_k) - (u_k, u_k) + (W_k u_k, A_k u_k) + \Phi_k$$
$$\le (Xu_k, u_k) - \|u_k\|^2 (1 - \|W_k\|) - \|\Phi_k\|.$$

According to (95):

$$\|\Phi_k(x)\| \le \|X\|(2\|A_k\|\|G_k(x)\|\|x\| + \|G_k(x)\|^2 \le \|X\|(2\|A_k\|v_k + v_k^2)\|x\|^2)$$

and:

$$\|W_k\| \le \|X\|(2\|A\|\|A - A_k\| + \|A - A_k\|^2).$$

Consequently,

$$\|\Phi_k(x)\| + \|W_k x\| \le \|X\|(2\|A\|\|A - A_k\| + \|A - A_k\|^2 + v_k\|A_k\| + v_k^2)\|x\|^2 \le \gamma\|x\|^2.$$

From (98), it follows:

$$(Xu_{k+1}, u_{k+1}) \le (Xu_k, u_k) - \gamma(u_k, u_k) \le (Xu_k, u_k)(1 - \frac{\gamma}{\|X\|}).$$

Hence, (97) follows, as claimed. □

Since $X \ge I$, X is invertible and:

$$\frac{1}{\|X^{-1}\|}(u_k, u_k) = \frac{1}{\|X^{-1}\|}(X^{-1}Xu_k, u_k) \le (Xu_k, u_k).$$

From the latter lemma with $\varrho = \infty$, we have:

$$(u_k, u_k) \le \|X^{-1}\|\|X\|(1 - \frac{1-\gamma}{\|X\|})^k (u_0, u_0) \ \ (u_0 \in \mathcal{H}),$$

and thus:

$$\|u_k\| \le (\|X^{-1}\|\|X\|)^{1/2}(1-\frac{1-\gamma}{\|X\|})^{k/2}\|u_0\| \ \ (k=1,2,...). \tag{99}$$

Theorem 6. *Let Condition (95) and there be an $A \in \mathcal{B}(\mathcal{H})$ with $r_s(A) < 1$ satisfying (96). In addition, let:*

$$\|u_0\| < \frac{\varrho}{(\|X^{-1}\|\|X\|)^{1/2}}. \tag{100}$$

Then, the solution to (94) admits the estimate (99).

Proof. In the case $\varrho = \infty$, the result is due to the latter lemma. Let $\varrho < \infty$. By the Urysohn theorem ([20], p. 15), there is a scalar-valued function ψ_ϱ defined on $\mathcal{H}$, such that:

$$\psi_\varrho(w) = 1 \ \ (w \in \mathcal{H}, \|w\| < \varrho) \text{ and } \psi_\varrho(w) = 0 \ \ (\|w\| \ge \varrho).$$

Put $G_k(\varrho, w) = \psi_\varrho(w)G_k(w)$ and consider the equation:

$$v_{k+1} = A_k v_k + G_k(\varrho, v_k), \ \ v_0 = u_0. \tag{101}$$

Besides, (95) yields the condition:

$$\|G_k(\varrho, w))\| \le \nu_k\|w\| \ \ (w \in \mathcal{H}; k \ge 0).$$

Thanks to the latter lemma, a solution v_k of Equation (101) satisfies (99). According to (100), $\|v_k\| \le (\|X^{-1}\|\|X\|)^{1/2}\|u_0\| < \varrho$ $(k = 1,2,...)$. Therefore, solutions of (101) and (94) under (102) coincide. This proves the required result. □

Definition 2. *The zero solution to (94) is said to be exponentially stable if there are constants $m_0 > 0, m_1 > 0$ and $m_2 \in (0,1)$, such that the solution u_k to (94) satisfies the inequality, $\|u_k\| \le m_1 m_2^k \|u_0\|$ $(k = 1,2,...)$, provided $\|u_0\| < m_0$.*

Corollary 3. *Under the hypothesis of Theorem 10.1, the zero solution to (94) is exponentially stable.*

Definition 3. *We will say that Equation (1) is quasi-linear, if:*

$$\lim_{w \to 0} \|G_j(w)\|/\|w\| = 0 \tag{102}$$

uniformly in $j \ge 0$.

Corollary 4. *Let (94) be quasi-linear and there be an $A \in \mathcal{B}(\mathcal{H})$ with $r_s(A) < 1$ satisfying the inequality:*

$$\|X\|[2\|A\|\|A - A_j\| + \|A - A_j\|^2] < 1 \ \ (j = 0,1,2...).$$

Then, the zero solution to (94) is exponentially stable.

Indeed, according to (102),

$$\|G_j(w)\| \le \hat{\nu}(\varrho)\|w\| \ \ (w \in \omega(\varrho))$$

with a $\hat{\nu}(\varrho) \to 0$ as $\varrho \to 0$. Therefore, for a sufficiently small ϱ, we have Condition (95) with $\hat{\nu}(.)$ instead of ν_k. Now, Theorem 10.1 yields the required result.

Funding: This research received no external funding.

Conflicts of Interest: The author declares no conflicts of interest.

References

1. Eisner, T. *Stability of Operators and Operator Semigroups*; Operator Theory: Advances and Applications; Birkhäuser Verlag: Basel, Switzerland, 2010; Volume 209.
2. Gil', M.I. *Difference Equations in Normed Spaces. Stability and Oscillations*; North-Holland, Mathematics Studies; Elsevier: Amsterdam, The Netherlands, 2007; Volume 206.
3. Gil', M.I. *Operator Functions and Operator Equations*; World Scientific: Hackensack, NJ, USA, 2018.
4. Huy, N.T.; Ha, V.T.N. Exponential dichotomy of difference equations in l_p-phase spaces on the half-line. *Adv. Differ. Equ.* **2006**, *2006*, 58453. [CrossRef]
5. Ngoc, P.H.A.; Naito, T. New characterizations of exponential dichotomy and exponential stability of linear difference equations. *J. Differ. Equ. Appl.* **2005**, *11*, 909–918. [CrossRef]
6. Pötzsche, C. *Geometric Theory of Discrete Nonautonomous Dynamical Systems*; Lecture Notes in Mathematics; Springer: Berlin, Germany, 2010; Volume 2002.
7. Preda, P.; Pogan, A.; Preda, C. Discrete admissibility and exponential dichotomy for evolution families. *Dyn. Contin. Discrete Impuls. Syst. Ser. A Math. Anal.* **2005**, *12*, 621–631.
8. Russ, E. Dichotomy spectrum for difference equations in Banach spaces. *J. Differ. Equ. Appl.* **2017**, *23*, 574–617. [CrossRef]
9. Sasu, A.L. Exponential dichotomy and dichotomy radius for difference equations. *J. Math. Anal. Appl.* **2008**, *344*, 906–920. [CrossRef]
10. Sasu, B.; Sasu, A.L. Exponential dichotomy and (l_p, l_q)-admissibility on the half-line. *J. Math. Anal. Appl.* **2006**, *316*, 397–408. [CrossRef]
11. Sasu, A.L.; Sasu, B. On the dichotomic behavior of discrete dynamical systems on the half-line. *Discrete Contin. Dyn. Syst.* **2013**, *33*, 3057–3084. [CrossRef]
12. Babutia, M.G.; Megan, M.; Popa, I.-L. On (h,k)-dichotomies for nonautonomous linear difference equations in Banach spaces. *Int. J. Differ. Equ.* **2013**, *2013*, 761680. [CrossRef]
13. Agarwal, R.P.; Thompson, H.B.; Tisdell, C.C. Difference equations in Banach spaces. *Comput. Math. Appl.* **2003**, *45*, 1437–1444. [CrossRef]
14. Megan, M.; Ceausu, T.; Tomescu, M.A. On exponential stability of variational nonautonomous difference equations in Banach spaces. *Ann. Acad. Rom. Sci. Ser. Math. Appl.* **2012**, *4*, 20–31.
15. Megan, M.; Ceausu, T.; Tomescu, M.A. On polynomial stability of variational nonautonomous difference equations in Banach spaces. *Int. J. Anal.* **2013**, *2013*, 407958. [CrossRef]
16. Bay, N.S.; Phat, V.N. Stability analysis of nonlinear retarded difference equations in Banach spaces. *Comput. Math. Appl.* **2003**, *45*, 951–960. [CrossRef]
17. Medina, R. New conditions for the exponential stability of pseudo-linear difference equations in Banach spaces. *Abstr. Appl. Anal.* **2016**, *2016*, 5098086. [CrossRef]
18. Daleckii, J.L.; Krein, M.G. *Stability of Solutions of Differential Equations in Banach Space*; Translations of Mathematical Monographs; American Mathematical Society: Providence, RI, USA, 1974; Volume 43.
19. Demidenko, G.V.; Bondar, A.A. Exponential dichotomy of systems of linear difference equations with periodic coefficients. *Sib. Math. J.* **2016**, *57*, 117–124. [CrossRef]
20. Dunford, N.; Schwartz, J.T. *Linear Operators, Part I*; Interscience: New York, NY, USA; London, UK, 1963.

axioms

MDPI

Article

Note on Limit-Periodic Solutions of the Difference Equation $x_{t+1} - [h(x_t) + \lambda]x_t = r_t, \lambda > 1$

Jan Andres [1,*] **and Denis Pennequin** [2]

[1] Department of Mathematical Analysis and Applications of Mathematics, Faculty of Science, Palacký University, 17. listopadu 12, 771 46 Olomouc, Czech Republic
[2] Centre PMF, Laboratory SAMM, Université Paris I Panthéon—Sorbonne, 90, Rue de Tolbiac, 75 634 Paris CEDEX 13, France; pennequi@univ-paris1.fr
* Correspondence: jan.andres@upol.cz; Tel.: +420-585-634-602

Received: 7 January 2019; Accepted: 1 February 2019; Published: 5 February 2019

Abstract: As a nontrivial application of the abstract theorem developed in our recent paper titled "Limit-periodic solutions of difference and differential systems without global Lipschitzianity restricitions", the existence of limit-periodic solutions of the difference equation from the title is proved, both in the scalar as well as vector cases. The nonlinearity h is not necessarily globally Lipschitzian. Several simple illustrative examples are supplied.

Keywords: limit-periodic solutions; difference equations; exponential dichotomy; strong nonlinearities; effective existence criteria; population dynamics

MSC: 39A11; 39A24; 42A75

1. Introduction

As far as we know, in spite of the intensive studies of limit-periodic (especially Schrödinger-type) operators (see, e.g., [1,2], and the references therein), the results about the existence of limit-periodic solutions to nonlinear differential and difference equations are very rare (see e.g., [3–8]). For some related periodicity and almost-periodicity problems and applications, see, e.g., [9–12], and the references therein. As already pointed out in [5], since the space of limit-periodic sequences is (unlike for limit-periodic functions) Banach (cf. [4,13]), the existence criteria for limit-periodic solutions of difference equations are significantly simpler than those for differential equations. Nevertheless, an obstruction related to the absence of global Lipschitzianity restrictions, imposed on the given right-hand sides of equations under consideration, remains also in the discrete case. For the recently investigated continuous case, see [6] and the references therein.

Hence, the aim of the present note is to obtain, by means of our technique (see [5], Theorem 3.2, resp. Corollary 3.3), which we state below in the form of Proposition 2, the effective solvability criteria of the equation from the title. Its scalar and vector cases will be treated separately. Let us note that, in particular cases, the equation from the title can describe discrete population models for a single species. For instance, if $h(x) := -\mu x + \mu - \lambda$, $\mu > 0$, then we get the *forced logistic equation*. For more details, see e.g., ([14], Chapter 1, [15], Chapter 2).

2. Preliminaries and Auxiliary Results

At first, we will recall the notion of a limit-periodic sequence and its basic properties.

Definition 1. *A sequence $\underline{x} := \{x_t\} \in (\mathbb{R}^n)^{\mathbb{Z}}$, where $\mathbb{R}$ and $\mathbb{Z}$ denote respectively the sets of reals and integers, is called limit-periodic if there exists a family of periodic sequences $\underline{x}^k := \{x_t^k\}$, $k \in \mathbb{N}$ ($\mathbb{N}$ denotes the set of positive integers), such that $\lim_{k\to\infty} x_t^k = x_t$, uniformly w.r.t. $t \in \mathbb{Z}$.*

It is well known (see e.g., [5]) that Definition 1 is equivalent to the following definition of a semi-periodic sequence (cf. [13]).

Definition 2. *A sequence* $\underline{x} := \{x_t\} \in (\mathbb{R}^n)^{\mathbb{Z}}$ *is called semi-periodic if*

$$\forall \varepsilon > 0,\ \exists T \in \mathbb{N},\ \forall m \in \mathbb{Z},\ \forall k \in \mathbb{Z}, \quad |x_{k+mT} - x_k| \leq \varepsilon.$$

Remark 1. *Since the uniform and Stepanov norms, namely* $\|\cdot\|_\infty$ *and* $\|\cdot\|_{S_T^p}$, *where*

$$\|\underline{x}\|_\infty := \sup_{t\in\mathbb{Z}} |x_t|,$$

$$\|\underline{x}\|_{S_T^p} := \sup_{m\in\mathbb{Z}} \left(\frac{1}{T+1} \sum_{t=m}^{m+T} |x_t|^p \right)^{\frac{1}{p}}, \ T \in \{0\} \cup \mathbb{N},\ p \geq 1,$$

were shown in ([16], Proposition 1 and Remark 2) to be equivalent, both Definitions 1 and 2 can be easily reformulated in terms of Stepanov limit-periodic and Stepanov semi-periodic sequences by means of the Stepanov norm $\|\cdot\|_{S_T^p}$, $p \geq 1$.

Summing up, it will be convenient to recall the following proposition (cf. [5], Proposition 2.12).

Proposition 1. *The following properties of a sequence* $\{x_t\} \in (\mathbb{R}^n)^{\mathbb{Z}}$ *are equivalent:*

(i) $\{x_t\}$ *is uniformly limit-periodic,*
(ii) $\{x_t\}$ *is (Stepanov)* S_T^p*-limit-periodic,*
(iii) $\{x_t\}$ *is uniformly semi-periodic,*
(iv) $\{x_t\}$ *is (Stepanov)* S_T^p*-semi-periodic.*

Moreover, the set of all (Stepanov) limit-periodic, resp. (Stepanov) semi-periodic sequences $\{x_t\} \in (\mathbb{R}^n)^{\mathbb{Z}}$, *endowed with the uniform or Stepanov norms* $\|\cdot\|_\infty$ *or* $\|\cdot\|_{S_T^p}$, $p \geq 1$, *is a Banach space.*

Now, let us proceed to the difference system

$$x_{t+1} = f(x_t) + p_t, \tag{1}$$

where $f \in C^1(\mathbb{R}^n, \mathbb{R}^n)$ and $\{p_t\} \in (\mathbb{R}^n)^{\mathbb{Z}}$ is a (Stepanov) limit-periodic sequence. Let us also consider the associated one-parameter family of systems

$$x_{t+1} = f(x_t) + p_t^N, \tag{2}$$

where the class of T_k-periodic ($T_k > 0$) sequences$\{p_t^N\} \in (\mathbb{R}^n)^{\mathbb{Z}}$, $N \in \mathbb{N}$, converges uniformly to $\{p_t\}$.

Our technique in ([5], Corollary 3.7) will be stated here in the form of the following proposition.

Proposition 2. *Assume still that*

(i) for each fixed N, system (2) admits a T_k*-periodic solution* $\{x_t^N\}$,
(ii) $\sup_{N\in\mathbb{N}} \|\{x_t^N\}\|_\infty < \infty$,
(iii) if A_t^N *is the Jacobian matrix of f at* x_t^N, *then there exists a non-singular solution of the homogeneous system*

$$y_{t+1} = A_t^N y_t, \tag{3}$$

which satisfies the exponential dichotomy, for all sufficiently large values of N, with common constants K and α, *characterizing the exponential dichotomy.*

Then, system (1) possesses a uniformly limit-periodic solution.

Let us recall the definition of an exponential dichotomy for Equation (3). Introducing the *resolvent* $R\colon \mathbb{Z}^2 \to \mathcal{M}_n$, where $\mathcal{M}_n$ denotes the space of real $n \times n$ matrices, and

$$R(t,s) := \begin{cases} A_{t-1} \dots A_s, & \text{for } t > s, \\ A_s^{-1} \dots A_{t-1}^{-1}, & \text{for } t < s, \\ I_n, & \text{for } t = s, \end{cases}$$

where I_n is the identity matrix, it has the semi-group property. Taking $\Phi_t := R(t,0)$, we say that Equation (3) satisfies an *exponential dichotomy* if there exists a projection P ($P^2 = P$) and constants $K > 0$, $\alpha \in (0,1)$ such that

$$\begin{cases} |\Phi_t P \Phi_s^{-1}| \leq K\alpha^{t-s}, & \text{for } t \geq s, \\ |\Phi_t (I_n - P)\Phi_s^{-1}| \leq K\alpha^{-(t-s)}, & \text{for } t \leq s. \end{cases} \tag{4}$$

Let us note that $\Phi_t P \Phi_s^{-1} y$ is the t-iterated image of the projection by P of the solution of Equation (3) such that $y_s = y$. In the stable case, $P = I_n$, and so $\Phi_t P \Phi_s^{-1} = R(t,s)$, when $t \geq s$, and $\Phi_t(I_n - P)\Phi_s^{-1} = 0$, when $t \leq s$.

Remark 2. *For some alternative definitions of an exponential dichotomy for Equation (3), see, e.g., [17–19]. In particular, Palmer gives in [18] a finite-time condition for an exponential dichotomy. In fact, all these conditions were formulated for a more general class of (uniformly) almost-periodic homogeneous systems (3).*

On this basis, we can define the associated *Green function* $G\colon \mathbb{Z}^2 \to \mathcal{M}_n$, where (see, e.g., [5], and the references therein)

$$G(t,s) := \Phi_t \left(P l_{t \geq s} + (I_n - P) l_{t \leq s} \right) \Phi_s^{-1} = \begin{cases} \Phi_t P \Phi_s^{-1}, & \text{for } t > s, \\ I_n, & \text{for } t = s, \\ \Phi_t (I_n - P) \Phi_s^{-1}, & \text{for } t < s, \end{cases} \tag{5}$$

and

$$l_{t \geq s} := \begin{cases} 1, & \text{for } t \geq s, \\ 0, & \text{for } t < s. \end{cases}$$

3. Limit-Periodic Solutions: Scalar Case

Consider the equation from the title in the scalar case ($n = 1$), i.e.,

$$x_{t+1} - [h(x_t) + \lambda]\, x_t = r_t, \tag{6}$$

where $\lambda > 1$, $h \in C^1(\mathbb{R}, \mathbb{R})$ and $\underline{r} = \{r_t\} : \mathbb{Z} \to \mathbb{R}$ is a (Stepanov or) uniformly limit-periodic sequence (see Definition 1 and Proposition 1).

At first, let us deal with the case, when $\underline{r}$ is periodic. Since we would like to obtain a periodic solution for Equation (6), we associate to it its Schauder-like parametrization, namely

$$x_{t+1} - [h(q_t) + \lambda]\, x_t = r_t, \tag{7}$$

where $\underline{q} \in Q_D := \{\underline{p} \in \mathbb{R}^{\mathbb{Z}}, \|\underline{p}\|_\infty \leq D\}$.

Consider still the homogeneous equation, obtained by the linearization of Equation (6) at $\{q_t\}$, i.e.,

$$x_{t+1} - \left[h'(q_t) q_t + h(q_t) + \lambda\right] x_t = 0. \tag{8}$$

Let us assume that there exists a constant $D > 0$ such that

$$\forall x \in [-D, D]\colon h(x) \geq 0 \text{ and } h'(x)x + h(x) \geq 0, \tag{9}$$

jointly with

$$\|\underline{r}\|_\infty \leq \frac{\lambda - 1}{\lambda + 1} D. \tag{10}$$

We are ready to formulate the first main theorem (for the scalar case), when applying Proposition 2.

Theorem 1. *Let $\lambda > 1$ and let there exist a constant $D > 0$ such that condition (9) holds for $h \in C^1(\mathbb{R}, \mathbb{R})$, jointly with condition (10) for a (Stepanov or) uniformly limit-periodic sequence $\{r_t\}\colon \mathbb{Z} \to \mathbb{R}$. Then, Equation (6) admits a uniformly limit-periodic solution $\underline{z}$ satisfying*

$$\|\underline{z}\|_\infty \leq \frac{\lambda + 1}{\lambda - 1} \|\underline{r}\|_\infty. \tag{11}$$

Proof. Observe that, under the assumption (9), the homogeneous Equation (8) exhibits an exponential dichotomy with constants $K = 1$ and $\alpha = 1/\lambda$ because

$$|h'(q_t)q_t + h(q_t) + \lambda| = h'(q_t)q_t + h(q_t) + \lambda \geq \lambda,$$

by which

$$|x_t| = \frac{1}{|h'(q_t)q_t + h(q_t) + \lambda|} |x_{t+1}| \leq \frac{1}{\lambda} |x_{t+1}|.$$

Moreover, Equation (7) admits a unique entirely bounded solution $\underline{u}$ which takes the form

$$u_t = \sum_{\ell \in \mathbb{Z}} G_q(t, \ell) r_{\ell - 1},$$

where G_q is the Green function for Equation (7), where $r_t = 0$, (see Formula (5)). By the standard calculations, we obtain that (see e.g., [5])

$$\|\underline{u}\|_\infty \leq K \frac{1 + \alpha}{1 - \alpha} \|\underline{r}\|_\infty = \frac{\lambda + 1}{\lambda - 1} \|\underline{r}\|_\infty.$$

If $\underline{r}$ is T_k-periodic, then so must be $\underline{u}$ (see e.g., [20], Theorem 2.6).

For each $k \in \mathbb{N}$, we introduce

$$Q_k := \left\{ \underline{p} \in \mathbb{R}^{\mathbb{Z}},\ \underline{p} \text{ is } T_k\text{-periodic and } \left\|\underline{p}\right\|_\infty \leq D \right\},$$

and the operator $\mathcal{T}_k\colon Q_k \to \mathbb{R}^{\mathbb{Z}}$, where

$$\mathcal{T}_k(\underline{q}) := \sum_{\ell \in \mathbb{Z}} G_q(t, \ell) r_{\ell - 1}.$$

One can easily check that this operator is continuous and compact ($Q_k \cap \mathbb{R}^{\mathbb{Z} \cap [0, T_k]}$ is compact), and such that $\mathcal{T}_k(Q_k) \subset Q_k$, provided condition (10) holds.

Thus, for a given $\underline{r}$, we take D such that $D \geq \frac{\lambda+1}{\lambda-1} \|\underline{r}\|_\infty$, and the set Q_k as above. Applying the well known Brouwer fixed point theorem, $\mathcal{T}_k$ possesses a fixed point $\varphi_k \in Q_k$, which represents a T_k-periodic solution of Equation (6), where $\underline{r}$ is T_k-periodic. Moreover, $\sup_{k \in \mathbb{N}} \|\varphi_k\|_\infty \leq D$.

Now, let us proceed to a limit-periodic sequence $\underline{r}$. According to Definition 1, it is a limit of a family of periodic sequences $\{\underline{r_k}\}$. We take $D > \frac{\lambda+1}{\lambda-1} \sup_{k \in \mathbb{N}} \|\underline{r_k}\|_\infty$. Since we have the exponential dichotomy with the same constants for each perturbated system, all the assumptions of Proposition 2 are satisfied. Thus, we obtain the existence of a limit-periodic solution $\underline{z}$ of Equation (6) satisfying the inequality (11), which completes the proof. □

Remark 3. *Taking $H(x) := h'(x)x + h(x)$, we can see that if h is even (resp. odd), then H must be also. Thus, if h is odd, then the only situation in order function H could satisfy the assumption (9) occurs, provided $h(x) = 0$, for each $x \in [0, D]$, which does not have much meaning. On the other hand, if h is even, then H must be too, and it is sufficient to satisfy the inequality in condition (9) on $(0, D]$. One can easily check that the function h satisfies this assumption on $(0, D]$ if and only if the function $x \mapsto xh(x)$ (≥ 0) is non-decreasing. We can see then that the assumption (9) is in fact a local non-decreasing property.*

We can give some illustrative examples in order condition (9) to be satisfied with an implicit or explicit D.

Example 1 (with an implicit D). *Assume that there is an expansion of h arround 0,*

$$h(x) := a_n x^n + o(x^n),\ a_n > 0,$$

where n is a (strictly) positive even number. In such a case, we have $h'(x)x + h(x) = x^n\,(na_n + o(1))$, with $na_n > 0$. Unfortunately, an implicit character of D does not lead to an effective solvability criterium here.

Example 2 (D can be any positive real number). *Let h be a polynomial of an even degree, namely*

$$h(x) := \sum_{k=0}^{N} a_k x^{2k},$$

where $a_k \geq 0$, for every k.

More generally, we can consider the case of the following even Taylor expansion:

$$h(x) := \sum_{k=0}^{+\infty} a_k x^{2k},$$

where $a_k \geq 0$, for every k. Then, any D (strictly) smaller than the radius of convergence of the series exists. For instance, we can consider the function

$$h(x) := \frac{1}{1 - x^2},$$

with $D \in (0, 1)$. In this case, condition (10) can be reduced to

$$\|\underline{r}\|_\infty < \frac{\lambda - 1}{\lambda + 1},$$

provided $D = 1 - \varepsilon$ holds with a sufficiently small $\varepsilon > 0$.

4. Limit-Periodic Solutions: Vector Case

In this section, we will consider again Equation (6), but this time in a vector case.

Hence, let $\lambda > 1$, $h \in C^1(\mathbb{R}^n, \mathbb{R})$ and $\underline{r} = \{r_t\} : \mathbb{Z} \to \mathbb{R}^n$ be a (Stepanov or) uniformly limit-periodic sequence. As before, we associate to this equation its Schauder-like parametrization (7), and the homogeneous equation

$$x_{t+1} - [(\nabla h(q_t) \cdot x_t) q_t + h(q_t) x_t + \lambda x_t] = 0, \tag{12}$$

where this time $\underline{q} \in Q_D := \{\underline{p} \in (\mathbb{R}^n)^{\mathbb{Z}}, \|\underline{p}\|_\infty \leq D\}$. We also introduce, for each $q \in \mathbb{R}^n$, the (continuous) linear mapping $L_q \colon \mathbb{R}^n \to \mathbb{R}^n$, where

$$L_q \colon x \mapsto (\nabla h(q) \cdot x) q + h(q) x. \tag{13}$$

Letting $B_D := \{x \in \mathbb{R}^n, \|x\| \leq D\}$, we will assume that one of the following assumptions holds. Let us note that the first one is a vector analogy of the scalar case. Moreover, it is equivalent to impose this assumption either for all $x \in \mathbb{R}^n$, or on the ball B_D:

$$\exists D > 0, \forall x \in \mathbb{R}^n, \forall q \in B_D, \quad h(q) \geq 0 \text{ and } L_q(x) \cdot x \geq 0, \tag{14}$$

$$\exists \beta > 1, \exists D > 0, \forall x \in \mathbb{R}^n, \forall q \in B_D, \quad h(q) \geq 0 \text{ and } \|L_q(x) + \lambda x\| \geq \beta\|x\|. \tag{15}$$

Assume still that

$$\|\underline{r}\|_\infty \leq \frac{\lambda - 1}{\lambda + 1} D, \text{ or (in the case of condition (15)) } \|\underline{r}\|_\infty \leq \frac{\beta - 1}{\beta + 1} D. \tag{16}$$

We are ready to formulate the second main theorem (for the vector case).

Theorem 2. *Let $\lambda > 1$ and let there exist a constant $D > 0$ such that either condition (14) or condition (15), with a suitable constant $\beta > 1$, hold for L_q defined in the mapping (13), where $h \in C^1(\mathbb{R}^n, \mathbb{R})$. Let condition (16) still hold for a (Stepanov or) uniformly limit-periodic sequence $\{r_t\} : \mathbb{Z} \to \mathbb{R}^n$. Then, Equation (6) admits a uniformly limit-periodic solution $\underline{z}$ satisfying*

$$\|\underline{z}\|_\infty \leq \frac{\lambda + 1}{\lambda - 1}\|\underline{r}\|_\infty, \textit{ resp. } \|\underline{z}\|_\infty \leq \frac{\beta + 1}{\beta - 1}\|\underline{r}\|_\infty.$$

Proof. Under the assumptions (14) or (15), the homogeneous Equation (12) exhibits an exponential dichotomy. Indeed, in the first case, for any solution, we have

$$\|x_{t+1}\| \cdot \|x_t\| \geq x_{t+1} \cdot x_t \geq L_q(x_t) \cdot x_t + \lambda x_t \cdot x_t \geq \lambda\|x_t\|^2.$$

Thus, we receive the exponential dichotomy with constants $K = 1$ and $\alpha = 1/\lambda$. In the second case, a simple calculation leads to the exponential dichotomy with constants $K = 1$ and $\alpha = 1/\beta$. To consider both situations together, let us replace λ by β in the first case. The unique entirely bounded solution $\underline{u}$ of Equation (7) satisfies this time the inequality

$$\|\underline{u}\|_\infty \leq \frac{\beta + 1}{\beta - 1}\|\underline{r}\|_\infty.$$

If $\underline{r}$ is T_k-periodic, then so must be $\underline{u}$ (see again [20], Theorem 2.6).

Proceeding in a quite analogous way as in the scalar case in Section 3, we can prove the existence of a T_k-periodic solution φ_k of Equation (6), where $\underline{r}$ is T_k-periodic and such that $\sup_{k \in \mathbb{N}} \|\varphi_k\|_\infty \leq D$, provided condition (16) holds.

The claim follows, when applying again Proposition 2. □

Although the second inequality in condition (14) is linear with respect to h, we will show that Example 2 cannot be directly extended in a vector way, even in the case of monoms, which justifies considering the vector case separately.

Let us consider the monome

$$h(x) := c\prod_{j=1}^{n} x_j^{\alpha_j},$$

where $c > 0$, and each α_j is even. For any positive D, take $q_i = \frac{D}{\sqrt{n}}$, for all $i = 1, \ldots, n-1$, and $q_n = \epsilon\frac{D}{\sqrt{n}}$, with $\epsilon \in (0, 1)$. Then, $q = (q_1, \ldots, q_n) \in B_D$. Now, let us compute $L_q(x) \cdot x$, for $x = (1, \ldots, 1, -\theta)$. It is a quadratic polynomial with respect to θ, whose discriminant Δ takes in terms of ϵ the form $\Delta = \frac{a}{\epsilon^2} + b + c\epsilon^2$. Thus, for a sufficiently small ϵ, the discriminant Δ is positive, demonstrating that $L_q(x) \cdot x$ can admit negative values.

On the other hand, in the following illustrative example, we will be able to obtain a suitable local condition for $h(0) > 0$, even with an explicit D.

Example 3 (condition (14)). *Let us consider*

$$h(x) := C + \prod_{j=1}^{n} x_j^{\alpha_j},$$

where $C > 0$, and each α_j is even. Observe that h is everywhere positive and such that

$$L_q(x) \cdot x \geq [-\|\nabla h(q)\| \cdot \|q\| + h(q)] \|x\|^2.$$

Hence, in order to satisfy condition (14), it is enough to obtain the inequality

$$\|\nabla h(q)\| \cdot \|q\| \leq h(q),$$

and since $h(q) \geq C$, it is still enough to have

$$\|\nabla h(q)\| \cdot \|q\| \leq C.$$

A basic majorization of the i-th component of $\nabla h(q)$, under the constraint $\|q\| \leq D$, is $\alpha_i D^{\sum_j \alpha_j - 1}$. From this, for $\|q\| \leq D$, we obtain

$$\|\nabla h(q)\|.\|q\| \leq D^{\sum_j \alpha_j} \sqrt{\sum_j \alpha_j^2}.$$

Thus, it is sufficient to choose

$$D = \left(\frac{C^2}{\sum_j \alpha_j^2} \right)^{\frac{1}{2\sum_j \alpha_j}}.$$

We have not obviously made an optimal majorization of $\|\nabla h(q)\|$ in order to obtain a simple and transparent condition. In other words, our estimation can be certainly improved for obtaining a larger D.

Let us deduce a slightly more effective condition for $n = 2$ (again not an optimal one). In this case,

$$\|\nabla h(q)\| = |q_1|^{\alpha_1 - 1} |q_2|^{\alpha_2 - 1} \sqrt{\alpha_1 q_2^2 + \alpha_2^2 q_1^2} \leq |q_1|^{\alpha_1 - 1} |q_2|^{\alpha_2 - 1} \max\{\alpha_1, \alpha_2\} \|q\| \leq c(\alpha) D^{\alpha_1 + \alpha_2 - 1},$$

where

$$c(\alpha) := \frac{(\alpha_1 - 1)^{\frac{\alpha_1 - 1}{2}} (\alpha_2 - 1)^{\frac{\alpha_2 - 1}{2}}}{(\alpha_1 + \alpha_2 - 2)^{\frac{\alpha_1 + \alpha_2 - 2}{2}}} \max\{\alpha_1, \alpha_2\}.$$

Thus,

$$\|\nabla h(q)\| \cdot \|q\| \leq c(\alpha) D^{\alpha_1 + \alpha_2},$$

and we can choose $D = \left(\frac{C}{c(\alpha)} \right)^{\frac{1}{\alpha_1 + \alpha_2}}$.

Example 4 (condition (15)). *Let us turn to the ball B_D with a fixed D. We assume a priori that $h(q) \geq 0$ holds in a neighbourhood of 0. For instance, let $h(0) > 0$. Furthermore, suppose that, for some $c > 0$ and $p > 0$, we have*

$$\|\nabla h(q)\| \leq c\|q\|^p.$$

Then, $|h(q)| \leq h(0) + \frac{c}{p+1} \|q\|^{p+1}$, and subsequently

$$\|L_q(x)\| \leq \left[h(0) + c\frac{p+2}{p+1} \|q\|^{p+1} \right] \|x\|.$$

Thus, let us still suppose that $h(0) > 1$ and $\beta \in (1, h(0))$. For (an explicit value of) D, we get for any $q \in B_D$ that

$$\|L_q(x) + \lambda x\| \leq \left[\lambda - \left(h(0) + c\frac{p+2}{p+1}\|q\|^{p+1}\right)\right] \|x\|.$$

Assume finally that $\lambda - h(0) > 1$, and take any $\beta \in (1, \lambda - h(0))$. We have

$$\left(\lambda - \left(h(0) + c\frac{p+2}{p+1}\|q\|^{p+1}\right) > \beta\right) \Leftrightarrow \left(\|q\| \leq \left(\frac{p+1}{p+2}\frac{\lambda - h(0) - \beta}{c}\right)^{\frac{1}{p+1}}\right).$$

After all, we can take

$$D \leq \left(\frac{p+1}{p+2}\frac{\lambda - h(0) - \beta}{c}\right)^{\frac{1}{p+1}}$$

in order to satisfy condition (15). By the optimization with respect to β, we can readily check that any D satisfying

$$D < \left(\frac{p+1}{p+2}\frac{\lambda - h(0) - 1}{c}\right)^{\frac{1}{p+1}}$$

can be chosen for it. The last step is to specify D such that $h(q) \geq 0$, for every $q \in B_D$.

5. Conclusions

Under the assumptions of Theorems 1 and 2, the obtained limit-periodic solutions are obviously also almost-periodic. On the other hand, if the forcing terms $\{r_t\}$ in Equation (6) are almost-periodic (or, in particular, quasi-periodic), then one should proceed in a different manner in order to get an almost-periodic (resp. quasi-periodic) solution. However, if the forcing terms $\{r_t\}$ in Equation (6) are at the same time limit-periodic and quasi-periodic, then they become simply periodic (see [4], Theorem 2 and [5], Remark 4). In this very special case, the existence criteria for periodic solutions can be significantly improved. Concretely, conditions (9), (14), and (15) can be reduced into $h(x) \geq 0$, for $x \in B_D$.

Observe that, in the special case of the limit-periodically forced logistic equation (i.e., $h(x) := -\mu x + \mu - \lambda$, $\mu > 0$), namely

$$x_{t+1} + \mu(x_t - 1)x_t = r_t,$$

condition (9) takes the form

$$\begin{aligned}&\exists D > 0 \text{ such that, for some } \lambda \in (1, \mu),\\ &\forall x \in [-D, D]: \; -2\mu x + \mu - \lambda \geq 0.\end{aligned}$$

Condition (10) is the same as above, i.e.,

$$\|\underline{r}\|_\infty \leq \frac{\lambda - 1}{\lambda + 1} D.$$

One can readily check that they can be satisfied for $D = \frac{\mu - \lambda}{2\mu}$, $\mu > \lambda > 1$, and

$$\|\underline{r}\|_\infty \leq \frac{\lambda - 1}{\lambda + 1} \cdot \frac{\mu - \lambda}{2\mu}.$$

For example, taking $\mu = 3.5$ and $\lambda = 2$, we have $D = \frac{3}{14}$, and subsequently $\|\underline{r}\|_\infty \leq \frac{1}{14}$.

Let us finally note that if condition (9) holds on the whole line, like e.g., for $h(x) := \frac{\pi}{2} + \arctan x$, then Equation (6) admits a limit-periodic solution for any limit-periodic forcing $\{r_t\}$. In the special case of $h(x) := \frac{\pi}{2} + \arctan x$, the second inequality in condition (9) namely holds because

$$[h'(x)x + h(x)]' = \frac{2}{x^4 + 2x^2 + 1} > 0,$$

by which $h'(x)x + h(x)$ is strictly increasing, jointly with

$$\lim_{x \to -\infty} [h'(x)x + h(x)] = \lim_{x \to -\infty} \left[\frac{x}{1 + x^2} + \frac{\pi}{2} + \arctan x \right] = 0.$$

More generally, a sufficient condition for satisfying condition (9), for all $x \in \mathbb{R}$, takes the form:

$$h(x) \geq 0, \ [h(x)x]'' \geq 0 \text{ and } \lim_{x \to -\infty} [h(x)x]' \geq 0.$$

In the vector case, the same is true, provided condition (14) or condition (15) holds, for all $q \in \mathbb{R}^n$.

Author Contributions: The contributions of both authors (J.A. and D.P.) are equal. All the main results and illustrative examples were developed together.

Funding: This research was funded by Grant Agency of Palacký University in Olomouc grant number IGA_PrF_2018_024 "Mathematical Models".

Conflicts of Interest: The authors declare no conflict of interest.

References

1. Damanik, D.; Fillman, J. Spectral properties of limit-periodic operators. *arXiv* **2018**, arXiv:1802.05794.
2. Damanik, D.; Gan, Z. Spectral properties of limit-periodic Schrödinger operators. *Commun. Pure Appl. Anal.* **2011**, *10*, 859–871. [CrossRef]
3. Alonso, A.I.; Obaya, R.; Ortega, R. Differential equations with limit-periodic forcings. *Proc. Am. Math. Soc.* **2003**, *131*, 851–857. [CrossRef]
4. Andres, J.; Pennequin, D. Semi-periodic solutions of difference and differential equations. *Bound. Value Prob.* **2012**, *2012*, 141. [CrossRef]
5. Andres, J.; Pennequin, D. Limit-periodic solutions of difference and differential systems without global Lipschitzianity restrictions. *J. Differ. Equ. Appl.* **2018**, *24*, 955–975. [CrossRef]
6. Andres, J.; Pennequin, D. Existence, localization and stability of limit-periodic solutions to differential equations involving cubic nonlinearities. *Topol. Meth. Nonlinear Anal.* **2019**, in press.
7. Seifert, G. Almost periodic solutions for limit periodic systems. *SIAM J. Appl. Math.* **1972**, *22*, 38–44. [CrossRef]
8. Tarallo, M.; Zhou, Z. Limit periodic upper and lower solutions in a generic case. *Discr. Cont. Dyn. Syst.* **2018**, *38*, 293–309. [CrossRef]
9. Akhmet, M.U. Almost periodic solutions of differential equations with piecewise constant argument of generalized type. *Nonlinear Anal. Hybrid Syst.* **2008**, *2*, 456–467. [CrossRef]
10. Akhmet, M.U.; Kivilcim, A. Periodic motions generated from non-autonomous grazing dynamics. *Commun. Nonlinear Sci. Numer. Simul.* **2017**, *49*, 48–62. [CrossRef]
11. Wang, C.; Agarwal, R.P. Almost periodic dynamics for impulsive delay neural networks of a general type on almost periodic time scales. *Commun. Nonlinear Sci. Numer. Simul.* **2016**, *36*, 238–251. [CrossRef]
12. Wang, C.; Agarwal, R.P. Almost periodic solution for a new type of neutral impulsive stochastic Lasota–Wazewska timescale model. *Appl. Math. Lett.* **2017**, *70*, 58–65. [CrossRef]
13. Berg, I.D.; Wilansky, A. Periodic, almost-periodic, and semiperiodic sequences. *Mich. Math. J.* **1962**, *9*, 363–368. [CrossRef]
14. Farkas, M. *Dynamical Models in Biology*; Academic Press: San Diego, CA, USA, 2001.
15. Murray, J.D. *Mathematical Biology*; Springer: Berlin, Germany, 1993.

16. Andres, J.; Pennequin, D. On Stepanov almost-periodic oscillations and their discretizations. *J. Differ. Equ. Appl.* **2012**, *18*, 1665–1682. [CrossRef]
17. Bodine, S.; Lutz, D.A. *Asymptotic Integration of Differential and Difference Equations*; LNM 2129; Springer: Berlin, Germany, 2015.
18. Palmer, K.J. A finite-time condition for exponential dichotomy. *J. Differ. Equ. Appl.* **2011**, *17*, 221–234. [CrossRef]
19. Alonso, A.I.; Hong J.; Obaya, R. Exponential dichotomy and trichotomy for difference equations. *Comput. Math. Appl.* **1999**, *38*, 41–49. [CrossRef]
20. Dannan, F.; Elaydi, S.; Liu, P. Periodic solutions of difference equations. *J. Differ. Equ. Appl.* **2000**, *6*, 203–232. [CrossRef]

axioms

MDPI

Article

Bäcklund Transformations for Nonlinear Differential Equations and Systems

Tatyana V. Redkina, Robert G. Zakinyan, Arthur R. Zakinyan *, Olesya B. Surneva and Olga S. Yanovskaya

Institute of Mathematics and Natural Sciences, North Caucasus Federal University, 1 Pushkin Street, 355009 Stavropol, Russia; TVR59@mail.ru (T.V.R.); zakinyan@mail.ru (R.G.Z.); surnevao@mail.ru (O.B.S.); olenka_yan@mail.ru (O.S.Y.)

* Correspondence: zakinyan.a.r@mail.ru; Tel.: +7-918-7630-710

Received: 5 February 2019; Accepted: 7 April 2019; Published: 11 April 2019

Abstract: In this work, new Bäcklund transformations (BTs) for generalized Liouville equations were obtained. Special cases of Liouville equations with exponential nonlinearity that have a multiplier that depends on the independent variables and first-order derivatives from the function were considered. Two- and three-dimensional cases were considered. The BTs construction is based on the method proposed by Clairin. The solutions of the considered equations have been found using the BTs, with a unified algorithm. In addition, the work develops the Clairin's method for the system of two third-order equations related to the integrable perturbation and complexification of the Korteweg-de Vries (KdV) equation. Among the constructed BTs an analog of the Miura transformations was found. The Miura transformations transfer the initial system to that of perturbed modified KdV (mKdV) equations. It could be shown on this way that, considering the system as a link between the real and imaginary parts of a complex function, it is possible to go to the complexified KdV (cKdV) and here the analog of the Miura transformations transforms it into the complexification of the mKdV.

Keywords: Bäcklund transformation; Clairin's method; generalized Liouville equation; Miura transformation; Korteweg-de Vries equation

1. Introduction

Currently, nonlinear partial differential equations are widely used to describe the so-called "fine processes", such as propagation of nonlinear waves in dispersive media [1]. Due to the complexity of different nonlinear equations, no common method of their solution exists. For the integrable systems, efficient methods have been developed, such as the inverse scattering method [2,3], Hirota method [4], Painlevé method [5], Bäcklund transformation [6], a method of mapping and deformation [3], nonlocal symmetry method [7,8], etc.

In the classical works [2,6] the Bäcklund transformations (BTs) were considered for the couple of differential second order partial differential equations and presented in form of a system of relations and containing independent variables, functions of the said equations, and their first-order derivatives. The BTs allow to obtain not only couples of equations but, if the solution of one of them is known, obtain the solution of the other one.

BT plays an important role in the integrable systems because it reveals the inner relations between different integrable properties, such as determination of the point symmetries [9,10], the presence of the Hamiltonian structure [11–13].

Lots of research has recently been conducted in this area. For example, determining the complementary symmetries and obtaining the Miura transformations for the hierarchy of the Kadomtsev–Petviashvili (KP) equation and modified KP, including for the discrete analog [14,15];

in [16] the new BTs relative to the residual symmetry of the (2 + 1)-dimensional Bogoyavlenskij equation [17] have been investigated; construction of new auto Bäcklund transformations for the Lagrange system and of Henon-Heiles system of equations in parabolic coordinates [18]; it has been shown that the calibration conditions in the theory of the relativistic string, which allow using the d'Alembert equation instead of the nonlinear Liouville equation, are direct consequences of the BT relating the solutions of these equations [19].

In Reference [20] it is shown how pseudo constants of the Liouville-type equations can be exploited as a tool for construction of the Bäcklund transformations. In Reference [21] it is proven that contact-nonequivalent three-dimensional linearly degenerate second-order equations that are Lax-integrable are related to each other by the corresponding Bäcklund transformations.

This work describes how new BTs for the Liouville generalized equations are obtained. The second and third sections deal with the special cases of the Liouville equation with exponential nonlinearity that have a multiplier that depends upon the independent variables and first-order derivatives from the function, and the three-dimensional case. The BTs construction is based on the method proposed by Clairin and has at such approach a clear geometric sense. The solutions of the considered equations have been found using the BTs, with a unified algorithm.

The fourth section contains the development of Clairin's method for the system of two third-order equations related to the integrable perturbation and complexification of the KdV (cKdV) equation [22]. An essential point for these dynamic systems of equations is that the application of special conditions to the differential forms may lead to different dynamic systems.

Among the constructed BTs an analog of the Miura transformations was found in Section five. The Miura transformations transfer the initial system to that of perturbed modified KdV (mKdV) equations. In this way, we were able to show that when considering the system as a relation between the real and imaginary parts of a complex function, we can pass to the cKdV, and the analog of the Miura transformations transforms it into the complexification of mKdV.

2. Bäcklund Transformations for Special Cases of Liouville Equations

Theorem 1. *Partial differential equation*

$$z_{\xi\eta} = f_1(\xi)f_2(\eta)e^z \tag{1}$$

and wave equation $w_{\xi\eta} = 0$ *are related by the Bäcklund transformation of the form:*

$$\frac{\partial w}{\partial \xi} = be^{\frac{w+z}{2}}\sqrt{f_1(\xi)f_2(\eta)} + \frac{\partial z}{\partial \xi} + \frac{f_1'(\xi)}{f_1(\xi)}, \ \frac{\partial w}{\partial \eta} = -\frac{1}{b}e^{-\frac{w-z}{2}}\sqrt{f_1(\xi)f_2(\eta)} - \frac{\partial z}{\partial \eta} - \frac{f_2'(\eta)}{f_2(\eta)} \tag{2}$$

where b is an arbitrary parameter, $f_1(\xi), f_2(\eta)$ *are arbitrary functions of one variable,* $w(\xi,\eta)$ *and* $z(\xi,\eta)$ *are functions of two variables.*

The Proof uses the cross differentiation of the Equations (2) and then summing or finding the difference of the resulting expression.

Corollary 1. *If the wave equation*$w_{\xi\eta} = 0$ *has the solution*

$$w(\xi,\eta) = \theta(\eta) + \vartheta(\xi) \tag{3}$$

then Equation (1) has the solution

$$z(\xi,\eta) = \vartheta(\xi) - \theta(\eta) - \ln\left[|f_2(\eta)f_1(\xi)|\left(\frac{1}{2b}\int e^{-\theta(\eta)}d\eta + \frac{b}{2}\int e^{\vartheta(\xi)}d\xi\right)^2\right] \tag{4}$$

where $\vartheta(\xi)$, $\theta(\eta)$ are arbitrary functions.

Proof of Corollary 1. Substitute solution (3) into transformations (2), and get the system of equations for the function $z(\xi,\eta)$

$$\frac{\partial\vartheta}{\partial\xi}=be^{\frac{\theta(\eta)+\vartheta(\xi)+z}{2}}\sqrt{f_1(\xi)f_2(\eta)}+\frac{\partial z}{\partial\xi}+\frac{f_1'(\xi)}{f_1(\xi)},\ \frac{\partial\theta}{\partial\eta}=-\frac{1}{b}e^{-\frac{\theta(\eta)+\vartheta(\xi)-z}{2}}\sqrt{f_1(\xi)f_2(\eta)}-\frac{\partial z}{\partial\eta}-\frac{f_2'(\eta)}{f_2(\eta)}. \quad (5)$$

Seek for the solution of the system in form $z(\xi,\eta)=2\ln\varphi(\xi,\eta)$, then (5) takes the form of two Bernoulli equations. Their solutions have the forms

$$\varphi_1=\frac{1}{\sqrt{f_1(\xi)}}\frac{2e^{\frac{1}{2}\vartheta(\xi)}}{be^{\frac{\theta(\eta)}{2}}\sqrt{f_2(\eta)}\int e^{\vartheta(\xi)}d\xi+\psi_1(\eta)},\varphi_2=\frac{1}{\sqrt{f_2(\eta)}}\frac{2e^{-\frac{1}{2}\theta(\eta)}}{b^{-1}e^{-\frac{\vartheta(\xi)}{2}}\sqrt{f_1(\xi)}\int e^{-\theta(\eta)}d\eta+\psi_2(\xi)} \quad (6)$$

where $\psi_1(\eta)$, $\psi_2(\xi)$ are arbitrary functions.

Compare the resulting solutions (6) and determine the condition at which they coincide, then functions $\psi_2(\xi)$ and $\psi_1(\eta)$ must be predetermined as follows

$$\psi_1(\eta)=\frac{1}{b}e^{\frac{\theta(\eta)}{2}}\sqrt{f_2(\eta)}\int e^{-\theta(\eta)}d\eta,\psi_2(\xi)=be^{\frac{-\vartheta(\xi)}{2}}\sqrt{f_1(\xi)}\int e^{\vartheta(\xi)}d\xi \quad (7)$$

obtaining the solution of system (5) in the form

$$\varphi(\xi,\eta)=2e^{\frac{\vartheta(\xi)-\theta(\eta)}{2}}(f_1(\xi)f_2(\eta))^{-\frac{1}{2}}\left(\frac{1}{b}\int e^{-\theta(\eta)}d\eta+b\int e^{\vartheta(\xi)}d\xi\right)^{-1}$$

and the solution of Equation (1) in the form (4).

Clairin has proposed a method of Bäcklund transformations construction for the hyperbolical form of nonlinear equations. This procedure will be applied to the equation

$$\tilde{z}_{\xi\eta}=e^{\tilde{z}}\left(B_1\tilde{z}_\xi+B_2\tilde{z}_\eta\right),\ B_1,B_2-const \quad (8)$$

where $\tilde{z}(\xi,\eta)$ is a function of two variables, and the Bäcklund transformation will be constructed. □

Theorem 2. *Bäcklund transformations of the form:*

$$\begin{aligned} w_{\xi\xi}&=-\tfrac{1}{2}B_2e^{\tilde{z}}w_\xi+w_\xi\tfrac{\partial\tilde{z}}{\partial\xi},\\ w_{\xi\eta}&=\tfrac{1}{2}B_1e^{\tilde{z}}w_\xi+\tfrac{w_\xi}{2}\tfrac{\partial\tilde{z}}{\partial\eta}, \end{aligned} \quad (9)$$

relate the two equations, (8), and

$$B_2(w^2)_{\xi\eta}+4B_1w_\xi^2=0 \quad (10)$$

where B_1,B_2 are arbitrary constants, and $w(\xi,\eta)$, $z(\xi,\eta)$ are functions of two variables.

The Proof is similar to that of Theorem 1.

Corollary 2. *If Equation (10) has the solution*

$$w=2B_1\eta-B_2\xi, \quad (11)$$

then Equation (8) has the solution

$$\tilde{z}=-\ln\left|C+B_1\eta-\frac{B_2}{2}\xi\right|,C-const. \quad (12)$$

Proof of Corollary 2. Use the found transformations (9) and substitute the known solution (11), the system takes the form

$$B_2\frac{\partial \widetilde{z}}{\partial \eta} = -2B_1\frac{\partial \widetilde{z}}{\partial \xi}, 0 = B_1 e^{\widetilde{z}} + \frac{\partial \widetilde{z}}{\partial \eta}, \tag{13}$$

where, from the first linear partial differential equation, find the relation between the independent variables ξ, η, and, from the second equation of the system (13) determine the form of the function $\widetilde{z} = -\ln|C + 0.5t|$, $t = 2B_1\eta - B_2\xi$, $C - const$. The result is the solution of Equation (8) in the form (12). □

Corollary 3. *If Equation (10) has the solution*

$$w = e^{\frac{\lambda}{2B_2}\eta - \frac{\lambda}{2B_1}\xi}, \ \lambda - const, \tag{14}$$

then Equation (8) has the solution

$$\widetilde{z} = \frac{\lambda(2B_1\eta - B_2\xi)}{2B_1B_2} - \ln\left|1 + B_1B_2Ce^{\frac{\lambda(2B_1\eta - B_2\xi)}{2B_1B_2}}\right| + \ln|C\lambda|, C - const. \tag{15}$$

The Proof is similar to that of Corollary 2.

Corollary 4. *If Equation (8) has the solution* $\widetilde{z} = B_1\eta - B_2\xi$, *then Equation (10) has the solution*

$$w(\xi, \eta) = -\frac{2}{B_2}\exp\left(\frac{1}{2}e^{B_1\eta - B_2\xi} - \frac{B_1}{2}\eta\right). \tag{16}$$

Proof of Corollary 4. Use the Bäcklund transformations (9) and substitute the available solution $\widetilde{z} = B_1\eta - B_2\xi$, and get the system of equations, that can be integrated by the relevant variables

$$\begin{aligned} \ln w_\xi &= \tfrac{1}{2}e^{B_1\eta - B_2\xi} - B_2\xi + \psi_1(\eta), \\ \ln w_\xi &= \tfrac{1}{2}e^{B_1\eta - B_2\xi} + \tfrac{B_1}{2}\eta + \psi_2(\xi), \end{aligned} \tag{17}$$

where $\psi_1(\eta)$, $\psi_2(\xi)$ are the integration constants. Complete the definition of functions $\psi_1(\eta)$ and $\psi_2(\xi)$, so that the resulting values of the right parts of system (17) coincide. This is possible if $\psi_1(\eta) = 0,5B_1\eta$, $\psi_2(\xi) = -B_2\xi$. As a result, the value

$$w_\xi = \exp\left(\frac{1}{2}e^{B_1\eta - B_2\xi} - B_2\xi + \frac{B_1}{2}\eta\right)$$

is determined. Integration by variable ξ yields

$$w(\xi, \eta) = \phi(\eta) - \frac{2}{B_2}e^{\frac{1}{2}e^{B_1\eta - B_2\xi} - \frac{B_1}{2}\eta},$$

where $\phi(\eta)$ is an arbitrary function. For a greater certainty of $\phi(\eta)$, substitute the found function into Equation (10). The equality will be fulfilled identically if

$$2\phi'(\eta) + B_1\phi(\eta)e^{B_1\eta - B_2\xi} + B_1\phi(\eta) = 0.$$

The obtained equation depends upon variable ξ, which must not happen, hence, assume $\phi(\eta) = 0$, then the desired function has the form (16). □

3. Bäcklund Transformations for Three-Dimensional Liouville Equation

Theorem 3. *Nonlinear partial differential equation*

$$v_{\eta\xi} + \frac{c}{\gamma^2} e^v \left(3\gamma v_\eta + v_\zeta + \gamma v_\xi\right) = 0 \tag{18}$$

is linked to the nonlinear equation

$$\varphi_\eta \left[\gamma\varphi_\xi + \varphi_\zeta + 3\gamma\varphi_\eta\right] = \varphi_{\xi\eta}, \tag{19}$$

by the Bäcklund transformations of the form:

$$\begin{aligned} &\gamma\varphi_\xi + 3\gamma\varphi_\eta + \varphi_\zeta = \gamma v_\xi, \\ &\gamma\left[\varphi_{\eta\eta} + \varphi_\eta^2\right] + ce^v\left[\tfrac{c}{\gamma}e^v + v_\eta\right] = -2ce^v\varphi_\eta, \\ &\gamma\left[\varphi_{\xi\eta} + \varphi_\xi\varphi_\eta\right] + \varphi_{\zeta\eta} + \varphi_\zeta\varphi_\eta - \gamma v_\xi\varphi_\eta - \tfrac{c}{\gamma}e^v[3ce^v - \gamma v_\xi - v_\zeta] = 6ce^v\varphi_\eta, \end{aligned} \tag{20}$$

wherec, γ *are arbitrary constants, and* $\varphi(\xi, \eta, \zeta)$, $v(\xi, \eta, \zeta)$ *are functions of three variables* ξ, η, ζ.

Proof of Theorem 3. Shows that system (20) leads to Equation (18). For this differentiate the first equality of relation (20) by variable η

$$\gamma\varphi_{\xi\eta} + 3\gamma\varphi_{\eta\eta} + \varphi_{\zeta\eta} = \gamma v_{\xi\eta} \tag{21}$$

and determine from the second and third equalities the second order derivatives $\varphi_{\xi\eta}$, $\varphi_{\eta\eta}$, $\varphi_{\zeta\eta}$, then, having substituted their values into (21), gives

$$\left[\gamma v_\xi - 3\gamma\varphi_\eta - \gamma\varphi_\xi - \varphi_\zeta\right]\varphi_\eta - \frac{c}{\gamma}e^v\left(\gamma v_\xi + v_\zeta + 3\gamma v_\eta\right) = \gamma v_{\xi\eta}. \tag{22}$$

By reason of the first equality of system (20), the coefficient at function φ_η becomes zero and there remains the equality that relates the only function $v(\xi, \eta, \zeta)$:

$$-se^v\left(\gamma v_\xi + v_\zeta + 3\gamma v_\eta\right) = \gamma^2 v_{\xi\eta}.$$

Then, try to get rid of function $v(\xi, \eta, \zeta)$ in the initial system of transformations (20). In the second equation of the system separate the combination of functions $\varphi_\eta + c\gamma^{-1}e^v$, so that the equality takes the form

$$\left(\varphi_\eta + \frac{c}{\gamma}e^v\right)_\eta + \left(\frac{c}{\gamma}e^v + \varphi_\eta\right)^2 = 0. \tag{23}$$

In the third equation of system (20), substitute the value γv_ξ from the first equality (20), then, after having grouped the elements together

$$\left(\frac{c}{\gamma}e^v + \varphi_\eta\right)_\xi - 3\left(\frac{c}{\gamma}e^v + \varphi_\eta\right)^2 + \frac{1}{\gamma}\left(\frac{c}{\gamma}e^v + \varphi_\eta\right)_\zeta = 0. \tag{24}$$

Having separated the total derivatives, rewrite the first and third equations as

$$\begin{aligned} &\left(\tfrac{\partial}{\partial\xi} + 3\tfrac{\partial}{\partial\eta} + \tfrac{1}{\gamma}\tfrac{\partial}{\partial\zeta}\right)\varphi = v_\xi, \\ &\left(\tfrac{\partial}{\partial\xi} + 3\tfrac{\partial}{\partial\eta} + \tfrac{1}{\gamma}\tfrac{\partial}{\partial\zeta}\right)\left(\tfrac{c}{\gamma}e^v + \varphi_\eta\right) = 0. \end{aligned} \tag{25}$$

Obviously if $c\gamma^{-1}e^{v} + \varphi_\eta = C,\ C \neq 0$ is assumed, such form is not a solution of Equation (23), hence, the two situations are possible:

$$\mathbf{1.}\ C = 0,\ \ \mathbf{2.}\ ce^{v} + \gamma\varphi_\eta = z(\xi, \eta, \zeta) \tag{26}$$

$z(\xi, \eta, \zeta)$ is some function. The simplest is that if $C = 0$ is assumed, then, for function v, get $v = \ln(-c\gamma^{-1}\varphi_\eta)$. As a result of the substitution, the first equality transforms into the nonlinear form (19). □

Corollary 5. *If nonlinear partial differential Equation (18) has the solution*

$$v(\xi, \eta, \zeta) = c\gamma^{-1}\eta + f(\xi - \gamma\zeta) - 3c\zeta \tag{27}$$

where $f(\xi - \gamma\zeta)$ is an arbitrary function of the combined variable $\xi - \gamma\zeta$, then Equation (19) has the solution in the form

$$\varphi(\xi, \eta, \zeta) = \frac{a\xi + \gamma b\zeta}{a+b} f'(\xi - \gamma\zeta) - \exp\left(\frac{c}{\gamma}\eta + f(\xi - \gamma\zeta) - 3c\zeta\right) + r(\xi - \gamma\zeta), \tag{28}$$

where $\varphi(\xi, \eta, \zeta),\ v(\xi, \eta, \zeta)$ are functions of three variables ξ, η, ζ, $r(\xi, \eta, \zeta)$ is an arbitrary function, a, b, c are arbitrary constants.

Proof of Corollary 5. Use Bäcklund transformations (20). Perform this substitution of function $v(\xi, \eta, \zeta)$ (27) into (20); obviously, the last two equations of the system will be fulfilled identically if

$$\varphi_\eta = -\frac{c}{\gamma}\exp\left(\frac{c}{\gamma}\eta + f(\xi - \gamma\zeta) - 3c\zeta\right). \tag{29}$$

Having integrated the last equality get the sought for a function of the form

$$\varphi(\xi, \eta, \zeta) = q(\xi, \zeta) - \exp\left(\frac{c}{\gamma}\eta + f(\xi - \gamma\zeta) - 3c\zeta\right) \tag{30}$$

where $q(\xi, \zeta)$ is an arbitrary function. For greater certainty, use the remaining first equality of the system (20), then

$$q_\xi(\xi, \zeta) + \gamma^{-1}q_\zeta(\xi, \zeta) = f'(\xi - \gamma\zeta). \tag{31}$$

As in the resulting linear Equation (31), one of the first integrals coincides with the form of the argument of the function of the right part, write the solution in the form

$$q(\xi, \zeta) = g(\xi, \zeta) f'(\xi - \gamma\zeta) \tag{32}$$

with the unknown function $g(\xi, \zeta)$, which is obtained from the linear equation obtained after substitution into (31),

$$g(\xi, \zeta) = \frac{1}{a+b}(a\xi + \gamma b\zeta) + r_1(\xi - \gamma\zeta) \tag{33}$$

which is determined with accuracy to the summand of the form $r_1(\xi - \gamma\zeta),\ a,\ b$ are arbitrary parameters simultaneously not equal to zero. Now put together the resulting values of the functions (30), (32), and (33); this yields the sought for solution (28). □

Corollary 6. *Equations (18) and (19) have a solution in the form*

$$F = F_1(3\gamma\zeta - \eta) + F_2(\xi - \gamma\zeta) \tag{34}$$

where $F_1,\ F_2$ are arbitrary functions.

Get back to the above rationale and consider the second case (26). It may be shown that Equation (18) relates to a more complex equation. For this, make in (23) change 2 in (26)

$$\gamma z_\eta + z^2 = 0. \tag{35}$$

It can be seen that this equality may be integrated

$$z(\xi,\eta,\zeta) = \frac{\gamma}{\eta + \psi(\xi,\zeta)} \tag{36}$$

where $\psi(\xi,\zeta)$ is an arbitrary function. Substitute the found function into the last equality (25), this yields the equation for the function $\psi(\xi,\zeta)$

$$[\eta + \psi(\xi,\zeta)]^{-2}[\gamma\psi_\xi(\xi,\zeta) + 3\gamma + \psi_\zeta(\xi,\zeta)] = 0. \tag{37}$$

The common solution (37) will be written in the form $F(\xi - \gamma\zeta, 3\gamma\zeta + \psi) = 0$, where F is an arbitrary function. Consider the partial solution in the form of a linear relation in relation to the second combined variable

$$\psi(\xi,\zeta) = f(\xi - \gamma\zeta) - 3\gamma\zeta \tag{38}$$

with an arbitrary form of the function f. Hence, (36) takes the form

$$z(\xi,\eta,\zeta) = \frac{\gamma}{\eta + f(\xi - \gamma\zeta) - 3\gamma\zeta}. \tag{39}$$

From (26), find the function $v(\xi,\eta,\zeta)$

$$v(\xi,\eta,\zeta) = \ln\left|\frac{\gamma}{c}\left(\frac{1}{\eta + f(\xi - \gamma\zeta) - 3\gamma\zeta} - \varphi_\eta\right)\right| \tag{40}$$

then the first equality of system (25) takes the form

$$\gamma\varphi_{\xi\eta} + \frac{\gamma f'(\xi - \gamma\zeta)}{[\eta + f(\xi - \gamma\zeta) - 3\gamma\zeta]^2} = \left(\varphi_\eta - \frac{1}{\eta + f(\xi - \gamma\zeta) - 3\gamma\zeta}\right)\left[\gamma\varphi_\xi + \varphi_\zeta + 3\gamma\varphi_\eta\right]. \tag{41}$$

Theorem 4. *Nonlinear partial differential Equation (18) relates to the class of nonlinear Equations (41) by Bäcklund transformations (20), where $f(\xi - \gamma\zeta)$ is an arbitrary function of the combined variable $\xi - \gamma\zeta$.*

The solution of Equation (41) may be obtained having assumed

$$\varphi_\eta = \frac{1}{\eta + f(\xi - \gamma\zeta) - 3\gamma\zeta}$$

then

$$\varphi = \ln\left|\eta + f(\xi - \gamma\zeta) - 3\gamma\zeta\right| + q(\xi,\zeta) \tag{42}$$

where $q(\xi,\zeta)$ is an arbitrary function.

Corollary 7. *Function (42), where $f(\xi - \gamma\zeta)$ and $q(\xi,\zeta)$ are arbitrary functions, is a solution of Equation (41).*

Use the fact that, according to theorem 2, Equation (18) and family of Equations (41) are related by Bäcklund transformations (20), and see how the trivial solution $v(\xi,\eta,\zeta) = C - const$ of the first equation may serve to construct a solution for the family (41).

Corollary 8. *Family of nonlinear partial differential Equations (41) has the solution*

$$\varphi(\xi,\eta,\zeta) = \ln\left|3\gamma\zeta - \eta + f(\xi-\gamma\zeta)\right| + ac\gamma^{-1}(3\gamma\zeta-\eta) + f_2(\xi-\gamma\zeta) \tag{43}$$

where $f_2(\xi-\gamma\zeta)$ is an arbitrary function of the combined variable $\xi-\gamma\zeta$, a is an arbitrary constant.

Proof of Corollary 8. Substitute function $v(\xi,\eta,\zeta)= C - const$ to system (20), then the first equality of system (20) yields

$$\varphi(\xi,\eta,\zeta) = F(3\gamma\zeta-\eta, \xi-\gamma\zeta) \tag{44}$$

with an arbitrary function F. Denote the first component derivative as $F'_{(1)}$ and the second component derivative as $F'_{(2)}$. Substitute (44) into the remaining two equations of the system (20) (for compaction: $e^C = a > 0$)

$$\gamma\left[F''_{(1)(1)} + (F'_{(1)})^2\right] + a^2\tfrac{c^2}{\gamma} = 2caF'_{(1)},$$
$$-F''_{(1)(2)} - F'_{(2)}F'_{(1)} - 3F''_{(1)(1)} + F''_{(1)(2)} - [3F'_{(1)} - F'_{(2)}]F'_{(1)} - 3a^2\tfrac{c^2}{\gamma^2} = -\tfrac{1}{\gamma}6caF'_{(1)}.$$

It is easily seen that both equalities reduce to the single equation $\gamma\left[F''_{(1)(1)} + (F'_{(1)})^2\right] + a^2c^2\gamma^{-1} = 2caF'_{(1)}$, whose solution has the form $F = \ln\left|3\gamma\zeta-\eta+f_1(\xi-\gamma\zeta)\right| + ac\gamma^{-1}(3\gamma\zeta-\eta) + f_2(\xi-\gamma\zeta)$, and $f_2(\xi-\gamma\zeta)$ plays the role of the integration constant.

As the resulting solution must comply with a whole class of equalities (41) differing from each other by the function $f(\xi-\gamma\zeta)$, the arbitrary functions $f_1(\xi-\gamma\zeta)$, $f_2(\xi-\gamma\zeta)$ relate to the defined function $f(\xi-\gamma\zeta)$. The check leads to the necessity to assume $f_1(\xi-\gamma\zeta) = -f(\xi-\gamma\zeta)$, then solution (41) has the form (43). □

4. Bäcklund Transformations for System of Two Third-Order Equations

We will develop the ideas of Clairin [5] and try to construct differential relations that transform the defined system of two equations on the function $u(x,t)$, $w(x,t)$ of the form

$$u_t + u_{xxx} - 12\mu w w_x - 6uu_x = 0, w_t - 2w_{xxx} + 6uw_x = 0 \tag{45}$$

into a certain unknown system on the function $f(x,t)$, $r(x,t)$ of the same order.

As the initial system describes the relation of two functions of two variables x, t, to define the transition from one system to another one, it is necessary to define two couples characterizing the differential transformations from the independent variables x and t. Assuming that the considered system (45) is of third-order for variable x, and of first-order for variable t, and to construct (45) the cross differentiation is used, the differential relationships of the first order should be defined from variable t, and those of the second order should be defined from variable x:

$$\begin{aligned} \tfrac{\partial^2 r}{\partial x^2} &= F_1(u,w,f,r,u_x,w_x,f_x,r_x), & \tfrac{\partial r}{\partial t} &= H_1(u,w,f,r,u_x,w_x,f_x,r_x), \\ \tfrac{\partial^2 f}{\partial x^2} &= F_2(u,w,f,r,u_x,w_x,f_x,r_x), & \tfrac{\partial f}{\partial t} &= H_2(u,w,f,r,u_x,w_x,f_x,r_x). \end{aligned} \tag{46}$$

To define the explicit form of transformation, functions F_1, F_2, and H_1, H_2 must be found. The condition of integrability (equality of mixed second order derivatives) requires functions (46) to comply with the relationship

$$\frac{\partial^3 r}{\partial x^2 \partial t} = \frac{\partial^3 r}{\partial t \partial x^2}, \quad \frac{\partial^3 f}{\partial x^2 \partial t} = \frac{\partial^3 f}{\partial t \partial x^2} \tag{47}$$

where all the functions $u, w, f, r, u_x, w_x, f_x, r_x$ depend upon the variables x, t. Taking into account (46),

$$\begin{gathered} \frac{\partial^3 r}{\partial x^2 \partial t} = \frac{\partial F_1}{\partial u} u_t + \frac{\partial F_1}{\partial w} w_t + \frac{\partial F_1}{\partial f} f_t + \ldots + \frac{\partial F_1}{\partial r_x} r_{xt}, \\ \frac{\partial^3 r}{\partial t \partial x^2} = \frac{\partial}{\partial u}\Big(\frac{\partial H_1}{\partial u} u_x + \frac{\partial H_1}{\partial w} w_x + \ldots + \frac{\partial H_1}{\partial r_x} r_{xx}\Big) u_x + \ldots + \frac{\partial}{\partial r_x}\Big(\frac{\partial H_1}{\partial u} u_x + \frac{\partial H_1}{\partial w} w_x + \ldots + \frac{\partial H_1}{\partial r_x} r_{xx}\Big) r_{xx}, \end{gathered} \tag{48}$$

similarly for functions f. Equaling the right parts of the obtained equalities, and using (46) to exclude $r_t, f_t, r_{xx}, f_{xx}, r_{xt}, f_{xt}$, finally get the condition of consistency, which must lead to system (45).

System (45) has the exponential nonlinearity of the first order $(u_t, u_{xxx}, w_t, w_{xxx})$ and second order $(ww_x, uu_x, w_x u)$, while each summand in (48) is a product of two or three co-multipliers. To make the condition of consistency (47) yield the considered system (45) and without terms of higher than second power it is necessary to assume that functions $F_j, H_j,\ j = 1,2$ are of linear structure in relation to variables $u,\ u_x, w,\ w_x$:

$$\begin{aligned} F_j &= F_{j1}(f,r,f_x,r_x)u + F_{j2}(f,r,f_x,r_x)u_x + \\ &\quad + F_{j3}(f,r,f_x,r_x)w + F_{j4}(f,r,f_x,r_x)w_x + F_{j5}(f,r,f_x,r_x), \\ H_j &= H_{j1}(f,r,f_x,r_x)u + H_{j2}(f,r,f_x,r_x)u_x + \\ &\quad + H_{j3}(f,r,f_x,r_x)w + H_{j4}(f,r,f_x,r_x)w_x + H_{j5}(f,r,f_x,r_x). \end{aligned} \tag{49}$$

When composing the condition of consistency (48) at differentiation F_j by variable t, summands occur with the co-multipliers u_{xt}, w_{xt} that are absent from the initial system (45) and cannot be replaced or compensated, hence, it is necessary to set the coefficients

$$F_{j2}(f,r,f_x,r_x) = 0, \quad F_{j4}(f,r,f_x,r_x) = 0. \tag{50}$$

As a result, the condition of consistency (47) takes the form

$$\begin{gathered} \frac{\partial F_{j1}}{\partial t} u + F_{j1} u_t + \frac{\partial F_{j3}}{\partial t} w + F_{j3} w_t + \frac{\partial F_{j5}}{\partial t} = \frac{\partial^2 H_{j1}}{\partial x^2} u + 2\frac{\partial H_{j1}}{\partial x} u_x + H_{j1} u_{xx} + \frac{\partial^2 H_{j2}}{\partial x^2} u_x + 2\frac{\partial H_{j2}}{\partial x} u_{xx} + \\ + H_{j2} u_{xxx} + \frac{\partial^2 H_{j3}}{\partial x^2} w + 2\frac{\partial H_{j3}}{\partial x} w_x + H_{j3} w_{xx} + \frac{\partial^2 H_{j4}}{\partial x^2} w_x + 2\frac{\partial H_{j4}}{\partial x} w_{xx} + H_{j4} w_{xxx} + \frac{\partial^2 H_{j5}}{\partial x^2}, \end{gathered} \tag{51}$$

where

$$\begin{gathered} \frac{\partial F_{jk}}{\partial t} = \frac{\partial F_{jk}}{\partial f} H_2 + \frac{\partial F_{jk}}{\partial r} H_1 + \frac{\partial F_{jk}}{\partial f_x} H_{2x} + \frac{\partial F_{jk}}{\partial r_x} H_{1x}, \\ \frac{\partial H_{jk}}{\partial x} = \frac{\partial H_{jk}}{\partial f} f_x + \frac{\partial H_{jk}}{\partial r} r_x + \frac{\partial H_{jk}}{\partial f_x} F_2 + \frac{\partial H_{jk}}{\partial r_x} F_1, \\ \frac{\partial^2 H_{jk}}{\partial x^2} = \frac{\partial H_{jk}}{\partial f} F_2 + \frac{\partial H_{jk}}{\partial r} F_1 + \frac{\partial H_{jk}}{\partial f_x} F_{2x} + \frac{\partial H_{jk}}{\partial r_x} F_{1x} + \frac{\partial^2 H_{jk}}{\partial f^2} f_x^2 + \frac{\partial^2 H_{jk}}{\partial r^2} r_x^2 + \frac{\partial^2 H_{jk}}{\partial f_x^2} F_2^2 + \\ + \frac{\partial^2 H_{jk}}{\partial r_x^2} F_1^2 + 2\Big(\frac{\partial^2 H_{jk}}{\partial f \partial r} f_x r_x + \frac{\partial^2 H_{jk}}{\partial f \partial f_x} f_x F_2 + \frac{\partial^2 H_{jk}}{\partial f \partial r_x} f_x F_1 + \frac{\partial^2 H_{jk}}{\partial r \partial r_x} r_x F_1 + \frac{\partial^2 H_{jk}}{\partial r \partial f_x} r_x F_2 + \frac{\partial^2 H_{jk}}{\partial f_x \partial r_x} F_2 F_1\Big). \end{gathered} \tag{52}$$

Functions $u(x,t),\ w(x,t)$ are known, while the form of system (51) is determined by the equalities (45). The terms with multipliers $u_t, w_t, u_{xxx}, w_{xxx}$ cannot occur during substitutions $F_j, H_j,\ \ j = 1,2$ and their first order derivatives F_{jx}, H_{jx} (only second-order derivatives from x may occur), hence, comparing the coefficients for the couple u_t, u_{xxx} and w_t, w_{xxx} in formulas (45) it is necessary to assume

$$F_{j1} = -H_{j2}, \quad 2F_{j3} = H_{j4}. \tag{53}$$

Taking into account (53), equality (51) takes the form

$$\begin{gathered} \Big(\frac{\partial F_{j1}}{\partial t} - \frac{\partial^2 H_{j1}}{\partial x^2}\Big) u + F_{j1}(u_t + u_{xxx}) + \Big(\frac{\partial F_{j3}}{\partial t} - \frac{\partial^2 H_{j3}}{\partial x^2}\Big) w + F_{j3}(w_t - 2w_{xxx}) + \frac{\partial F_{j5}}{\partial t} - \frac{\partial^2 H_{j5}}{\partial x^2} = \\ = \Big(2\frac{\partial H_{j1}}{\partial x} - \frac{\partial^2 F_{j1}}{\partial x^2}\Big) u_x + \Big(H_{j1} - 2\frac{\partial F_{j1}}{\partial x}\Big) u_{xx} + 2\Big(\frac{\partial H_{j3}}{\partial x} + \frac{\partial^2 F_{j3}}{\partial x^2}\Big) w_x + \Big(H_{j3} + 4\frac{\partial F_{j3}}{\partial x}\Big) w_{xx}. \end{gathered} \tag{54}$$

System (45) has no terms not containing u_{xx} or w_{xx}. Hence, differentiate (54) by variable u_{xx} (correspondingly, by w_{xx}), and obtain the relation that must be fulfilled identically

$$\begin{aligned}&\Big(\frac{\partial F_{j1}}{\partial f_x}F_{21}+\frac{\partial F_{j1}}{\partial r_x}F_{11}\Big)u+\Big(\frac{\partial F_{j3}}{\partial f_x}F_{21}+\frac{\partial F_{j3}}{\partial r_x}F_{11}\Big)w+\frac{\partial F_{j5}}{\partial f_x}F_{21}+\frac{\partial F_{j5}}{\partial r_x}F_{11}+H_{j1}=\\&=2\Big(\frac{\partial F_{j1}}{\partial f}f_x+\frac{\partial F_{j1}}{\partial r}r_x+\frac{\partial F_{j1}}{\partial f_x}F_2+\frac{\partial F_{j1}}{\partial r_x}F_1\Big),\end{aligned}\tag{55}$$

similarly for w_{xx}:

$$\begin{aligned}&\Big(\frac{\partial F_{j1}}{\partial f_x}F_{23}+\frac{\partial F_{j1}}{\partial r_x}F_{13}\Big)u+\Big(\frac{\partial F_{j3}}{\partial f_x}F_{23}+\frac{\partial F_{j3}}{\partial r_x}F_{13}\Big)w+\frac{\partial F_{j5}}{\partial f_x}F_{23}+\frac{\partial F_{j5}}{\partial r_x}F_{13}=\\&=\frac{1}{2}H_{j3}+2\Big(\frac{\partial F_{j3}}{\partial f}f_x+\frac{\partial F_{j3}}{\partial r}r_x+\frac{\partial F_{j3}}{\partial f_x}F_2+\frac{\partial F_{j3}}{\partial r_x}F_1\Big).\end{aligned}\tag{56}$$

As the equalities have functions u, w and do not have similar summands, the coefficients at these functions must return to zero, hence, (55), (56) separate into system $j = 1, 2$:

$$\begin{aligned}&\frac{\partial F_{j1}}{\partial f_x}F_{21}+\frac{\partial F_{j1}}{\partial r_x}F_{11}=0,\quad \frac{\partial F_{j3}}{\partial f_x}F_{21}+\frac{\partial F_{j3}}{\partial r_x}F_{11}=2\frac{\partial F_{j1}}{\partial f_x}F_{23}+2\frac{\partial F_{j1}}{\partial r_x}F_{13},\\&\frac{\partial F_{j5}}{\partial f_x}F_{21}+\frac{\partial F_{j5}}{\partial r_x}F_{11}+H_{j1}=2\Big(\frac{\partial F_{j1}}{\partial f}f_x+\frac{\partial F_{j1}}{\partial r}r_x+\frac{\partial F_{j1}}{\partial f_x}F_{25}+\frac{\partial F_{j1}}{\partial r_x}F_{15}\Big),\\&\frac{\partial F_{j1}}{\partial f_x}F_{23}+\frac{\partial F_{j1}}{\partial r_x}F_{13}=2\frac{\partial F_{j3}}{\partial f_x}F_{21}+2\frac{\partial F_{j3}}{\partial r_x}F_{11},\quad \frac{\partial F_{j3}}{\partial f_x}F_{23}+\frac{\partial F_{j3}}{\partial r_x}F_{13}=0,\\&\frac{\partial F_{j5}}{\partial f_x}F_{23}+\frac{\partial F_{j5}}{\partial r_x}F_{13}=\frac{1}{2}H_{j3}+2\Big(\frac{\partial F_{j3}}{\partial f}f_x+\frac{\partial F_{j3}}{\partial r}r_x+\frac{\partial F_{j3}}{\partial f_x}F_{25}+\frac{\partial F_{j3}}{\partial r_x}F_{15}\Big).\end{aligned}$$

To make the first, second, fourth, and fifth equalities be fulfilled identically, assume F_{j1}, F_{j3} independent of functions f_x, r_x. Note that here the simplest variant is selected. Other relations between functions F_{j1}, F_{j3} are possible as well. The introduced assumptions are not final and may be changed when constructing transformations in the event when, at the next steps, incompatible systems or terms that cannot be eliminated occur. The third and sixth equalities yield

$$H_{j1}=2\Big(\frac{\partial F_{j1}}{\partial f}f_x+\frac{\partial F_{j1}}{\partial r}r_x\Big)-\frac{\partial F_{j5}}{\partial f_x}F_{21}-\frac{\partial F_{j5}}{\partial r_x}F_{11},\tag{57}$$

$$H_{j3}=2\frac{\partial F_{j5}}{\partial f_x}F_{23}+2\frac{\partial F_{j5}}{\partial r_x}F_{13}-4\Big(\frac{\partial F_{j3}}{\partial f}f_x+\frac{\partial F_{j3}}{\partial r}r_x\Big).\tag{58}$$

As a result of the performed analysis, functions (4.5) were transformed into the form

$$\begin{aligned}&F_j=F_{j1}(f,r)u+F_{j3}(f,r)w+F_{j5}(f,r,f_x,r_x),\\&H_j=\Big(2\Big[\frac{\partial F_{j1}}{\partial f}f_x+\frac{\partial F_{j1}}{\partial r}r_x\Big]-\frac{\partial F_{j5}}{\partial f_x}F_{21}-\frac{\partial F_{j5}}{\partial r_x}F_{11}\Big)u-F_{j1}(f,r)u_x+2F_{j3}(f,r)w_x+\\&+\Big(2\frac{\partial F_{j5}}{\partial f_x}F_{23}+2\frac{\partial F_{j5}}{\partial r_x}F_{13}-4\Big[\frac{\partial F_{j3}}{\partial f}f_x+\frac{\partial F_{j3}}{\partial r}r_x\Big]\Big)w+H_{j5}(f,r,f_x,r_x).\end{aligned}\tag{59}$$

Continue examining equality (54). See with what coefficient the term with the multiplier uu_x, point (1) (point (2): ww_x, point (3): uw_x), enters the condition of consistency (54); for this, differentiate (54) twice, first by variable u (by w in (2), and by u in (3)), then by variable u_x (in (2), (3) by variable w_x). During the manipulations, interrelated equations are obtained, hence, describe their construction separately.

1. After differentiation of (54) in relation to multiplier uu_x, the following summands remain

$$\frac{\partial^3 F_{j1}}{\partial t\partial u_x\partial u}u+\frac{\partial^2 F_{j1}}{\partial t\partial u_x}-\frac{\partial^3 H_{j1}}{\partial x^2\partial u_x}+\frac{\partial^3 F_{j3}}{\partial t\partial u_x\partial u}w+\frac{\partial^3 F_{j5}}{\partial t\partial u_x\partial u}-2\frac{\partial^2 H_{j1}}{\partial x\partial u}+\frac{\partial^3 F_{j1}}{\partial x^2\partial u},\tag{60}$$

where, taking into account (59), derivatives are transformed into a simpler form

$$\begin{aligned} &\frac{\partial^2 H_{jk}}{\partial x \partial u} = \frac{\partial^2 H_{jk}}{\partial x \partial u_x} = \frac{\partial H_{jk}}{\partial f_x} F_{21} + \frac{\partial H_{jk}}{\partial r_x} F_{11}, \quad \frac{\partial^3 F_{jk}}{\partial t \partial u_x \partial u} = 0, \\ &\frac{\partial^3 F_{jk}}{\partial x^2 \partial u} = -\frac{\partial^2 F_{jk}}{\partial t \partial u_x} = \frac{\partial F_{jk}}{\partial f} F_{21} + \frac{\partial F_{jk}}{\partial r} F_{11}, \quad j = 1,2, \quad k = 1,3. \end{aligned} \tag{61}$$

As a result of the performed differentiation of the condition of consistency get the coefficient (60), that will be at the multiplier uu_x. As such term occurs in system (45), the expression (60) must not be identically equal to zero but must be proportional to the coefficient F_{j1}, with which the terms $u_t + u_{xxx}$ enter. The coefficient of proportionality conforms to the coefficient of term uu_x in system (45) and equals −6. As a result, after substitution (57), expression (60) yields equation

$$\left(2\frac{\partial F_{j1}}{\partial f} - \frac{\partial^2 F_{j5}}{\partial f_x^2} F_{21} - 2\frac{\partial^2 F_{j5}}{\partial r_x \partial f_x} F_{11}\right) F_{21} + \left(2\frac{\partial F_{j1}}{\partial r} - \frac{\partial^2 F_{j5}}{\partial r_x^2} F_{11}\right) F_{11} = 2F_{j1}. \tag{62}$$

2. Perform similar actions in relation to the term ww_x. In relationship (54) the following summands remain

$$\frac{\partial^3 F_{j1}}{\partial t \partial w_x \partial w} u + \frac{\partial^2 F_{j3}}{\partial t \partial w_x} - \frac{\partial^3 H_{j3}}{\partial x^2 \partial w_x} + \frac{\partial^3 F_{j3}}{\partial t \partial w_x \partial w} w + \frac{\partial^3 F_{j5}}{\partial t \partial w_x \partial w} - 2\frac{\partial^2 H_{j3}}{\partial x \partial w} - 2\frac{\partial^3 F_{j3}}{\partial x^2 \partial w}. \tag{63}$$

where, taking into account (59), derivatives are transformed into a simpler form

$$\begin{aligned} &\frac{\partial^2 H_{jk}}{\partial x \partial w} = \frac{\partial^3 H_{jk}}{\partial x^2 \partial w_x} = \frac{\partial H_{jk}}{\partial f_x} F_{23} + \frac{\partial H_{jk}}{\partial r_x} F_{13}, \quad \frac{\partial^3 F_{jk}}{\partial t \partial w_x \partial w} = 0, \\ &\frac{\partial^2 F_{jk}}{\partial t \partial w_x} = 2\frac{\partial^3 F_{jk}}{\partial x^2 \partial w} = 2\frac{\partial F_{jk}}{\partial f} F_{23} + 2\frac{\partial F_{jk}}{\partial r} F_{13}, \quad j = 1,2, \quad k = 1,3. \end{aligned} \tag{64}$$

Expression (63) must not be identically equal to zero but must be proportional to F_{j1} with the coefficient of proportionality corresponding to the term ww_x in system (45) and equal to -12μ. As a result, after substitution (58), expression (63) yields equation

$$\left(\frac{\partial^2 F_{j5}}{\partial f_x^2} F_{23} + 2\frac{\partial^2 F_{j5}}{\partial r_x \partial f_x} F_{13} - 2\frac{\partial F_{j3}}{\partial f}\right) F_{23} + \left(\frac{\partial^2 F_{j5}}{\partial r_x^2} F_{13} - 2\frac{\partial F_{j3}}{\partial r}\right) F_{13} = 2\mu F_{j1}. \tag{65}$$

3. After differentiation (54) with multiplier uw_x the following non-zero summands remain

$$\frac{\partial^3 F_{j1}}{\partial t \partial w_x \partial u} u + \frac{\partial^2 F_{j1}}{\partial t \partial w_x} - \frac{\partial^3 H_{j1}}{\partial x^2 \partial w_x} + \frac{\partial^3 F_{j3}}{\partial t \partial w_x \partial u} w + \frac{\partial^3 F_{j5}}{\partial t \partial w_x \partial u} - 2\frac{\partial^2 H_{j3}}{\partial x \partial u} - 2\frac{\partial^2 F_{j3}}{\partial x^2 \partial u}. \tag{66}$$

Specifying the form of the derivatives using the earlier found form (53), rewrite the remaining coefficients (66) and equate $6F_{j3}$

$$\begin{aligned} &2\frac{\partial F_{j1}}{\partial f} F_{23} + 2\frac{\partial F_{j1}}{\partial r} F_{13} - \frac{\partial H_{j1}}{\partial f_x} F_{23} - \frac{\partial H_{j1}}{\partial r_x} F_{13} - 2\frac{\partial H_{j3}}{\partial f_x} F_{21} - \\ &-2\frac{\partial H_{j3}}{\partial r_x} F_{11} - 2\frac{\partial F_{j3}}{\partial f} F_{21} - 2\frac{\partial F_{j3}}{\partial r} F_{11} = 6F_{j3}, \end{aligned} \tag{67}$$

or, after substitution of the earlier found functions (57), (58):

$$2\frac{\partial F_{j3}}{\partial f} F_{21} + 2\frac{\partial F_{j3}}{\partial r} F_{11} - \left(\frac{\partial^2 F_{j5}}{\partial f_x^2} F_{21} + \frac{\partial^2 F_{j5}}{\partial r_x \partial f_x} F_{11}\right) F_{23} - \left(\frac{\partial^2 F_{j5}}{\partial f_x \partial r_x} F_{21} + \frac{\partial^2 F_{j5}}{\partial r_x^2} F_{11}\right) F_{13} = 2F_{j3}. \tag{68}$$

Now it is necessary to solve the system of six quasilinear partial second order differential equations (62), (65), (68), $j = 1, 2$

$$
\begin{aligned}
&2\left(\tfrac{\partial F_{j1}}{\partial f}F_{21} + \tfrac{\partial F_{j1}}{\partial r}F_{11}\right) - \left(F_{21}\tfrac{\partial}{\partial f_x} + F_{11}\tfrac{\partial}{\partial r_x}\right)^2 F_{j5} = 2F_{j1},\\
&\left(F_{23}\tfrac{\partial}{\partial f_x} + F_{13}\tfrac{\partial}{\partial r_x}\right)^2 F_{j5} - 2\left(\tfrac{\partial F_{j3}}{\partial f}F_{23} + \tfrac{\partial F_{j3}}{\partial r}F_{13}\right) = 2\mu F_{j1},\\
&2\left(\tfrac{\partial F_{j3}}{\partial f}F_{21} + \tfrac{\partial F_{j3}}{\partial r}F_{11}\right) - \left(F_{23}\tfrac{\partial}{\partial f_x} + F_{13}\tfrac{\partial}{\partial r_x}\right)\left(F_{21}\tfrac{\partial}{\partial f_x} + F_{11}\tfrac{\partial}{\partial r_x}\right)F_{j5} = 2F_{j3}.
\end{aligned}
\tag{69}
$$

In the resulting system (69) the summands $F_{j1f}F_{21} + F_{j1r}F_{11}, F_{j3f}F_{23} + F_{j3r}F_{13}$, and $F_{j3f}F_{21} + F_{j3r}F_{11}$ have occurred that depend only upon variables f, r, and operators of second order differentiation by variables f_x, r_x, for which the dependence upon variables f, r is parametric. Obviously, the system decomposes into two subsystems determining the dependence upon variables f_x, r_x:

$$
\begin{aligned}
&\left(F_{21}\tfrac{\partial}{\partial f_x} + F_{11}\tfrac{\partial}{\partial r_x}\right)^2 F_{j5} = 0, \quad \left(F_{23}\tfrac{\partial}{\partial f_x} + F_{13}\tfrac{\partial}{\partial r_x}\right)^2 F_{j5} = 0,\\
&\left(F_{23}\tfrac{\partial}{\partial f_x} + F_{13}\tfrac{\partial}{\partial r_x}\right)\left(F_{21}\tfrac{\partial}{\partial f_x} + F_{11}\tfrac{\partial}{\partial r_x}\right)F_{j5} = 0,
\end{aligned}
\tag{70}
$$

and the dependence upon variables f, r:

$$
\frac{\partial F_{j1}}{\partial f}F_{21} + \frac{\partial F_{j1}}{\partial r}F_{11} = F_{j1}, \quad \frac{\partial F_{j3}}{\partial f}F_{23} + \frac{\partial F_{j3}}{\partial r}F_{13} = -\mu F_{j1}, \quad \frac{\partial F_{j3}}{\partial f}F_{21} + \frac{\partial F_{j3}}{\partial r}F_{11} = F_{j3}. \tag{71}
$$

It can be seen that both systems (70), (71) are over-determined, hence, we will not search for their solutions here (they may exist; this variant has not been examined). The second possibility is when the action of the second order differential operators on function F_{j5} yields the expression, dependent only upon variables f, r. This is possible if F_{j5} has quadratic dependence upon variables f_x, r_x; write it in the form:

$$
F_{j5} = s_{j1}(f,r)f_x^2 + s_{j2}(f,r)f_x r_x + s_{j3}(f,r)r_x^2 + s_{j4}(f,r)f_x + s_{j5}(f,r)r_x + s_{j6}(f,r). \tag{72}
$$

In this case, system (69) takes the form:

$$
\begin{aligned}
&F_{21}\tfrac{\partial F_{j1}}{\partial f} + F_{11}\tfrac{\partial F_{j1}}{\partial r} - s_{j1}F_{21}^2 - F_{21}F_{11}s_{j2} - s_{j3}F_{11}^2 = F_{j1},\\
&s_{j1}F_{23}^2 + F_{23}F_{13}s_{j2} + s_{j3}F_{13}^2 - F_{23}\tfrac{\partial F_{j3}}{\partial f} - F_{13}\tfrac{\partial F_{j3}}{\partial r} = \mu F_{j1},\\
&2\left(F_{21}\tfrac{\partial F_{j3}}{\partial f} + F_{11}\tfrac{\partial F_{j3}}{\partial r}\right) - F_{23}[2s_{j1}F_{21} + F_{11}s_{j2}] - F_{13}[F_{21}s_{j2} + 2s_{j3}F_{11}] = 2F_{j3}.
\end{aligned}
\tag{73}
$$

The first equation yields the system, relating two functions F_{11}, F_{21}. Select the simplest solutions (such an approach is justified because the Bäcklund transformations must, if possible, be of simple form)

$$
F_{21} = 0, \quad F_{11} = a - const \tag{74}
$$

then, it must be additionally assumed

$$
s_{23} = 0, \quad s_{13} = -a^{-1}. \tag{75}
$$

Taking into account (74), (75), the remaining equalities take the form

$$\begin{aligned} &s_{11}F_{23}^2 + s_{12}F_{23}F_{13} - \tfrac{1}{a}F_{13}^2 - \left(F_{23}\tfrac{\partial}{\partial f} + F_{13}\tfrac{\partial}{\partial r}\right)F_{13} = \mu a, \qquad && 2a\tfrac{\partial F_{13}}{\partial r} = as_{12}F_{23}, \\ &s_{21}F_{23}^2 + s_{22}F_{23}F_{13} - \left(F_{23}\tfrac{\partial}{\partial f} + F_{13}\tfrac{\partial}{\partial r}\right)F_{23} = 0, && \tfrac{\partial F_{23}}{\partial r} = \tfrac{2+as_{22}}{2a}F_{23}. \end{aligned} \tag{76}$$

Select, if possible, simpler solutions; for this suppose that F_{13}, F_{23} depend upon f and do not depend upon r, then

$$s_{12} = 0, \quad s_{22} = -2a^{-1} - const. \tag{77}$$

Only two first-order differential equations remain

$$s_{11}F_{23}^2 - \frac{1}{a}F_{13}^2 - F_{23}\frac{\partial F_{13}}{\partial f} = \mu a, \qquad s_{21}F_{23}^2 - \frac{2}{a}F_{23}F_{13} - F_{23}\frac{\partial F_{23}}{\partial f} = 0 \tag{78}$$

whose solutions may be varied. Let

$$F_{13} = 0, \quad s_{11} = \mu a e^{-2f}, \qquad s_{21} = 1 \qquad e^f = F_{23}. \tag{79}$$

As a result, formulas (59) are transformed into the form

$$\begin{aligned} &F_1 = au + a\mu e^{-2f} f_x^2 - \tfrac{1}{a} r_x^2 + S_1, \qquad F_2 = e^f w + f_x^2 - \tfrac{2}{a} f_x r_x + S_2, \\ &H_1 = (2r_x - as_{15})u - au_x + 2\left(2a\mu e^{-f} f_x + e^f s_{14}\right)w + H_{15}(f, r, f_x, r_x), \\ &H_2 = [2f_x - as_{25}]u + 2e^f\left(s_{24} - \tfrac{2}{a} r_x\right)w + 2e^f w_x + H_{25}(f, r, f_x, r_x), \end{aligned} \tag{80}$$

where, for compactness of entry

$$\begin{aligned} S_1 &= s_{14}(f, r) f_x + s_{15}(f, r) r_x + s_{16}(f, r), \\ S_2 &= s_{24}(f, r) f_x + s_{25}(f, r) r_x + s_{26}(f, r). \end{aligned}$$

Return to the condition of consistency (54)

$$\begin{aligned} &a(u_t + u_{xxx}) + \tfrac{\partial F_{15}}{\partial t} - \tfrac{\partial^2 H_{15}}{\partial x^2} = \tfrac{\partial^2 (H_{11}u + H_{13}w)}{\partial x^2}, \\ &e^f(w_t - 2w_{xxx}) + \tfrac{\partial F_{25}}{\partial t} - \tfrac{\partial^2 H_{25}}{\partial x^2} = \tfrac{\partial^2 (H_{21}u + H_{23}w)}{\partial x^2} + e^f\left[4f_x w_{xx} + 2\left(f_x^2 + F_2\right)w_x - H_2 w\right], \end{aligned} \tag{81}$$

and find the dependence upon u^2 (1) (w^2, step (2), uw, step (3)); for this differentiate by u^2 (1) (by variable w^2 at step (2), and by variable uw at step (3)). By reason of the only linear dependence F_j, H_j, H_{jkx} in relation to function u (52), the condition (81) will, after differentiation by u^2 taking the form

$$\frac{\partial^2 F_{j5}}{\partial t \partial (u^2)} - \frac{\partial^3 H_{j5}}{\partial x^2 \partial (u^2)} = \frac{\partial}{\partial (u^2)}\left(\frac{\partial^2 H_{j1}}{\partial x^2}u\right) + \frac{\partial^3 H_{j3}}{\partial x^2 \partial (u^2)}w, \quad j = 1, 2, \tag{82}$$

that yields

$$\frac{\partial^2 H_{15}}{\partial r_x^2} = \frac{\partial s_{15}}{\partial r}, \qquad \frac{\partial^2 H_{25}}{\partial r_x^2} = \frac{\partial s_{25}}{\partial r}. \tag{83}$$

4. Perform the second step of the algorithm. According to (52), and, taking into account the linear character of $F_j,\ H_j,\ H_{jkx}$ in relation to function w, (81) will, after differentiation, take the form:

$$\begin{aligned}\frac{\partial^2 F_{15}}{\partial t\partial(w^2)}-\frac{\partial^3 H_{15}}{\partial x^2\partial(w^2)}&=\frac{\partial}{\partial(w^2)}\left(\frac{\partial^2 H_{13}}{\partial x^2}w\right)+\frac{\partial^3 H_{11}}{\partial x^2\partial(w^2)}u,\\ \frac{\partial^2 F_{25}}{\partial t\partial(w^2)}-\frac{\partial^3 H_{25}}{\partial x^2\partial(w^2)}&=\frac{\partial}{\partial(w^2)}\left(\frac{\partial^2 H_{23}}{\partial x^2}w\right)+\frac{\partial^3 H_{21}}{\partial x^2\partial(w^2)}u-e^f H_{23}.\end{aligned}\tag{84}$$

After transformations, the relationship (83) takes the form

$$4a\mu s_{15}-e^{2f}\frac{\partial^2 H_{15}}{\partial f_x^2}=2e^{2f}\left(s_{14}+\frac{\partial s_{14}}{\partial f}\right)+4a\mu s_{24},\quad 4a\mu s_{25}-e^{2f}\frac{\partial^2 H_{25}}{\partial f_x^2}=-\frac{4}{a}e^{2f}s_{14}.\tag{85}$$

5. The condition of consistency (81) for the values $j=1,2$ yields the system:

$$\begin{aligned}\frac{\partial^2 F_{15}}{\partial t\partial(uw)}-\frac{\partial^3 H_{15}}{\partial x^2\partial(uw)}&=\frac{\partial}{\partial(uw)}\left(\frac{\partial^2 H_{11}}{\partial x^2}u+\frac{\partial^2 H_{13}}{\partial x^2}w\right),\\ \frac{\partial^2 F_{25}}{\partial t\partial(uw)}-\frac{\partial^3 H_{25}}{\partial x^2\partial(uw)}&=\frac{\partial}{\partial(uw)}\left(\frac{\partial^2 H_{21}}{\partial x^2}u+\frac{\partial^2 H_{23}}{\partial x^2}w\right)-e^f H_{21}.\end{aligned}\tag{86}$$

Using the earlier found form of coefficients and their dependence upon variables $r_x,\ r,\ f_x,\ f$, obtain from (85) two new differential equations:

$$\begin{aligned}&2a\frac{\partial^2 H_{15}}{\partial f_x\partial r_x}-a\frac{\partial s_{15}}{\partial f}+4s_{14}+2a\frac{\partial s_{14}}{\partial r}+4a^2\mu e^{-2f}s_{25}=0,\\ &2a\frac{\partial^2 H_{25}}{\partial f_x\partial r_x}+4s_{24}-a\frac{\partial s_{25}}{\partial f}+as_{25}+2a\frac{\partial s_{24}}{\partial r}-4s_{15}=0.\end{aligned}\tag{87}$$

Equalities (83), (85), (87) do not contain in explicit form the variables $f_x,\ r_x$; this allows to suppose that functions $H_{j5}=0,\ s_{jk}=0,\ j=1,2,\ k=4,5,6$. Perform a check having returned to equalities (4.37), where

$$\frac{\partial F_{15}}{\partial t}=-2a\mu e^{-2f}f_x^2H_2+2a\mu e^{-2f}f_xH_{2x}-\frac{2}{a}r_xH_{1x},\quad \frac{\partial F_{25}}{\partial t}=2\left(f_x-\frac{1}{a}r_x\right)H_{2x}-\frac{2}{a}f_xH_{1x},$$

$$\frac{\partial H_{11}}{\partial x}=2F_1,\quad \frac{\partial H_{13}}{\partial x}=4a\mu e^{-f}\left(F_2-f_x^2\right),\quad \frac{\partial H_{21}}{\partial x}=2F_2,\quad \frac{\partial H_{23}}{\partial x}=-\frac{4}{a}e^f\left(r_xf_x+F_1\right),$$

$$\frac{\partial^2 H_{11}}{\partial x^2}=2F_{1x},\quad \frac{\partial^2 H_{13}}{\partial x^2}=4a\mu e^{-f}\left(F_{2x}+f_x^3-3f_xF_2\right),\quad \frac{\partial^2 H_{21}}{\partial x^2}=2F_{2x},$$

$$\frac{\partial^2 H_{23}}{\partial x^2}=-\frac{4}{a}e^f\left(r_xF_2+F_{1x}+r_xf_x^2+2f_xF_1\right),\quad \frac{\partial F_{23}}{\partial t}=e^fH_2,\quad \frac{\partial^2 F_{23}}{\partial x^2}=e^fF_2+e^ff_x^2,$$

after substitution

$$a(u_t+u_{xxx})=6au_xu+12a\mu w_xw,\quad w_t-2w_{xxx}=-6w_xu.$$

It can be seen that equalities coincide, hence, a Bäcklund transformation of the form (80), where $s_{jk}=0,\ H_{j5}=0,\ j=1,2,\ k=4,5,6$ has been found.

Theorem 5. *Nonlinear systems of partial differential equations (45) and*

$$\begin{aligned}r_t&=2a^{-2}r_x^3-r_{xxx}+6\mu e^{-2f}f_x\left(af_{xx}-af_x^2+f_xr_x\right),\\ f_t&=6a^{-1}f_x\left(r_{xx}-a^{-1}r_x^2-af_{xx}\right)+2f_x^3[1-\mu e^{-2f}]+2f_{xxx},\end{aligned}\tag{88}$$

are interrelated by the Bäcklund transformations of the form:

$$\begin{aligned} r_{xx} &= au + a\mu e^{-2f} f_x^2 - a^{-1} r_x^2, & f_{xx} &= e^f w + f_x^2 - 2a^{-1} f_x r_x, \\ r_t &= 2r_x u - a u_x + 4a\mu e^{-f} f_x w, & f_t &= 2 f_x u - 4a^{-1} e^f r_x w + 2e^f w_x, \end{aligned} \tag{89}$$

where $u(x,t)$, $w(x,t)$, $f(x,t)$, $r(x,t)$ differentiable functions of two independent variables, $a \neq 0$, μ are arbitrary non-zero parameters.

Another form of transformation may be obtained, as well. For this, return in the procedure of examination, to the moment that determines the form, i.e., to system (78). Such an approach has been implemented in Reference [17].

5. Analog of Miura Transformations

We demonstrate how the results obtained in the previous section can be used. Use the earlier obtained Bäcklund transformation (89) and substitute the functions $f(x,t)$, $r(x,t)$ by the functions $g(x,t)$, $v(x,t)$:

$$\left(e^{-f(x,t)}\right)_x = g(x,t), \quad r_x(x,t) = v(x,t), \tag{90}$$

To perform the complete substitution with the new functions, the second couple of equalities (89) must be previously differentiated by variable x. The substitution yields the following relation

$$\begin{aligned} v_x &= au + a\mu g^2 - a^{-1} v^2, & g_x &= -2a^{-1} g v - w, \\ v_t &= \tfrac{\partial}{\partial x}(2vu - 4a\mu g w - a u_x), & g_t &= 2\tfrac{\partial}{\partial x}\left(g u + \tfrac{2}{a} v w - w_x\right). \end{aligned} \tag{91}$$

The first line yields the explicit form of functions $u(x,t)$, $w(x,t)$ via the two other functions:

$$u = a^{-1} v_x + a^{-2} v^2 - \mu g^2, \qquad w = -2a^{-1} g v - g_x. \tag{92}$$

Supposing that $g(x,t) = v(x,t)$, the resulting relation has terms similar to the known Miura transformation ($q = v^2 - i v_x$) [22], which determines the conformity between the KdV equation and the modified KdV equation, hence, (92) may be considered a certain analog of this transformation.

Substitute (92) into the equalities of the second line (91), and get the system of two equations

$$v_t = \left(2a^{-2} v^3 - v_{xx} + 6\mu g^2 v + 6a\mu g g_x\right)_x, \quad g_t = 2\left[3a^{-2} g\left(a v_x - v^2\right) - \mu g^3 + g_{xx}\right]_x, \tag{93}$$

each of which is a perturbation of modified KdV equation.

Theorem 6. *Systems of partial differential Equations (45) and (93) are related by transformations (92).*

Proof of Theorem 6. Substitute (92) into (90). Transform the first equation and separate the total derivatives

$$\begin{aligned} u_t + u_{xxx} - 12\mu w w_x - 6u u_x &= \tfrac{1}{a}\tfrac{\partial}{\partial x}\left[v_t + v_{xxx} - 6\tfrac{1}{a^2} v^2 v_x\right] - \tfrac{6}{a}\mu \tfrac{\partial^2}{\partial x^2}\left[g^2 v + a g g_x\right] + \\ &+ \tfrac{2}{a^2} v\left[v_t + \left(v_{xx} - \tfrac{2}{a^2} v^3 - 6\mu g^2 v - 6a\mu g g_x\right)_x\right] + 2\mu g\left[2\left(g_{xx} - \mu g^3 + \tfrac{3}{a^2} g\{a v_x - v^2\}\right)_x - g_t\right]. \end{aligned}$$

In the resulting equality, the linear operator $(a^{-1}\partial_x + 2a^{-2} v)$ may be removed:

$$\begin{aligned} u_t + u_{xxx} - 12\mu w w_x - 6u u_x = {} & \left(\tfrac{1}{a}\tfrac{\partial}{\partial x} + \tfrac{2}{a^2} v\right)\left[v_t + \left(v_{xx} - \tfrac{2}{a^2} v^3 - 6\mu g^2 v - 6a\mu g g_x\right)_x\right] + \\ & + 2\mu g\left[2\left(g_{xx} - \mu g^3 + \tfrac{3}{a^2} g\{a v_x - v^2\}\right)_x - g_t\right]. \end{aligned} \tag{94}$$

Do the same with the second equality of system (45) and factor out the operator $(\partial_x + 2a^{-1}v)$.

$$w_t - 2w_{xxx} + 6uw_x = \left(\tfrac{2}{a}v + \tfrac{\partial}{\partial x}\right)\left[2\left(g_{xx} - \mu g^3 + \tfrac{3}{a}gv_x - \tfrac{3}{a^2}gv^2\right)_x - g_t\right] + \\ + \tfrac{2}{a}g\left(\left[\tfrac{2}{a^2}v^3 - v_{xx} + 6\mu v g^2 + 6a\mu g g_x\right]_x - v_t\right). \tag{95}$$

If functions $u(x,t),\ w(x,t)$ are solutions of system (45) and $u(x,t) \neq 0,\ w(x,t) \neq 0$, then, at $g(x,t) \neq 0,\ v(x,t) \neq 0$, it follows from (94) and (95) that functions $g(x,t),\ v(x,t)$ are solutions of system (93). □

Corollary 9. *Complexification of Korteweg-de Vries equation*

$$q_t = 3(\overline{q} - q)q_x + 6\overline{q}_x q - 0,5(3\overline{q} - q)_{xxx}, \tag{96}$$

and

$$s_t = \left[s\left(3\overline{s}^2 - s^2\right) + 3(s - \overline{s})\overline{s}_x + 0,5(s - 3\overline{s})_{xx}\right]_x, \tag{97}$$

are related by transformation

$$q = \overline{s}_x + \overline{s}^2, \tag{98}$$

where $q(x,t),\ s(x,t)$ are complex functions of independent variables x,t.

The pattern of proof fully coincides with the proof of the theorem above, where $u(x,t) = \mathrm{Re}q(x,t)$, $w(x,t) = \mathrm{Im}s(x,t),\ v(x,t) = \mathrm{Re}s(x,t),\ g(x,t) = \mathrm{Im}s(x,t)$, is supposed to contain parameters $a = 1$, $\mu = 1$.

Assuming in equality (97) that $s(x,t)$ is a real function, get a routinely modified KdV equation $s_t = 6s^2 s_x - s_{xxx}$, hence, (97) may be considered as a modification of the KdV equation complexification [22].

In the classic case, the resulting transformations can be used to build exact solutions. Let us show that the found relation (91) of the two systems (45) and (93) allows us to do this. We take, as the solution of system (45), the following trivial functions

$$w(x,t) = 0,\ u(x,t) = \beta - const. \tag{99}$$

Using (91) and integrating, we obtain the solution of system (93) in the form of traveling waves:

$$g(x,t) = \left[\sqrt{C_1^2 + \tfrac{\mu}{\beta}}\,\mathrm{ch}\left(C_2 - 2\sqrt{\beta}x - 4\beta\sqrt{\beta}t\right) - C_1\right]^{-1},$$
$$v(x,t) = -a\sqrt{C_1^2 + \tfrac{\mu}{\beta}}\,\mathrm{sh}\left(C_2 - 2\sqrt{\beta}x - 4\beta\sqrt{\beta}t\right)\left[\sqrt{C_1^2 + \tfrac{\mu}{\beta}}\,\mathrm{ch}\left(C_2 - 2\sqrt{\beta}x - 4\beta\sqrt{\beta}t\right) - C_1\right]^{-1},$$

where C_1 and C_2 are the arbitrary integration constants. At $C_1 = 0$ we obtain classical solutions:

$$g(x,t) = \sqrt{\frac{\beta}{\mu}}\,\mathrm{ch}^{-1}\left(C_2 - 2\sqrt{\beta}x - 4\beta\sqrt{\beta}t\right), \quad v(x,t) = -a\,\mathrm{th}\left(C_2 - 2\sqrt{\beta}x - 4\beta\sqrt{\beta}t\right).$$

6. Conclusions

1. In this work, new Bäcklund transformations (BTs) have been obtained for the particular cases of Liouville equations with the exponential nonlinearity that has a multiplier dependent upon independent variables and first-order derivatives from the function.
2. BT for three-dimensional Liouville equation has been constructed.
3. A solution of coupled pairs of equations using BT has been found.
4. Clairin's method for the system of two third-order partial differential equations has been generalized and algorithm for construction of BTs for these dynamic systems has been demonstrated.

5. Non-uniqueness of differential relations has been shown because the application of special conditions to differential forms leads to different dynamic systems.
6. Analog of Miura transformations that relates the initial system to the system of perturbed modified KdV equations has been determined.
7. Natural transition of KdV to mKdV using Miura transformations has been received from the relation of cKdV and complexification of mKdV with an analog of Miura transformations, supposing that the function is real.

Author Contributions: Conceptualization, methodology, investigation, writing—original draft preparation, T.V.R.; investigation, writing—original draft preparation, R.G.Z.; validation, writing—review and editing, A.R.Z.; validation, formal analysis, O.B.S.; formal analysis, O.S.Y.

Funding: This research received no special funding.

Conflicts of Interest: The authors declare no conflict of interest.

References

1. Mitrinović, D.S.; Kečkić, J.D. *Jednačine Matematičke Fizike*; Nauka: Beograd, Serbia, 1994.
2. Gardner, C.S.; Greene, J.M.; Kruskal, M.D.; Miura, R.M. Method for solving the Korteweg-de Vries equation. *Phys. Rev. Lett.* **1967**, *19*, 1095–1097. [CrossRef]
3. Ablowitz, M.J.; Clarkson, P.A. *Solitons, Nonlinear Equations and Inverse Scattering*; Cambridge University Press: Cambridge, UK, 1991.
4. Hirota, R. Exact solution of the Korteweg-de Vries equation for multiple collisions of solitons. *Phys. Rev. Lett.* **1971**, *27*, 1192–1194. [CrossRef]
5. Lamb, G.L. *Elements of Soliton Theory*; John Wiley & Sons: New York, NY, USA, 1980.
6. Miura, R.M. Korteweg-de Vries equation and generalizations. I. A remarkable explicit nonlinear transformation. *J. Math. Phys.* **1968**, *9*, 1202–1204. [CrossRef]
7. Zhao, Z. Bäcklund transformations, rational solutions and soliton-cnoidal wave solutions of the modified Kadomtsev–Petviashvili equation. *Appl. Math. Lett.* **2019**, *89*, 103–110. [CrossRef]
8. Zhao, Z.; Chen, Y.; Han, B. Lump soliton, mixed lump stripe and periodic lump solutions of a (2 + 1)-dimensional asymmetrical Nizhnik–Novikov–Veselov equation. *Mod. Phys. Lett. B* **2017**, *31*, 1750157. [CrossRef]
9. Veerakumar, V.; Daniel, M. Modified Kadomtsev–Petviashvili (MKP) equation and electromagnetic soliton. *Math. Comput. Simul.* **2003**, *62*, 163–169. [CrossRef]
10. Sun, Z.Y.; Gao, Y.T.; Yu, X.; Meng, X.H.; Liu, Y. Inelastic interactions of the multiple-front waves for the modified Kadomtsev–Petviashvili equation in fluid dynamics, plasma physics and electrodynamics. *Wave Motion* **2009**, *46*, 511–521. [CrossRef]
11. Chang, J.H. Soliton interaction in the modified Kadomtsev–Petviashvili-(II) equation. *Appl. Anal.* **2018**. [CrossRef]
12. Ren, B. Interaction solutions for mKP equation with nonlocal symmetry reductions and CTE method. *Phys. Scr.* **2015**, *90*, 065206. [CrossRef]
13. Zhao, Z.; Han, B. Residual symmetry, Bäcklund transformation and CRE solvability of a (2 + 1)-dimensional nonlinear system. *Nonlinear Dyn.* **2018**, *94*, 461–474. [CrossRef]
14. Chen, J.C.; Xin, X.P.; Chen, Y. Nonlocal symmetries of the Hirota-Satsuma coupled Korteweg-de Vries system and their applications: Exact interaction solutions and integrable hierarchy. *J. Math. Phys.* **2014**, *55*, 053508. [CrossRef]
15. Hu, X.R.; Lou, S.Y.; Chen, Y. Explicit solutions from eigenfunction symmetry of the Korteweg-de Vries equation. *Phys. Rev. E* **2012**, *85*, 056607. [CrossRef] [PubMed]
16. Cheng, J.; He, J. Miura and auto-Backlund transformations for the discrete KP and mKP hierarchies and their constrained cases. *Commun. Nonlinear Sci. Numer. Simul.* **2019**, *69*, 187–197. [CrossRef]
17. Bogoyavlenskij, O.I. Breaking solitons in (2 + 1)-dimensional integrable equations. *Russ. Math. Surv.* **1990**, *45*, 1–86. [CrossRef]
18. Tsiganov, A.V. Backlund transformations and divisor doubling. *J. Geom. Phys.* **2018**, *126*, 148–158. [CrossRef]

19. Barbashov, B.M.; Nesterenko, V.V. Bäcklund transformation for the Liouville equation and gauge conditions in the theory of a relativistic string. *Theor. Math. Phys.* **1983**, *56*, 180–191. [CrossRef]
20. Demskoi, D. On application of Liouville type equations to constructing Bäcklund transformations. *J. Nonlinear Math. Phys.* **2007**, *14*, 147–156. [CrossRef]
21. Morozov, O.I.; Pavlov, M.V. Bäcklund transformations between four Lax-integrable 3D equations. *J. Nonlinear Math. Phys.* **2017**, *24*, 465–468. [CrossRef]
22. Redkina, T.V. Some properties of the complexification of the Korteweg-de Vries equation. *Izv. Acad. Sci. USSR Ser. Math.* **1991**, *55*, 1300–1311.

axioms

MDPI

Article

First Order Coupled Systems With Functional and Periodic Boundary Conditions: Existence Results and Application to an SIRS Model

João Fialho [1,2,*] and **Feliz Minhós** [2,3]

[1] Department of Mathematics, British University of Vietnam, Ecopark Campus, 160000 Hung Yen, Hanoi, Vietnam
[2] Centro de Investigação em Matemática e Aplicações (CIMA), Instituto de Investigação e Formação Avançada, Universidade de Évora, Rua Romão Ramalho, 59, 7000-671 Évora, Portugal; fminhos@uevora.pt
[3] Departamento de Matemática, Escola de Ciências e Tecnologia, Universidade de Évora, 7000-812 Évora, Portugal
* Correspondence: joao.f@buv.edu.vn

Received: 6 January 2019; Accepted: 13 February 2019; Published: 16 February 2019

Abstract: The results presented in this paper deal with the existence of solutions of a first order fully coupled system of three equations, and they are split in two parts: 1. Case with coupled functional boundary conditions, and 2. Case with periodic boundary conditions. Functional boundary conditions, which are becoming increasingly popular in the literature, as they generalize most of the classical cases and in addition can be used to tackle global conditions, such as maximum or minimum conditions. The arguments used are based on the Arzèla Ascoli theorem and Schauder's fixed point theorem. The existence results are directly applied to an epidemic SIRS (Susceptible-Infectious-Recovered-Susceptible) model, with global boundary conditions.

Keywords: coupled nonlinear systems; functional boundary conditions; Schauder's fixed point theory; Arzèla Ascoli theorem; lower and upper solutions; first order periodic systems; SIRS epidemic model; mathematical modelling

MSC: 34B10; 34B15; 34B40

1. Introduction

In this paper two different problems are analyzed.

Part one is concerned with the study of a fully nonlinear coupled system of equations

$$\begin{cases} u_1'(t) = f_1\left(t, u_1(t), u_2(t), u_3(t)\right) \\ u_2'(t) = f_2\left(t, u_1(t), u_2(t), u_3(t)\right) \\ u_3'(t) = f_3\left(t, u_1(t), u_2(t), u_3(t)\right) \end{cases}, \tag{1}$$

$f_i : [a, b] \times \mathbb{R}^3 \to \mathbb{R}$ and $i = 1, 2, 3$ are L^1-Carathéodory functions, subject to the nonlinear functional boundary conditions

$$\begin{aligned} u_1(a) &= L_1\left(u_1, u_2, u_3, u_1(b), u_2(b), u_3(b), u_2(a), u_3(a)\right) \\ u_2(a) &= L_2\left(u_1, u_2, u_3, u_1(b), u_2(b), u_3(b), u_1(a), u_3(a)\right) \\ u_3(a) &= L_3\left(u_1, u_2, u_3, u_1(b), u_2(b), u_3(b), u_1(a), u_2(a)\right) \end{aligned} \tag{2}$$

where $L_i : (C[a,b])^3 \times \mathbb{R}^5 \to \mathbb{R},\ i = 1,2,3$, are continuous functions with properties later to be defined. The technique used for the functional problem is based on the Arzèla-Ascoli theorem and Schauder's fixed point theorem.

Part two, more precisely, Section 4, deals with the fully nonlinear couple system of Equation (1) coupled with the periodic boundary conditions

$$\begin{aligned} u_1(a) &= u_1(b) \\ u_2(a) &= u_2(b) \\ u_3(a) &= u_3(b). \end{aligned} \tag{3}$$

Given that the conditions on L_i, do not allow the problem (1)–(2) to cover the periodic case, a different approach for the problem (1)–(3) is required. In this case, in order to obtain the existence and location of periodic solutions, the upper and lower solutions method, along with some adequate local monotone assumptions on the nonlinearities, is used.

Mathematical modelling and applications are becoming increasingly popular nowadays. With the sudden outburst of keywords such as big data, data analytics and modelling, the quest for mathematical models is on high demand. In the area of mathematical modelling, systems of differential equations are a must, due to their high applicability in areas such as population dynamics [1–3], finance [4], medicine [5], biotechnology [6] and physics [7,8], and also examples treated in [9,10].

Nevertheless, in the literature available, the cases dealing with coupled systems of equations are not abundant. Such systems can be found in [11–13], however, in this paper the authors present a problem where both equations are coupled. In addition to that, to the best of our knowledge, it is the first time where coupled systems are considered with coupled functional boundary conditions.

This feature allows to generalize the classical boundary data in the literature, such as two-point or multi-point, nonlinear, nonlocal, integro-differential conditions, among others. Indeed, the functional part can deal with global boundary assumptions, such as minimum or maximum arguments, infinite multi-point data, and integral conditions on the several unknown functions. Functional problems, along with their features, can be seen in [14–19] and the references therein.

The methods and techniques applied in this paper can be easily adapted to coupled systems with n equations and variables. However, as the notation and writing appear to be heavy, and may avoid the clarity of the results, we prefer to prove our theoretical part for $n = 3$, which is adequate for our application.

This paper has the following structure: Section 2 contains some definitions and generic assumptions on the nonlinearities. Section 3 shows the main result for problems with functional boundary conditions. In Section 4 it is studied the periodic problem via lower and upper solutions technique together with some local growth conditions. The final section presents an application of (1)–(2) to an epidemic SIRS model to illustrate the applicability of the problem discussed and to show the potentialities of the functional boundary conditions, exploring global initial boundary conditions on the system.

2. Definitions and Assumptions

Throughout this work we consider the space of continuous functions in $[a,b]$, on the Banach space $E := (C[a,b])^3$, equipped with the norm

$$\|u\|_E = \max\left\{\|u_i\|, i = 1,2,3\right\}$$

where $\|u_i\| = \max_{t\in[a,b]} |u_i(t)|$.

The functional boundary functions verify the assumption:

Hypothesis 1. *$L_i : (C[a,b])^3 \times \mathbb{R}^5 \to \mathbb{R}$ are continuous functions. Moreover, each of the functions, $L_i(\eta, x, y)$ are uniformly continuous when (η, x, y) is bounded.*

The admissible nonlinearities will be L^1-Carathéodory functions, according the following definition:

Definition 1. *The functions $f_i : [a,b] \times \mathbb{R}^3 \to \mathbb{R}$, $i = 1, ..., n$, are L^1-Carathéodory if they verify:*

(*i*) *for each $(y_1, y_2, y_3) \in \mathbb{R}^3$, $t \mapsto f_i(t, y_1, y_2, y_3)$ are measurable on $[a,b]$, for $i = 1,2,3$;*
(*ii*) *for almost every $t \in [a,b]$, $(y_1, y_2, y_3) \mapsto f_i(t, y_1, y_2, y_3)$ are continuous on $\mathbb{R}^3$, for $i = 1,2,3$;*
(*iii*) *for each $L > 0$, there exists a positive function $\psi_{iL} \in L^1[a,b]$, $i = 1,2,3$, such that, for $\max\{\|y_i\|, i = 1,2,3\} < L$,*

$$|f_i(t, y_1(t), y_2(t), y_3(t))| \leq \psi_{iL}(t), \text{ a.e. } t \in [a,b], \ i = 1,2,3.$$

To demonstrate the final result, Schauder's fixed point theorem will be an important tool to guarantee the existence of fixed points for the operator to be defined:

Theorem 1. *([20]) Let Y be a nonempty, closed, bounded and convex subset of a Banach space X, and suppose that $P : Y \to Y$ is a completely continuous operator. Then P has at least one fixed point in Y.*

3. Main Result for Functional Problems

In this section, we present and prove the main existence result for (1)–(2), given by the following theorem:

Theorem 2. *If f_i are L^1-Carathéodory functions, for $i = 1,2,3$, and the continuous functions L_i, $i = 1,2,3$, verify (H1) and*

Hypothesis 2. *there exists $R > 0$ such that*

$$\max\left\{k_i + \int_a^t \psi_{iR}(s)\,ds, \ i = 1,2,3\right\} \leq R,$$

with

$$\begin{aligned} k_1 &:= \max\{L_1(u_1, u_2, u_3, u_1(b), u_2(b), u_3(b), u_2(a), u_3(a))\}, \\ k_2 &:= \max\{L_2(u_1, u_2, u_3, u_1(b), u_2(b), u_3(b), u_1(a), u_3(a))\}, \\ k_3 &:= \max\{L_3(u_1, u_2, u_3, u_1(b), u_2(b), u_3(b), u_1(a), u_2(a))\}, \end{aligned}$$

then the problem (1)–(2) has at least one solution $u = (u_1, u_2, u_3) \in (C[a,b])^3$.

Proof. Let us consider the integral system given by

$$\begin{cases} u_1(t) = L_1(u_1, u_2, u_3, u_1(b), u_2(b), u_3(b), u_2(a), u_3(a)) + \\ \qquad \int_a^t f_1(s, u_1(s), u_2(s), u_3(s))\,ds, \\ u_2(t) = L_2(u_1, u_2, u_3, u_1(b), u_2(b), u_3(b), u_1(a), u_3(a)) + \\ \qquad \int_a^t f_2(s, u_1(s), u_2(s), u_3(s))\,ds, \\ u_3(t) = L_3(u_1, u_2, u_3, u_1(b), u_2(b), u_3(b), u_1(a), u_2(a)) + \\ \qquad \int_a^t f_3(s, u_1(s), u_2(s), u_3(s))\,ds, \end{cases} \tag{4}$$

and the operator

$$T : (C\,[a,b])^3 \to (C\,[a,b])^3$$

defined by

$$T\,(u_1, u_2, u_3) = (T_1\,(u_1, u_2, u_3)\,, T_2\,(u_1, u_2, u_3)\,, T_3\,(u_1, u_2, u_3))\,, \tag{5}$$

where $T_i : (C\,[a,b])^3 \to C\,[a,b]\,, i = 1,2,3$, given by

$$\begin{aligned} T_1\,(u_1, u_2, u_3) &= L_1\,(u_1, u_2, u_3, u_1(b), u_2(b), u_3(b), u_2(a), u_3(a)) + \\ &\quad \textstyle\int_a^t f_1\,(s, u_1(s), u_2(s), u_3(s))\,ds, \\ T_2\,(u_1, u_2, u_3) &= L_2\,(u_1, u_2, u_3, u_1(b), u_2(b), u_3(b), u_1(a), u_3(a)) + \\ &\quad \textstyle\int_a^t f_2\,(s, u_1(s), u_2(s), u_3(s))\,ds, \\ T_3\,(u_1, u_2, u_3) &= L_3\,(u_1, u_2, u_3, u_1(b), u_2(b), u_3(b), u_1(a), u_2(a)) + \\ &\quad \textstyle\int_a^t f_3\,(s, u_1(s), u_2(s), u_3(s))\,ds, \end{aligned} \tag{6}$$

As the fixed points of T are fixed points of T_i, $i = 1,2,3$, and vice-versa, which are solutions of (1)–(2), the goal of this architecture will be to use the Arzèla-Ascoli theorem and Schauder's fixed point theorem to prove that the problem (1)–(2) has at least one solution.

For clarity, we consider several claims:

Claim 1. *$TD \subset D$, for some $D \subset (C\,[a,b])^3$ a bounded, closed and convex subset.*

Consider

$$D = \left\{ (u_1, u_2, u_3) \in (C\,[a,b])^3 : \|(u_1, u_2, u_3)\|_E \le k \right\},$$

with $k > 0$ such that

$$\max \left\{ k_i + \int_a^t \psi_{ik}\,(s)\,ds,\; i = 1,2,3 \right\} \le k.$$

Given $(u_1, u_2, u_3) \in D$, by Definition 1 (iii) and (Hypothesis 2), we have that for $(u_1, u_2, u_3) \in D$, $i = 1,2,3$,

$$|f_i\,(s, u_1(s), u_2(s), u_3(s))| \le \psi_{ik}\,(s)\,, \quad \text{a.e. } s \in [a,b]\,.$$

Therefore, $\|T\,(u_1, u_2, u_3)\|_E \le k$ and Claim 1 is proved.

Claim 2. *The operator T is completely continuous.*

To prove that the operator T is completely continuous it is sufficient to show that T is uniformly bounded and T is equicontinuous. Using the above arguments, it can be proved that T_i are uniformly bounded, for $i = 1,2,3$, and therefore T is uniformly bounded.

In order to show that the operator T is equicontinuous, let us consider $t_1, t_2 \in [a,b]$, such that, without any loss of generality, $t_1 < t_2$.

Then for T_i, $i = 1,2,3$, we have

$$\begin{aligned} &|T_i\,(u_1\,(t_2)\,, u_2\,(t_2)\,, u_3\,(t_2)) - T_i\,(u_1\,(t_1)\,, u_2\,(t_1)\,, u_3\,(t_1))| \\ \le\; &\int_{t_1}^{t_2} |f_i\,(s, u_1(s), u_2(s), u_3(s))|\,ds \le \int_{t_1}^{t_2} \psi_{iL}\,(s)\,ds \to 0, \end{aligned}$$

as $t_1 \to t_2$. So, each operator T_i is equicontinuous and, hence, the operator T is equicontinuous.

Therefore, by Arzèla-Ascoli's theorem, the operator T is compact and using Schauder's fixed point theorem, we obtain that T has a fixed point, that is, the problem (1)–(2) has at least a solution $u \in (C[a,b])^3$. □

Remark that (Hypothesis 2) implies that the periodic case is not covered by (2).

4. Existence and Localization Result for the Periodic Case

Consider now the system (1) with the periodic boundary conditions (3). As it is well known, in nonlinear differential equations, the periodic case is more delicate and requires a different approach for general nonlinearities.

The method to be used will apply lower and upper solutions technique, based on the definition:

Definition 2. *Consider the C^1-functions $\alpha_i, \beta_i : [a,b] \to \mathbb{R}$, $i = 1,2,3$. The triple $(\alpha_1, \alpha_2, \alpha_3)$ is a lower solution of the periodic problem (1), (3) if*

$$\begin{cases} \alpha_1'(t) \le f_1(t, \alpha_1(t), \alpha_2(t), \alpha_3(t)) \\ \alpha_2'(t) \le f_2(t, \alpha_1(t), \alpha_2(t), \alpha_3(t)) \\ \alpha_3'(t) \le f_3(t, \alpha_1(t), \alpha_2(t), \alpha_3(t)) \end{cases} \tag{7}$$

and

$$\begin{array}{c} \alpha_1(a) \le \alpha_1(b) \\ \alpha_2(a) \le \alpha_2(b) \\ \alpha_3(a) \le \alpha_3(b). \end{array} \tag{8}$$

The triple $(\beta_1, \beta_2, \beta_3)$ is an upper solution of the periodic problem (1), (3) if the reversed inequalities hold.

This method allows to obtain an existence and localization theorem:

Theorem 3. *Let $(\alpha_1, \alpha_2, \alpha_3)$ and $(\beta_1, \beta_2, \beta_3)$ be lower and upper solutions of (1), (3), respectively, such that $\alpha_i(t) \le \beta_i(t)$, $\forall t \in [a,b]$ and for $i = 1,2,3$.*

Define the set

$$A = \left\{ (t, u_1, u_2, u_3) \in [a,b] \times \mathbb{R}^3 : \alpha_i(t) \le u_i \le \beta_i(t), i = 1,2,3 \right\}$$

and assume that f_i are L^1-Carathéodory functions on A, for $i = 1,2,3$ verifying

$$f_1(t, x, \alpha_2(t), \alpha_3(t)) \le f_1(t, x, y, z) \le f_1(t, x, \beta_2(t), \beta_3(t)), \tag{9}$$

for $t \in [a,b]$, $\alpha_2(t) \le y \le \beta_2(t), \alpha_3(t) \le z \le \beta_3(t)$,

$$f_2(t, \alpha_1(t), y, \alpha_3(t)) \le f_2(t, x, y, z) \le f_2(t, \beta_1(t), y, \beta_3(t)),$$

for $t \in [a,b]$, $\alpha_1(t) \le x \le \beta_1(t), \alpha_3(t) \le z \le \beta_3(t)$,

$$f_3(t, \alpha_1(t), \alpha_2(t), z) \le f_3(t, x, y, z) \le f_3(t, \beta_1(t), \beta_2(t), z),$$

for $t \in [a,b]$, $\alpha_1(t) \le x \le \beta_1(t), \alpha_2(t) \le y \le \beta_2(t)$.

Then the problem (1), (3) has, at least, a solution $u = (u_1, u_2, u_3) \in (C[a,b])^3$ such that

$$\alpha_i(t) \le u_i(t) \le \beta_i(t), i = 1,2,3, \text{ for all } t \in [a,b].$$

Proof. For $i = 1, 2, 3$, define the truncature functions δ_i given by

$$\delta_i(t, u_i) = \begin{cases} \alpha_i(t) & \text{if} \quad u_i < \alpha_i(t) \\ u_i & \text{if} \quad \alpha_i(t) \leq u_i \leq \beta_i(t) \\ \beta_i(t) & \text{if} \quad u_i > \beta_i(t) \end{cases}$$

and consider the modified problem composed by the truncated and perturbed differential equations

$$\begin{cases} u_1'(t) + u_1(t) = f_1\left(t, \delta_1(t, u_1), \delta_2(t, u_2), \delta_3(t, u_3)\right) + \delta_1(t, u_1) \\ u_2'(t) + u_2(t) = f_2\left(t, \delta_1(t, u_1), \delta_2(t, u_2), \delta_3(t, u_3)\right) + \delta_2(t, u_2) \\ u_3'(t) + u_3(t) = f_3\left(t, \delta_1(t, u_1), \delta_2(t, u_2), \delta_3(t, u_3)\right) + \delta_3(t, u_3), \end{cases} \tag{10}$$

together with the boundary conditions (3).

As the linear and homogeneous problem associated to (10), (3) has only the null solution, then we can write (10), (3) in the integral form

$$\begin{cases} u_1(t) = \int_a^b G_1(t,s) \left[\begin{array}{c} f_1\left(s, \delta_1(s, u_1(s)), \delta_2(s, u_2(s)), \delta_3(s, u_3(s))\right) \\ +\delta_1(s, u_1(s)) \end{array} \right] ds \\ u_2(t) = \int_a^b G_2(t,s) \left[\begin{array}{c} f_2\left(s, \delta_1(s, u_1(s)), \delta_2(s, u_2(s)), \delta_3(s, u_3(s))\right) \\ +\delta_2(s, u_2(s)) \end{array} \right] ds \\ u_3(t) = \int_a^b G_3(t,s) \left[\begin{array}{c} f_3\left(s, \delta_1(s, u_1(s)), \delta_2(s, u_2(s)), \delta_3(s, u_3(s))\right) \\ +\delta_3(s, u_3(s)) \end{array} \right] ds, \end{cases} \tag{11}$$

where $G_i(t,s)$ are the Green functions corresponding to the problem

$$\begin{array}{rcl} u_i'(t) + u_i(t) & = & f_i(t) \\ u_i(a) & = & u_i(b), \end{array}$$

for $i = 1, 2, 3$.

Then the operator

$$T : (C[a,b])^3 \to (C[a,b])^3$$

given by

$$T(u_1, u_2, u_3) = \left(T_1(u_1, u_2, u_3), T_2(u_1, u_2, u_3), T_3(u_1, u_2, u_3)\right),$$

with $T_i : (C[a,b])^3 \to C[a,b]$, $i = 1, 2, 3$, defined as

$$T_1(u_1, u_2, u_3) = \int_a^b G_1(t,s) \left[\begin{array}{c} f_1\left(s, \delta_1(s, u_1(s)), \delta_2(s, u_2(s)), \delta_3(s, u_3(s))\right) \\ +\delta_1(s, u_1(s)) \end{array} \right] ds$$

$$T_2(u_1, u_2, u_3) = \int_a^b G_2(t,s) \left[\begin{array}{c} f_2\left(s, \delta_1(s, u_1(s)), \delta_2(s, u_2(s)), \delta_3(s, u_3(s))\right) \\ +\delta_2(s, u_2(s)) \end{array} \right] ds$$

$$T_3(u_1, u_2, u_3) = \int_a^b G_3(t,s) \left[\begin{array}{c} f_3\left(s, \delta_1(s, u_1(s)), \delta_2(s, u_2(s)), \delta_3(s, u_3(s))\right) \\ +\delta_3(s, u_3(s)) \end{array} \right] ds$$

is completely continuous in $(C[a,b])^3$.

By Schauder's fixed point theorem, we obtain that T has a fixed point, that is, the problem (10), (3) has at least a solution $u_* := (u_{*1}, u_{*2}, u_{*3}) \in (C[a,b])^3$.

To prove that this function u_* is a solution of the initial problem (1), (3), it will be enough to show that

$$\alpha_i(t) \leq u_{*i}(t) \leq \beta_i(t), i = 1,2,3, \text{ for all } t \in [a,b].$$

Suppose that the first inequality does not hold for $i = 1$ and some $t_1 \in [a,b]$, that is

$$u_{*1}(t_1) < \alpha_1(t_1).$$

Extend the function $u_{*1} - \alpha_1$ by periodicity and consider

$$t_0 := \inf\{t \in [t_1 - b + a, t_1] : \forall s \in]t, t_1], u_{*1}(s) < \alpha_1(s)\}.$$

Therefore, for $t \in [t_0, t_1]$ we have, by Definition 2 and (9),

$$\begin{aligned} & u'_{*1}(t) - \alpha'_1(t) \\ \geq\ & f_1(t, \alpha_1(t), \delta_2(t, u_2), \delta_3(t, u_3)) + \alpha_1(t) - u_{*1}(t) - f_1(t, \alpha_1(t), \alpha_2(t), \alpha_3(t)) \\ \geq\ & \alpha_1(t) - u_{*1}(t) > 0. \end{aligned}$$

So, $\alpha_1 - u_{*1}$ is increasing on $[t_0, t_1]$ and

$$0 > u_{*1}(t_1) - \alpha_1(t_1) > u_{*1}(t_0) - \alpha_1(t_0),$$

where $t_0 = t_1 - b + a$, which is in contradiction of the periodicity of the extension of $u_{*1} - \alpha_1$. So $\alpha_1(t) \leq u_{*1}(t)$, for all $t \in [a,b]$.

Applying similar arguments it can be proved that $u_{*1}(t) \leq \beta_1(t)$, for $t \in [a,b]$, and

$$\alpha_i(t) \leq u_{*i}(t) \leq \beta_i(t), i = 2,3, \text{ for all } t \in [a,b].$$

□

5. An Epidemic Model of an SIRS System With Nonlinear Incidence Rate and Interaction from Infectious to Susceptible Subjects

The existent literature has innumerous examples of applications of SIR models, namely in [21–23] where the population is divided into Susceptible (S), Infectious (I) and Recovery (R). However, in SIR models, recovered individuals are assumed to develop lifelong immunity, which for some diseases such as seasonal flu, influenza or venereal diseases is not necessarily true. A recovered individual becomes susceptible and possibly infected after some time. In this case, SIRS models, where recovered individuals lose immunity and become susceptible again, are far more adequate. Examples can be found in [24–29]. In [30], the authors develop and explore a mathematical model of an SIRS epidemic model where a transfer between infectious and susceptible rate is included, as shown in Figure 1.

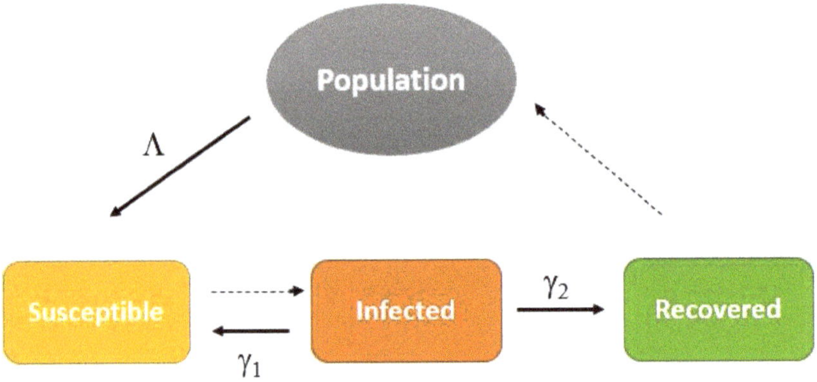

Figure 1. SIRS model diagram.

This approach was designed to model the cases where recovery cannot generate immunity for a long time. Infected individuals may recover after some treatments and go back directly to the susceptible class. In addition, a nonlinear incidence with the average number of new cases of a disease per unit time, $Sg(I)$, is included, as suggested in [31]. This nonlinear incidence replicates a more complex dynamic than the ones presented on bilinear or standard incidence models, and it seems to perform better when modelling more complex cholera cases, as shown in [31].

The model presented is then

$$\begin{cases} \frac{dS}{dt} = \frac{1}{100}\left(\Lambda - \mu S - Sg(I) + \gamma_1 I + \delta R\right), \\ \frac{dI}{dt} = \frac{1}{100}\left[Sg(I) - \left(\mu + \gamma_1 + \gamma_2 + \alpha\right) I\right], \\ \frac{dR}{dt} = \frac{1}{100}\left[\gamma_2 I - \left(\mu + \delta\right) R\right] \end{cases} \tag{12}$$

where:

- Λ represents the recruitment rate of susceptible individuals;
- μ is the natural death rate;
- γ_1 is the transfer rate from the infected class to the susceptible class;
- γ_2 is the transfer rate from the infected class to the recovered class;
- α is the disease-induced death rate;
- δ the immunity loss rate.

In this model Λ and μ are assumed to be positive and δ, γ_1, γ_2 and α are assumed to be nonnegative. As per the nonlinear incidence, $Sg(I)$, g is a real locally Lipschitz function on $\mathbb{R}^+$ with the following conditions, as presented in [32]:

- $g(0) = 0$ and $g(I) > 0$ for $I > 0$,
- $g(I)/I$ is continuous and monotonously increasing for $I > 0$ and $\lim_{I \to 0^+} \frac{g(I)}{I}$ exists as $\beta > 0$.

Unlike the previous models presented, in this paper we couple the system (12) with the boundary conditions, for $t \in [0, T]$ and $T > 0$,

$$S(0) = \max_{t\in[0,T]} I(t),$$
$$I(0) = \int_0^T S(T)S(s)ds, \tag{13}$$
$$R(0) = \int_0^T S(T)I(s)ds.$$

These functional boundary conditions fully reflect the considerations and conclusions presented in [30]. These types of boundary conditions have the following practical meaning:

- $S(0)$, the initial number of susceptible subjects, is equal to the maximum of the infected subjects;
- $I(0)$, the initial number of infected subjects, is a weighted average of the susceptible individuals, weighted by the final value of the susceptible S, at time T;
- $R(0)$, the initial number of individuals who recovered, is equal to a weighted average of the infected individuals, weighted by the final value of the susceptible, S, at time T.

Functional boundary conditions allow these assumptions to be examined in full, as operators can be considered as boundary conditions. These types of global conditions, that can include somewhat more abstract conditions, can only be contemplated via functional boundary conditions.

This model can therefore be presented in the form of (1)–(2), where $u_1 = S$, $u_2 = I$, $u_3 = R$, $a = 0$ and $b = T$,

$$f_1(t, u_1, u_2, u_3) = \Lambda - \mu u_1 - Sg(u_2) + \gamma_1 u_2 + \delta u_3,$$

$$f_2(t, u_1, u_2, u_3) = u_1 g(I) - (\mu + \gamma_1 + \gamma_2 + \alpha) u_2,$$

$$f_3(t, u_1, u_2, u_3) = \gamma_2 u_1 - (\mu + \delta) u_3,$$

$$L_1(u_1, u_2, u_3, u_1(T), u_2(T), u_3(T), u_2(0), u_3(0)) = \max_{t\in[0,T]} u_2 := k_1,$$

$$L_2(u_1, u_2, u_3, u_1(T), u_2(T), u_3(T), u_1(0), u_3(0)) = \int_0^T u_1(T)u_1(s)ds := k_2,$$

$$L_3(u_1, u_2, u_3, u_1(T), u_2(T), u_3(T), u_1(0), u_2(0)) = \int_0^T u_1(T)u_2(s)ds := k_3.$$

It is clear that the boundary conditions L_1, L_2 and L_3 satisfy (H1).

Moreover, f_1, f_2 and f_3 are L^1-Carathéodory functions such that, for $\max\{\|u_1\|, \|u_2\|, \|u_3\|\} < k$,

$$\begin{aligned} |f_1(t, u_1, u_2, u_3)| &\leq \Lambda + \mu k + k\beta + \gamma_1 k + \delta k := \psi_{1k}(t), \\ |f_2(t, u_1, u_2, u_3)| &\leq k\beta + (\mu + \gamma_1 + \gamma_2 + \alpha) k := \psi_{2k}(t), \\ |f_3(t, u_1, u_2, u_3)| &\leq \gamma_2 k + (\mu + \delta) k := \psi_{3k}(t). \end{aligned}$$

To satisfy (Hypothesis 2), one must verify that there exists $k > 0$, such that,

$$k \geqslant \max \left\{ \begin{array}{l} k_1 + \int_0^T (\Lambda + k(+\mu k + \beta + \gamma_1 + \delta)) \, ds, \\ k_2 + \int_0^T k(\beta + \mu + \gamma_1 + \gamma_2 + \alpha) \, ds, \\ k_3 + \int_0^T k(\gamma_2 + \mu + \delta) \, ds, \end{array} \right\}.$$

Let $T = 1, \Lambda = 2, \mu = 12.6, \gamma_1 = 2.7, \gamma_2 = 0.28, \alpha = 6.32, \delta = 0.38, k_1 = 2, k_2 = 0.27, k_3 = 5, \beta = 12$.

For $k \geq 0.06$, condition (Hypothesis 2) is verified and therefore, by Theorem 2, the system (12)–(13) has at least one solution $(S, I, R) \in (C[0,1])^3$, for the values considered.

6. Conclusions

In this paper, the authors show the existence of solution for a first order fully coupled system of three equations, involving two different cases. The first case, with coupled functional boundary conditions, is an existence result. The second case, with periodic boundary conditions, which is not covered by the first result, is an existence and location result. The extra information obtained in this second case is associated with the technique used, as it relies on the upper and lower solution method.

The application to an SIRS model, with global boundary conditions, underlines the key advantage and flexibility of the functional boundary conditions. The example shown illustrates not only the theorem proved in this paper, but it also provides guidance on how to arrange global conditions, in order to conform with the layout in (2).

As a matter of fact, a similar approach can be taken in several other models, allowing global conditions to be considered as boundary conditions, highly increasing the level of applicability of these models.

Author Contributions: The authors equally contributed to this work.

Funding: This project was supported by Fundação para a Ciência e a Tecnologia (FCT) via project UID/MAT/04674/2019.

Conflicts of Interest: The authors declare no conflict of interest.

References

1. Al-Moqbali, M.K.A.; Al-Salti, N.S.; Elmojtaba, I.M. Prey–Predator Models with Variable Carrying Capacity. *Mathematics* **2018**, *6*, 102. [CrossRef]
2. Gumus, O.A.; Kose, H. On the Stability of Delay Population Dynamics Related with Allee Effects. *Math. Comput. Appl.* **2012**, *17*, 56–67. [CrossRef]
3. Song, H.S.; Cannon, W.R.; Beliaev, A.S.; Konopka, A. Mathematical Modeling of Microbial Community Dynamics: A Methodological Review. *Processes* **2014**, *2*, 711–752. [CrossRef]
4. Osakwe, C.J.U. Incentive Compatible Decision Making: Real Options with Adverse Incentives. *Axioms* **2018**, *7*, 9. [CrossRef]
5. Deuflhard, P. Differential equations in technology and medicine: Computational concepts, adaptive algorithms, and virtual labs. In *Computational Mathematics Driven by Industrial Problems*; Springer: Berlin/Heidelberg, Germany, 2000; pp. 69–125.
6. Jang, S.S.; de la Hoz, H.; Ben-zvi, A.; McCaffrey, W.C.; Gopaluni, R.B. Parameter estimation in models with hidden variables: An application to a biotech process. *Can. J. Chem. Eng.* **2012**, *90*, 690–702. [CrossRef]
7. Akarsu, M.; Özbaş, Ö. Monte Carlo Simulation for Electron Dynamics in Semiconductor Devices. *Math. Comput. Appl.* **2005**, *10*, 19–26. [CrossRef]
8. Malinzi, J.; Quaye, P.A. Exact Solutions of Non-Linear Evolution Models in Physics and Biosciences Using the Hyperbolic Tangent Method. *Math. Comput. Appl.* **2018**, *23*, 35. [CrossRef]

9. Nieto, J.J. Periodic boundary value problems for first-order impulsive ordinary differential equations. *Nonlinear Anal.* **2002**, *51*, 1223–1232. [CrossRef]
10. Zhang, W.; Fan, M. Periodicity in a generalized ecological competition system governed by impulsive differential equations with delays. *Math. Comput. Model.* **2004**, *39*, 479–493. [CrossRef]
11. Agarwal, R.P.; O'Reagan, D. A coupled system of boundary value problems. *Appl. Anal.* **1998**, *69*, 381–385. [CrossRef]
12. Asif, N.A.; Talib, I.; Tunc, C. Existence of solutions for first-order coupled system with nonlinear coupled boundary conditions. *Bound. Val. Prob.* **2015**, *2015*, 134. [CrossRef]
13. Asif, N.A.; Khan, R.A. Positive solutions to singular system with four-point coupled boundary conditions. *J. Math. Anal. Appl.* **2012**, *386*, 848–861. [CrossRef]
14. Cabada, A.; Fialho, J.; Minhós, F. Extremal solutions to fourth order discontinuous functional boundary value problems. *Math. Nachr.* **2013**, *286*, 1744–1751. [CrossRef]
15. Cabada, A.; Pouso, R.; Minhós, F. Extremal solutions to fourth-order functional boundary value problems including multipoint condition. *Nonlinear Anal. Real World Appl.* **2009**, *10*, 2157–2170. [CrossRef]
16. Fialho, J.; Minhós, F. Higher order functional boundary value problems without monotone assumptions. *Bound. Val. Prob.* **2013**, *2013*, 81. [CrossRef]
17. Fialho, J.; Minhós, F. Multiplicity and location results for second order functional boundary value problems. *Dyn. Syst. Appl.* **2014**, *23*, 453–464.
18. Graef, J.; Kong, L.; Minhós, F. Higher order boundary value problems with ϕ -Laplacian and functional boundary conditions. *Comput. Math. Appl.* **2011**, *61*, 236–249. [CrossRef]
19. Graef, J.; Kong, L.; Minhós, F.; Fialho, J. On the lower and upper solution method for higher order functional boundary value problems. *Appl. Anal. Discret. Math.* **2011**, *5*, 133–146. [CrossRef]
20. Zeidler, E. *Nonlinear Functional Analysis and Its Applications, I: Fixed-Point Theorems*; Springer: New York, NY, USA, 1986.
21. Angstmann, C.N.; Henry, B.I.; McGann, A.V. A Fractional-Order Infectivity and Recovery SIR Model. *Fract. Fract.* **2017**, *1*, 11. [CrossRef]
22. Cui, Q.; Qiu, Z.; Liu, W.; Hu, Z. Complex Dynamics of an SIR Epidemic Model with Nonlinear Saturate Incidence and Recovery Rate. *Entropy* **2017**, *19*, 305. [CrossRef]
23. Secer, A.; Ozdemir, N.; Bayram, M. A Hermite Polynomial Approach for Solving the SIR Model of Epidemics. *Mathematics* **2018**, *6*, 305. [CrossRef]
24. Alexander, M.E.; Moghadas, S.M. Bifurcation analysis of an SIRS epidemic model with generalized incidence. *SIAM J. Appl. Math.* **2005**, *65*, 1794–1816. [CrossRef]
25. Chen, J. An SIRS epidemic model. *Appl. Math. J. Chin. Univ.* **2004**, *19*, 101–108. [CrossRef]
26. Hu, Z.; Bi, P.; Ma, W.; Ruan, S. Bifurcations of an SIRS epidemic model with nonlinear incidence rate. *Discret. Contin. Dyn. Syst. Ser. B* **2011**, *15*, 93–112. [CrossRef]
27. Liu, J.; Zhou, Y. Global stability of an SIRS epidemic model with transport-related infection. *Chaos Solitons Fract.* **2009**, *40*, 145–158. [CrossRef]
28. Teng, Z.; Liu, Y.; Zhang, L. Persistence and extinction of disease in non-autonomous SIRS epidemic models with disease-induced mortality. *Nonlinear Anal. Theory Methods Appl.* **2008**, *69*, 2599–2614. [CrossRef]
29. Jin, Y.; Wang, W.; Xiao, S. An SIRS model with a nonlinear incidence rate. *Chaos Solitons Fract.* **2007**, *34*, 1482–1497. [CrossRef]
30. Li, T.; Zhang, F.; Liu, H.; Chen, Y. Threshold dynamics of an SIRS model with nonlinear incidence rate and transfer from infectious to susceptible. *Appl. Math. Lett.* **2017**, *70*, 52–57. [CrossRef]
31. Capasso, V.; Serio, G. A generalization of the Kermack-McKendrick deterministic epidemic model. *Math. Biosci.* **1978**, *42*, 43–61. [CrossRef]
32. Li, J.; Yang, Y.; Xiao, Y.; Liu, S. A class of Lyapunov functions and the global stability of some epidemic models with nonlinear incidence. *J. Appl. Anal. Comput.* **2016**, *6*, 38–46. [CrossRef]

MDPI
St. Alban-Anlage 66
4052 Basel
Switzerland
Tel. +41 61 683 77 34
Fax +41 61 302 89 18
www.mdpi.com

Axioms Editorial Office
E-mail: axioms@mdpi.com
www.mdpi.com/journal/axioms

www.ingramcontent.com/pod-product-compliance
Lightning Source LLC
LaVergne TN
LVHW071948080526
838202LV00064B/6701